Agrobiodiversity

Integrating Knowledge for a Sustainable Future

Strüngmann Forum Reports

Julia R. Lupp, series editor

The Ernst Strüngmann Forum is made possible through the generous support of the Ernst Strüngmann Foundation, inaugurated by Dr. Andreas and Dr. Thomas Strüngmann.

Agrobiodiversity

Integrating Knowledge for a Sustainable Future

Edited by

Karl S. Zimmerer and Stef de Haan

Program Advisory Committee:

Conny J. M. Almekinders, Stephen B. Brush, Stef de Haan,
Timothy Johns, Julia R. Lupp, Yves Vigouroux,
and Karl S. Zimmerer

The MIT Press

Cambridge, Massachusetts
London, England

© 2019 Massachusetts Institute of Technology and
the Frankfurt Institute for Advanced Studies

Series Editor: J. R. Lupp
Editorial Assistance: A. Ducey-Gessner, M. Turner, C. Stephen
Photographs: N. Miguletz
Lektorat: BerlinScienceWorks

The book was set in TimesNewRoman and Arial.

Library of Congress Cataloging-in-Publication Data is available.
ISBN: 978-0-262-03868-3 (hardcover)
ISBN: 978-0-262-54969-1 (paperback)

Contents

The Ernst Strüngmann Forum

Science is a highly specialized enterprise—one that enables areas of enquiry to be minutely pursued, establishes working paradigms and normative standards, and supports rigor in experimental research. Some issues, however, do not fall neatly into the purview of a single disciplinary field and for these areas, specialization can actually hinder conceptualization and limit the generation of potential problem-solving approaches. The Ernst Strüngmann Forum was created to address such topics.

Founded on the tenets of scientific independence and the inquisitive nature of the human mind, the Ernst Strüngmann Forum is dedicated to the continual expansion of knowledge. Its activities promote interdisciplinary communication on high-priority issues encountered in basic science. Through its innovative communication process, the Ernst Strüngmann Forum provides a creative environment within which experts scrutinize high-priority issues from multiple vantage points.

This process begins with the identification of themes. By nature, a theme constitutes a problem area that transcends classic disciplinary boundaries—a topic of high-priority interest that requires concentrated, multidisciplinary perusal. Proposals are received from leading scientists active in their field and reviewed by an independent Scientific Advisory Board. Once approved, a steering committee is convened to refine the scientific parameters of the proposal and select participants. Approximately one year later, a central gathering, or Forum, is held to which circa forty experts are invited. Expansive discourse is employed to address the problem. Often, this necessitates reexamining long-established ideas and relinquishing previously held perspectives, yet when accomplished, novel insights begin to emerge. The resultant ideas and newly gained perspectives from the entire process are disseminated to the scientific community for further consideration and implementation.

Preliminary discussion on this topic began in 2014, when Karl Zimmerer approached me to discuss the possibility of convening a Forum on agrobiodiversity. Stef de Haan joined him in proposing the topic and, from September 25–27, 2015, the Program Advisory Committee (Conny Almekinders, Stephen Brush, Stef de Haan, Timothy Johns, Julia Lupp, Yves Vigouroux, and Karl Zimmerer) met to fine-tune the scientific framework for the Forum, which was held in Frankfurt am Main, Germany, from October 2–7, 2016.

This volume synthesizes the discourse that evolved between a diverse group of experts and is comprised of two types of contributions. Background information is provided on key aspects of the overall theme. These chapters, drafted before the Forum, have been peer reviewed and subsequently revised. In addition, Chapters 2, 6, 9, and 14 provide an overview of the working groups. These chapters are not consensus documents. Their intent was to summarize the discourse, to expose diverging opinions, and to highlight areas where future enquiry is needed.

An endeavor of this kind creates its own unique group dynamics and puts demands on everyone who participates. Each invitee played an active role and for their efforts, I am grateful to all. A special word of thanks goes to the members of the Program Advisory Committee, to the authors and reviewers of the background papers, as well as to the moderators of the individual working groups (Yves Vigouroux, Stephen Brush, Timothy Johns, and Glenn Davis Stone). The rapporteurs of the working groups (Kristin Mercer, Bert Visser, Anna Herforth, and Conny Almekinders) deserve special recognition, for to draft a report during the Forum and finalize it in the months thereafter is no simple matter. Finally, I extend my appreciation to Karl Zimmerer and Stef de Haan, whose commitment was essential to this 24th Ernst Strüngmann Forum.

A communication process of this nature relies on institutional stability and an environment that encourages free thought. The generous support of the Ernst Strüngmann Foundation, established by Dr. Andreas and Dr. Thomas Strüngmann in honor of their father, enables the Ernst Strüngmann Forum to pursue its work in the service of science. In addition, valuable partnerships with the following groups are gratefully acknowledged: the Scientific Advisory Board, which ensures the scientific independence of the Forum, and the Frankfurt Institute for Advanced Studies, which shares its intellectual setting with the Forum. Long-held views are never easy to put aside. Yet, when this is achieved, when the edges of the unknown begin to appear and the resulting gaps in knowledge are able to be identified, the act of formulating strategies to fill such gaps becomes a most invigorating activity. On behalf of everyone involved, I hope that this volume will advance understanding of agrobiodiversity and contribute to the sustainability of agriculture and food systems in the future.

Julia R. Lupp, Director
Ernst Strüngmann Forum
Frankfurt Institute for Advanced Studies (FIAS)
Ruth-Moufang-Str. 1, 60438 Frankfurt am Main, Germany
https://esforum.de/

List of Contributors

Aistara, Guntra A. Environmental Science and Policy, Central European University, 1051 Budapest, Hungary

Almekinders, Conny J. M. Knowledge, Technology and Innovation Group, Wageningen UR, 6706 KN Wageningen, The Netherlands

Andersen, Regine Fridtjof Nansen Institute (FNI), 1326 Lysaker, Norway

Baranski, Marci United States Department of Agriculture, Climate Change Program Office, Washington, D.C. 20250, U.S.A.

Benyei, Petra Institute of Environmental Science and Technology, Universitat Autònoma de Barcelona, 08193 Cerdanyola del Vallès, Barcelona, Spain

Borelli, Teresa Bioversity International, 00054 Maccarese (Fiumicino), Rome, Italy

Brush, Stephen B. Department of Human Ecology, University of California Davis, Davis, CA 95616, U.S.A.

Carney, Judith A. Department of Geography, University of California Los Angeles, Los Angeles, CA 90095-1524, U.S.A.

Castañeda-Álvarez, Nora P. Decision and Policy Analysis Area, International Center for Tropical Agriculture (CIAT), Cali, Colombia

Creed-Kanashiro, Hilary M. Instituto de Investigación Nutricional, Lima 12, Perú

de Haan, Stef International Center for Tropical Agriculture (CIAT), Agricultural Genetics Institute, Tu Liem, Hanoi, Vietnam

Giuliani, Alessandra School of Agricultural, Forest and Food Sciences (HAFL), Bern University of Applied Sciences, 3052 Zollikofen, Switzerland

Hanspach, Jan Faculty of Sustainability, Leuphana University Lüneburg, 21335 Lüneburg, Germany

Herforth, Anna Agriculture and Food Security Center, Columbia University Earth Institute, New York, NY 10025, U.S.A.

Hijmans, Robert J. Environmental Science and Policy, University of California Davis, Davis, CA 95616, U.S.A.

Hunter, Danny Bioversity International, 00054 Maccarese (Fiumicino), Rome, Italy

Jäger, Matthias International Center for Tropical Agriculture (CIAT), Cali, Colombia

Johns, Timothy   School of Human Nutrition, McGill University, Quebec H9X 3V9, Canada

Jones, Andrew D.   Department of Nutritional Sciences, School of Public Health, University of Michigan, Ann Arbor, MI 48109, U.S.A.

Kennedy, Gina   Bioversity International, 00054 Maccarese (Fiumicino), Rome, Italy

Khoury, Colin K.   International Center for Tropical Agriculture, (CIAT), Cali, Colombia

Krishna, Vijesh V.   International Maize and Wheat Improvement Center (CIMMYT), Texcoco 56237, Mexico

Lang, Timothy   Centre for Food Policy, City University, London EC1V 0HB, U.K.

Leclerc, Christian   CIRAD, UMR AGAP, 34398 Montpellier; and AGAP, University of Montpellier, CIRAD, INRA, Montpellier SupAgro, 34394 Montpellier, France

Maundu, Patrick   Kenya Resource Centre for Indigenous Knowledge (KENRIK), National Museums of Kenya, Nairobi, Kenya

McKey, Doyle   CEFE, Centre d'Ecologie Fonctionnelle et Evolutive, CNRS, University of Montpellier, University Paul Valéry Montpellier 3, EPHE, IRD, 34293 France

Mercer, Kristin L.   Department of Horticulture and Crop Science, Ohio State University, Columbus, OH 43202, U.S.A.

Nemogá, Gabriel   Indigenous Studies, University of Winnipeg, Winnipeg, MB R3B 2E9, Canada; and the Research Group on Biodiversity, Genetic Resources and Traditional Knowledge, Universidad Nacional de Colombia

Padmanabhan, Martina   Comparative Development and Cultural Studies, Focus Southeast Asia, Universität Passau, 94030 Passau, Germany

Powell, Bronwen   Department of Geography/African Studies Program, Pennsylvania State University, University Park, PA 16802, U.S.A.

Ramirez-Villegas, Julian   Decision and Policy Analysis Area, International Center for Tropical Agriculture, (CIAT), Cali, Colombia; and School of Earth and Environment, University of Leeds, U.K.

Raneri, Jessica E.   Department of Food Safety and Food Quality, Faculty of Bioscience Engineering, Ghent University, 9000 Ghent, Belgium

Reyes-García, Victoria   ICREA (Barcelona, Spain) and Institute of Environmental Science and Technology, Universitat Autònoma de Barcelona, 08193 Cerdanyola del Vallès, Barcelona, Spain

Sherwood, Stephen G.   Knowledge, Technology and Innovation Group, Wageningen UR, 6706 KN Wageningen, The Netherlands

Stone, Glenn Davis   Department of Anthropology, Washington University in St. Louis, St. Louis, MO 63130-4899, U.S.A.

Vanek, Steven J.   GeosyntheSES Laboratory, Department of Geography, Pennsylvania State University, PA 16802, and Department of Soil and Crop Sciences, Colorado State University, Fort Collins, CO 80523, U.S.A.

van Etten, Jacob   Information Services and Seed Supplies, Bioversity International (CGIAR), CATIE Campus, Turrialba, Costa Rica

van Loosen, Irene   International Center for Tropical Agriculture, (CIAT), Cali, Colombia

Vigouroux, Yves   Institut de Recherche pour le Développement (IRD), Montpellier, 34394, France, and Université de Montpellier, Montpellier, 34090, France

Visser, Bert   Centre for Genetic Resources, Wageningen University and Research Centre, 6708 PB Wageningen, The Netherlands

Zimmerer, Karl S.   Department of Geography, Pennsylvania State University, University Park, PA 16801, U.S.A.

1

Integrating Agrobiodiversity Knowledge for a Sustainable Future

Karl S. Zimmerer, Stef de Haan, and Julia R. Lupp

Introduction

If humanity is to achieve sustainable food systems in the twenty-first century, it must confront, comprehend, and respond to myriad interactions that transpire over time and across multiple scales. This requires us not only to recognize the individual elements and actors involved, but to understand the mechanisms that operate within and between the resulting interactions, the significance of how they are linked, and the competing interests and systemic tensions that exist and continue to emerge.

To meet this challenge, research and activities into the processes that impact agrobiodiversity have grown in importance over the last few decades. Given the wide-ranging scope of work, integrating knowledge from the distinct disciplines and fields of expertise has proven difficult. This Ernst Strüngmann Forum was convened to advance this process.

Dedicated to the expansion of knowledge in basic science, the Ernst Strüngmann Forum invites experts to partner with it to address problem areas confronted in research. It creates dialogues between multiple experts that are best likened to intellectual retreats—carefully crafted forms of interaction that promote synergy between diverse areas of science. The resulting, extended discourse serves to identify knowledge gaps, to explore novel ways of conceptualizing pressing issues, and to delineate trajectories for future research.

The dialogue on agrobiodiversity began when Karl Zimmerer and Stef de Haan approached Julia Lupp to discuss the possibility of putting together a Forum "to develop a new integrated scientific framework for understanding and advancing the current and future roles of agrobiodiversity in land use and food systems of the twenty-first century" (Zimmerer and de Haan, unpublished). Together they developed the initial proposal for formal submission to the Forum's Scientific Advisory Board. After a process of review and approval, Conny Almekinders, Stephen Brush, Timothy Johns, and Yves Vigouroux joined Zimmerer, de Haan, and Lupp on the Program Advisory Committee to refine the proposal and expand the primary goals for the Forum:

- To examine the linkages among key areas in agrobiodiversity
- To consolidate and advance the multidisciplinary foundations of science and scholarship needed in agrobiodiversity
- To develop an integrated scientific framework that will guide future work on sustainable food systems amid global change

To achieve these goals, the Ernst Strüngmann Forum invited experts from agriculture, agronomy, plant and animal breeding, anthropology, ecology, food systems and nutrition, geography, plant and biological sciences including genetics, political science, law, and sociology to attend the 24th Ernst Strüngmann Forum on "Agrobiodiversity in the 21st Century: Foundations and Integration for Sustainability." Working groups addressed the following themes:

- Group 1: Evolutionary ecology of genetic components, crop and livestock functions, and agroecology
- Group 2: Governance, including cultural and policy frameworks, at different geospatial scales
- Group 3: Whole-system approach to human health, nutrition, and disease
- Group 4: Socioecological interactions amid global change

Specific topics were introduced in advance through invited papers, and questions were proposed by the committee to initiate discussion at the Forum. With both as a backdrop, each group identified additional areas of inquiry, prioritized topics, and generated a discussion agenda. Interactions within and between the groups enabled a cross-fertilization of perspectives and ideas. Throughout, consensus was neither forced nor did preexisting biases or ideas drive the process. Instead, existing viewpoints were challenged and knowledge gaps exposed. The expansive discussions that resulted have been captured in "reports" (see Chapters 2, 6, 9, and 14).

This introductory chapter summarizes the individual chapters contained in this volume as well as the key outcomes from the Forum. Of particular note is the development of an integrated scientific framework—the response to the third goal—that emerged from the discourse. This framework is presented below to provide a way of conceptualizing the complexities surrounding agrobiodiversity, to support the integration of further knowledge, and to guide future research, scholarship, policy, and practice. In conclusion, we reflect on what lies ahead and issue a call for collective action: If sustainable food systems are to be achieved, new ideas need to be set into motion, with the requisite sensitivity to diverse interests and systemic tensions, informed by an integrated knowledge base.

The Dynamic Nature of Agrobiodiversity

Over the last century, crop genetic resources have declined significantly. In China, only 10% of the 10,000 varieties of wheat recorded in 1949 are now produced, and in the United States, more than 95% of the apple varieties known in 1900 are no longer harvested (Gepts 2006). Crop diversity plays an important role in food security amid a changeable environment and the urgent need for improved nutritional security, socioeconomic well-being, health, etc., and concerted action is needed immediately. Due to the diverse pressures that combine to cause loss or changes in range distribution, actions must be based on an integrated knowledge base that reflects the various actors and systems involved.

Broadly defined as the variation of crops and livestock in agriculture and food systems that result from and include heterogeneous (e.g., economic, ecological, institutional, sociocultural, and technological) factors, the concept of agrobiodiversity embodies dynamic processes between humans and nature on multiple organizational levels and spatial scales. It is simultaneously social and biological by nature, applicable to microbiomes, genes, species, habitats, and landscapes as well as to the historical, cultural, and social dimensions that frame the continuously evolving interactions between people and their environments. Utilizing a broad definition of agrobiodiversity enables us to integrate familiar biology, conservation, and ecology-centered meanings (e.g., CBD 2000; FAO 1999a; Jackson et al. 2007; Perrings et al. 2006) with sociocultural systems. It also presupposes the inclusion of new knowledge that may emerge as a result of pressures from globalization, global environmental change, or varied economic and sociocultural valuation. Implicit in this broad definition are three key areas.

The first refers to the *biodiversity* that is rooted in biology, genetics, and related fields of knowledge (e.g., taxonomy, conservation biology): organisms and communities (e.g., crops, vertebrates, trees, fish, insects, fungi, and other cultivated organisms) that pertain to domesticated and semidomesticated plants, wild biota used for food and health (Jacobsen et al. 2015; Reyes-García et al. 2006; Vandermeer et al. 1998), livestock, and the wild relatives of domesticates. This biodiversity must be understood across a spectrum of organismal groups and systems (Gepts et al. 2012), including

- alleles, genes, genomes, and microbiomes embedded in or constituting the actual domesticates and their wild relatives;
- crop and livestock varieties, landraces, breeds and or races that humans have historically selected and continue to create at the intraspecific level through traditional or modern practices;
- cultivated crop and livestock species that humans have domesticated through direct and indirect selection as well as the (un)conscious past and ongoing domestication of wild candidate flora and fauna; and

- crop and livestock wild relative (sub)species, including primary, secondary, and tertiary gene pools.

The second pertains to *interactions within and among habitats* that affect biodiversity (e.g., fields, farms, ecosystems, landscapes) (FAO 1999a). Many of the biological elements of such interactions, termed "associated agrobiodiversity" by Vandermeer et al. (1998), are extremely important to the functionality of agrobiodiversity and involve

- beneficial and nonbeneficial (micro)organisms (e.g., insects, arachnids, bacteria, fungi, viruses, and nematodes) associated with domesticates and farming systems;
- wide-ranging environments that are developed and managed by humans as diverse food systems: fish-, fungi- and insect-raising schemes; forestry, crop, and livestock agroecosystems; and a broad spectrum of farming ranging from swidden cultivation with extensive wild plant collecting to modern greenhouse hydroponics; and
- agricultural, pastoral, and forestry landscapes that contain the above-mentioned subsystems and traverse geographic spaces: from rural and peri-urban spaces to cities.

The third key area relates to the *human dimension of sociocultural relations* that affect and are integral to biological diversity in agriculture. This includes farm management, diverse belief and knowledge systems, cultural factors, collective processes such as seed exchange, and tourism associated with agricultural landscapes. Without these elements, biological and ecological subsystems of agrobiodiversity would not be able to exist or coevolve (Bellon et al. 2017; Brush 2000). In addition, variability in human resource management, skills, and knowledge is integral (Almekinders et al. 1995).

To support the integration of knowledge and areas of expertise, this volume embraces this broad definition of agrobiodiversity and holds that an inclusive view is imperative if diverse perspectives are to be incorporated to create greater understanding of the complexities inherent in agrobiodiversity. As knowledge is integrated, we should also be aware of the implicit use of knowledge systems as conceptual boundary objects. One example of a boundary object (Clark et al. 2016) would be the link created between the knowledge system of a particular group (e.g., scientists, resource management institutions, or NGOs) and the knowledge systems and practices of others (e.g., Indigenous, farmers, consumer groups). The objects and concepts that result from such connections constitute the "boundary" where communication occurs between different knowledge communities (Cash et al. 2003). The utilization of agrobiodiversity as a conceptual boundary object (Zimmerer 2015b) is thus relevant to the broad integration undertaken in this volume.

The individual chapters in this volume address wide-ranging issues related to environmental and socioeconomic changes, human nutrition, health, and governance, including policy, cultural, and economic practices. Below, the foci of the working groups are described, starting with the initial questions that were considered by the groups. An overview of the individual chapters follows, along with a summary of the key messages that emerged from the discussions.

Evolutionary Ecology, Agroecology, Conservation, and Cultural Interactions

- What are the complex genetic, evolutionary, and ecological interactions that underpin *in situ* conservation based on the continued cultivation of diversity in agroecosystems?
- How does the ongoing evolution of diversity, as practiced by individual farmers, function as an emergent adaptive mechanism in response to environmental change?
- What is the potential complementarity of *ex situ* and *in situ* approaches to genetic resources and how can this complementarity be strengthened?

Both natural and human factors cause agricultural systems to change. For millennia, humanity has relied on crops and biota that are semidomesticated and wild for the purposes of food, fiber, medicine, and fuel production. This reliance, in turn, has played a major role in shaping agrobiodiversity over time. Biotic and abiotic factors also impact agrobiodiversity, as do other organisms (e.g., pollinators, soil fauna, wild relatives). Understanding the interactions between crops and their wild relatives may reveal additional selection pressures at work across a range of ecosystems. Further, the phenotypic traits that crops express may help to clarify the ongoing process of crop evolution and domestication.

In their discussions, which aimed at understanding the genetic and functional dimensions of agrobiodiversity and associated knowledge, Kristin L. Mercer et al. (Chapter 2) stress that crop evolutionary agroecology must be viewed as a combined product of historical factors, local knowledge systems, and varied interactions with human society and associated biodiversity. Since each of these factors is affected by social and global change, the state of agrobiodiversity in any given environment must be considered to be in a state of flux. To gain a greater understanding of this complex, dynamic system, Mercer et al. propose research agendas to address the following areas:

- Quantify crop diversity and farmer knowledge that currently exist on the landscape to discern a baseline from which to understand future change.
- Increase understanding of the historical, evolutionary, and ecological factors that have led to current agrobiodiversity.
- Increase understanding of the drivers and effects of interactions between crops and their associated agrobiodiversity.

- Clarify the role of *in situ* conservation in farmers' fields and explore how *ex situ* collections can be better linked to *in situ* use of agrobiodiversity.
- Generate a theory of agrobiodiversity and project trajectories of agrobiodiversity capable of responding to social and environmental change.

To extend our understanding of how agrobiodiversity has evolved, population genetics can assume an important role. As discussed by Yves Vigouroux et al. (Chapter 3), population genetics offers the possibility of high-resolution, precise data as well as a robust way of monitoring spatial and temporal changes in crop–livestock populations through its ability to delineate trajectories of allele frequencies within and between given populations. Vigouroux et al. recognize that additional dimensions (e.g., space, time, stress, drivers, conservation approaches, biosystematic scale) are needed as well, and emphasize the importance of utilizing an integrated, multidisciplinary approach. They suggest that advances in modeling and the use of genomic markers present novel opportunities to evaluate, test, and increase our understanding of agrobiodiversity.

In Chapter 4, Steven J. Vanek discusses how crop and varietal diversity impacts the functioning and resilience of agroecosystems. He assesses impacts from pollination services, pests and disease, soil biota and soil nutrient cycling, as well as abiotic stress resistance. Vanek highlights the distinction between production characteristics related to plant phenotypes (provisioning services of ecosystems) and functional traits that support ecosystem services (supporting services of ecosystems), including the tendency toward trade-offs and the need to reconcile these to achieve resilience. He describes how these production and supporting services are linked to broader social and economic contexts and ecosystem resilience, and lists a number of questions to direct future efforts.

In Chapter 5, Nora P. Castañeda-Álvarez et al. characterize the processes that inhibit crop diversity and may lead to genetic erosion in crop resources. To mitigate the risk of loss, they suggest that *in situ* and *ex situ* conservation can be used in complementary ways. To predict possible changes in agrobiodiversity in the future, spatial analysis can assist both approaches and inform conservation action. Data availability, completeness, and quality are needed to secure effective spatial analysis of crop diversity. Castañeda-Álvarez et al. emphasize the role that modeling can play in assessing future responses to human and environmental events (e.g., floods, drought, variable rainfalls, land-use changes). Ultimately, the challenge is to expand spatial analyses and turn patterns of crop diversity into models that can explain how crop diversity is affected (positively or negatively) by different drivers and change scenarios.

Key messages that emerged are summarized as follows:

- Integrating *ex situ* and *in situ* approaches will strengthen knowledge of global genetic resources conservation, but has not yet been realized. Baseline quantification and characterization of agrobiodiversity and farmer knowledge is needed at different scales to track systematically and understand trajectories of change (Chapter 3). Regular gap analysis, timeline comparison, registration of unique diversity in both systems, and gap filling (*in situ* to *ex situ*) or redeployment (*ex situ* to *in situ*) are among the mechanisms available to link approaches (Chapter 5).

- Ecosystem-, organism-, and trait-level functionalities and services from agrobiodiversity need to be more systematically documented, evaluated, and valued. These functionalities and services (e.g., ongoing evolution, adaptive capacity, nutrient provision, yield stability, pest and disease regulation, relationships to cultural management practices) need to be better understood in different contexts and at multiple scales (Chapters 2, 4, and 10).
- Trajectories of past and future social and environmental change require information from geospatial and temporal scales to be fully integrated. To make predictions, a wider range of modeling approaches needs to be implemented. The scientific advancement and expanding tools offered by genomics and microbiomics provide multiple ways to unravel evolutionary pathways in response to cultural, migratory, environment, and management interactions (including climate change). Advances in modeling population genetics enable the testing of original hypotheses about the drivers that shape crop and livestock biodiversity (Chapters 2 and 3).
- The importance of culture and ethnicity in understanding processes involving selection, ongoing *in situ* conservation, and biogeographic distribution requires significantly more attention in agrobiodiversity science. The codistribution between farmers' cultural and crop–livestock genetic diversity has been described at different scales. Various expressions of culture and ethnicity, including different food systems and trait preferences, are recognized as persistent drivers of smallholder agrobiodiversity management. Nevertheless, integrative ethnographic, gender, and consumer behavioral research that would enable an analysis of the social processes underlying agrobiodiversity selection, (de) diversification, and migration is sparse compared to solely biological enquiries (Chapters 2 and 3).

Global Change and Socioecological Interactions

- How do agrobiodiversity use and conservation link to globally significant trends of urbanization and migration characteristic of high-agrobiodiversity regions?
- How do use (conservation) and disuse (genetic erosion) of agrobiodiversity intersect with the intensification of land use and food systems?
- What is included in agrobiodiversity beyond seeds (e.g., cultural and knowledge systems, consumer interest, cultural identities)?
- What are the driving values and visions?
- What are the prospects for intentional agrobiodiversity?
- How do key institutions constrain/facilitate adaptation to change?

In their discussions on socioecological interactions amid global change, Conny Almekinders et al. (Chapter 6) explore how agrobiodiversity has been used to improve human well-being under dynamic conditions by diverse groups and institutions. Their examination of different users (e.g., producers, consumers, and institutions) demonstrates how global change has impacted each group at the local level. In reviewing user responses, they found interesting instances where novel initiatives have been generated to link user groups and institutions, thus creating new collaborations and configurations diverse in nature, space, and scale. Such initiatives, Almekinders et al. propose, provide compelling evidence that socioecological interactions involved in agrobiodiversity can be positive and lead to increased human well-being amid global change. As efforts continue to resolve the many challenges posed by global change, this perspective holds promise that production and consumption can find complementary ways to interact.

In the conceptualization of interactions between climate and agrobiodiversity, it is important to recognize that different framings affect the types of questions addressed as well as the problem-solving approaches attempted. In Chapter 7, Jacob van Etten reviews how different scientific disciplines use distinct framings to explain these interactions and base their actions. Archaeological and environmental studies, for example, frame climate–agrobiodiversity interactions as part of a historical coevolutionary process, whereas agricultural and climate sciences focus more on genotype–environment interactions and diversification. Agrobiodiversity becomes critically important when agricultural development is framed as the central engine of economic growth to counteract loss due to climate change. Given the systemic nature, uncertainty, and intrinsic human values associated with climate change and agrobiodiversity management, van Etten stresses the need for integrated scientific approaches to address these complexities explicitly and to accommodate opposing sets of values.

In Chapter 8, Karl S. Zimmerer and Judith A. Carney review models and empirical studies that link demographic and spatial changes to socioecological interactions that involve agrobiodiversity at different spatial and temporal scales. Understanding the drivers of these global changes (e.g., human population changes, urbanization, economic and cultural globalization, spatial planning, food security, food sovereignty, historical, cultural, and social network considerations) is crucial as they shape agrobiodiversity outcomes. Zimmerer and Carney view engagement with policy communities as a high priority and recommend that researchers and organizations partner together to ensure that policy is informed by scientific analyses and scholarly understanding. They point out, however, that expanding the understandings of the varied processes involved in agrobiodiversity change will increase the complexity of research. Thus, they recommend that conceptual frameworks be developed to address this complexity and recommend promising areas for future research.

Key messages that emerged from this group are as follows:

- Global agricultural intensification has not resulted in a full-fledged wipeout of agrobiodiversity, yet the continued, predominant focus of the "new Green Revolutions" on efficiency and uniformity has severely limited the utilization of agrobiodiversity. The extent of agrobiodiversity in the farms and foods of smallholder and Indigenous farmers is more resilient and consistently conserved than many scientists predicted several decades ago. Farmers around the world continue to manage landraces and traditional breeds. In many areas, partial displacement or complete replacement of agrobiodiversity production and consumption has occurred. Nonetheless, farmers have found the means to adjust and incorporate agrobiodiversity into intensified food systems. Still, new initiatives for agricultural intensification, exemplified by the so-called New Green Revolution, the Next Green Revolution, and the initiative known as Alliance for a Green Revolution in Africa, continue to rely on a limited number of crop species and a few widely adapted varieties (Chapters 6 and 8).
- The globalization of the food industry and trade has made global food supply chains increasingly uniform, leading to increasingly standardized diets in terms of species and varietal usage. The worldwide spread of a "standard globalized diet"—one that is highly industrialized and often subsidized—is making cheap food increasingly accessible at the expense of the cultivation and consumption of local and regional agrobiodiversity. This, in turn, has fueled major global movements and alternative food systems concerned with food, health, and nutritional awareness as well as organizations that recognize the essential role of agrobiodiversity, local food culture, and inclusive value chains. Researching the interactions of these systems is important (Chapter 8).
- As a major driver of global change, urbanization poses both challenges and opportunities for agrobiodiversity, its producers, and consumers. Most people reside and work in urban areas, and the majority of global gross domestic product is produced in these spaces. Research into future agrobiodiversity use and conservation related to urbanization is needed to understand the complex interactions that exist and the potentially positive impacts that might be derived from feedback loops (Chapters 6 and 8).
- Climate change is another major global driver, one expected to exert both negative and positive selection pressures on agrobiodiversity. Deleterious impacts of global climate change are predicted to occur where the natural habitats of crop wild relatives and crop agroecologies impede viable ecological range displacement. Impacts from extreme weather, crop pests, and disease pose additional threats. Local environmental knowledge systems may also erode amid a rapidly changing environment (Chapter 7). Positive impacts on agrobiodiversity are potentially rooted in the capacities of diverse cropping systems, seed

networks, and cultivar management, although much research is still needed on actual socioecological adaptation and resilience capacities as well as vulnerability.

From Food and Human Diets to Nutrition, Health, and Disease

- What is the significance and role of agrobiodiversity in food-based approaches to assure nutrition security?
- What are the relations between agrobiodiversity and dietary diversity that support human well-being?
- How does agrobiodiversity interact with the main pillars of global food security: availability, access, stability, and utilization?
- What are the systemic and structural determinants of food preference, and how do they interact with cultural and social determinants?

In reviewing the complex relationships that exist between agrobiodiversity, food, and nutritional health, Anna Herforth et al. (Chapter 9) stress the need to move past the productionist paradigm, which has dominated agricultural and food policy since the middle of the twentieth century. Despite numerous adjustments being made, the productionist paradigm has been unable to improve global nutrition substantially or meet sustainable development goals: it is neither environmentally nor socially sustainable, and is thus unable to support economic sustainability. To promote nutritious, just, and sustainable food systems, Herforth et al. sketch out actions that support an alternative food narrative, where agrobiodiversity is viewed both as an essential element and key mechanism to the resolution of the world's food problems. They highlight the recent gains made as donors and policy makers adopt a more holistic view of food systems and practice in their attempts to balance nutrition security and the socioeconomic–environmental imprint of agriculture. They urge civil society organizations to give higher priority to farm and food systems that do not "mine" the earth, to restrict herbicide and agrichemical use, to push skills sharing and training that build on local knowledge, and to link agrobiodiversity to youth engagement, education, and revalued local identity. Further, they call for renewed public engagement, as change is needed at the consumer and food industry levels. Different messages are obviously required for different regions and social groups. Finally, Herforth et al. issue a direct appeal to the scientific community for clear, coherent, and evidence-based messages. Although the complexities involved in agrobiodiversity undoubtedly require ongoing research, they hold that enough is currently known to support the quest for sustainable, nutritious, and just food systems.

In Chapter 10, Andrew D. Jones et al. present key findings on the principal ways by which agrobiodiversity acts to influence human diet. Reviewing evidence of linkages between terrestrial agrobiodiversity (cultivated and wild harvested) and diet diversity and quality, they assess the research challenges that emerge when agrobiodiversity is linked to nutrition, and analyze diet diversity and quality indicators that would increase our understanding of these relationships. Jones et al. conclude with a set of policy recommendations directed toward global- and country-level policies, with the goal of producing more diverse foods and improving diet quality by mainstreaming biodiversity into overall development objectives.

Local choices about health (physical and mental) and food systems (production and consumption) do not occur in a void: they interact with and are affected by policies and economic trends that unfold across regional and global levels. Although evidence suggests that reduced agrobiodiversity is concomitant to dietary simplification and related health effects, complete understanding of this complex relationship is lacking. In Chapter 11, Victoria Reyes-García and Petra Benyei explore potential pathways at the local level (e.g., individuals, households, communities, or local landscapes) that link agrobiodiversity to physical (e.g., diet, nutrition) and mental health (e.g., how food culture and traditional agrobiodiversity management knowledge contribute to identity and self-esteem). They review social aspects related to the production and consumption of agrobiodiversity that promote health and well-being and contextualize how local solutions might fit into a broader political context. They stress the value of social support in attaining good physical and mental health, and argue that participation in social networks (e.g., seed exchanges or agrobiodiversity-based social networks) offers a range of supportive resources—emotional (e.g., nurturance), tangible (e.g., seeds), informational (e.g., advice), companionship (e.g., sense of belonging)—that relate back to physical and mental health.

Key messages that emerged are as follows:

- More of the same will result in more of the same: we must adjust our focus away from a small group of food species and varieties if we are to gain a broader understanding of agrobiodiversity as an integral part of true nutrition-sensitive agriculture. The consumption of cereals, starchy root crops, meat and dairy, oilseeds, and sugar has drastically increased during the last 50 years. Predominant crop and livestock sectors continue to receive most private and public investment, in terms of research and development. Offering a viable option and true alternative food narrative requires us to understand and use agrobiodiversity as a key to the world's current food system problems (Chapter 9).
- Understanding the impact of the full range of agrobiodiversity on diet diversity requires increased focus on intraspecific diversity, underutilized species, and wild foods. There are over 20,000 species of edible plants in the world yet fewer than twenty species now provide 90% of our food. Four staple crops—maize, potato, rice, and wheat—supply more than 60% of humanity's energy intake (FAO 2010b). Achieving healthy diets and diversified production requires (a) an expansion of

"underutilized" species and varieties of crops and animal breeds, semi-domesticates, and wild plants and animals that are nutritionally dense and (b) increased use of fruits, vegetables, pseudograins, nuts, minor roots, and tubers, among others (Chapters 9 and 10).

- The complex relationships between agrobiodiversity, human physical and mental health, and human well-being are poorly understood and under-researched. Evidence is growing that the overall reduction of agrobiodiversity in agroecosystems and value chains is concomitant to dietary simplification and negative health effects (Chapter 9). Yet, beyond the links between species and dietary diversity, exploration of other pathways linking agrobiodiversity to health and well-being remains uncommon. Social aspects related to the production and consumption of agrobiodiversity, as well as effects on overall well-being, need to be investigated and promoted using new integrative knowledge systems (Chapter 11).

- Healthy diets require agrobiodiversity; increased consumer demand and agrobiodiversity use can potentially provide a major stimulus for conservation. One of the main autonomous drivers triggering agrobiodiversity use, and thereby conservation, involves diverse cuisines. Human health is also directly and indirectly influenced by environmental health for which richness in agrobiodiversity, in turn, is essential. Thus, the consumption of agrobiodiversity can have positive effects on conservation and environmental health. Regional and farmer cuisines, enabling food environments that promote diversity, food literacy and awareness, high-value niche markets, certification schemes, and designation of origin are among the multiple options to create positive feedback loops from diets to conservation (Chapter 15).

Governance, Including Policy, Cultural, and Economic Frameworks

- How does agrobiodiversity interact with the current legal, policy, and political economic frameworks for food and agriculture?
- Given increased attention over the past two-plus decades, what major lessons can be derived for agrobiodiversity governance?
- What are the characteristic perils and promises related to the predominant macro-level market-based approaches to agrobiodiversity conservation?
- What is meant by "governance" beyond conventional policy/frameworks, including the balance between ownership, stewardship, and access via the market?
- How do power dynamics influence the management of agrobiodiversity across scales and systems?
- What are the challenges and potential for governance in rapidly changing food systems?

To date, the ability of Indigenous Peoples to determine governance of agrobio-diversity has not been fully recognized nor explicitly addressed in international policies or legislation. Gabriel Nemogá (Chapter 12) reviews the practices and politics of Indigenous Peoples relative to agrobiodiversity and proposes an inclusive, biocultural perspective of agrobiodiversity that accounts for the customs, worldviews, and rights of Indigenous Peoples. Nemogá discusses the epistemological and political barriers that currently exist and calls for a research agenda in support of a consistent policy for agrobiodiversity use and *in situ* conservation. He analyzes international and national policy and legal instruments that impact agrobiodiversity by Indigenous Peoples. Although the right of self-determination has been recognized on a global level for Indigenous Peoples, the contributions (past as well as present) and role that Indigenous Peoples play in agrobiodiversity governance have not been recognized at all levels of governance, global as well as domestic. Since the goal of overcoming hunger and malnutrition worldwide affects all peoples, Nemogá calls for proper recognition and protection of Indigenous Peoples, as their practice may contribute to robust approaches in agrobiodiversity governance.

Issues related to governance are often guided by estimates of countable and measurable objects: the number and diversity of heirloom seeds or landraces from a certain location, the frequency of seed exchange among actors, or the rates at which varieties disappear. Such variables provide information about conservation status at different scales as well as the dynamic, reciprocal roles and relationships that seeds and agrobiodiversity assume in local cultures and communities. In Chapter 13, Guntra A. Aistara explores the important cultural roles that seeds play in agrobiodiversity governance. She highlights how the practices, knowledge, and social networks through which farmers manage seeds are anchored in cultural memories and future visions of place. These places are further embedded in nested ecological, social, and political processes across scales. Aistara proposes that agrobiodiversity governance be studied as a set of nested but unequal relationships between people and their seeds, practices, and knowledge systems as well as to other people and species in their landscapes; also, within the broader politics of rural development and the cultural vision of place and landscape.

Furthering this discussion, Bert Visser et al. (Chapter 14) explore the multifaceted and highly dynamic realities of agrobiodiversity, which itself is the result of interactions between humans and nature, and is thus simultaneously both social and biological in nature. As carriers of major agrobiodiversity components, seeds are not mere material objects that exist outside of social relations; they are embedded sociobiological artifacts. Therefore, when addressing governance, we need to understand the limitations and political implications of the complementary and sometimes contradictory instrumental and relational perspectives. In many communities, agrobiodiversity constitutes a major part of the living environments of farmers and often plays a primary role in shaping both cultural identity and food systems. This situation is different in modern industrialized production systems, as farmers have become increasingly detached from the agrobiodiversity setting of their crops and animals. In addition, research and industry practices regarding the collection, taxonomic classification, and manipulation of seeds and plants has historically separated seeds and plants from the sociobiological context in which they were domesticated and the knowledge systems in which they functioned. This reality poses a dilemma. The multiple ways in which people relate to agrobiodiversity mirror myriad lifestyles, visions, cultures, and beliefs as well as the different social systems that help determine how resources are owned, exchanged, and distributed. These nuanced relationships reflect unique histories and ways of life, evoke unique questions, and necessitate a different type of research. To appreciate the potentialities of agrobiodiversity and the wealth of options for conservation and governance, the physical, biological, social, and cultural contexts must all be taken into account. Multiple worldviews must be managed and novel questions need to be raised and addressed. Researchers wishing to work with people with unique experience and value systems must not only be respectful of multiple and sometimes incompatible worldviews, they must be willing and able to represent competing worldviews as equally valid. Doing this will strengthen the unique plurality that historically gave rise to rich patterns of agrobiodiversity and promote relationship building and trust implicit in a highly multicultural, cosmopolitan world.

How have markets affected the governance of agrobiodiversity? In Chapter 15, Matthias Jäger et al. analyze the role of agricultural product markets in agrobiodiversity governance. The expansion of these markets globally over the past two decades has generally promoted the simplification of agricultural and food systems, thus reducing diversity within crop and animal species. Farmers who continue to conserve on-farm agrobiodiversity provide valuable public goods, in terms of food security and environmental sustainability. However, because the market does not compensate farmers for conserving high levels of agrobiodiversity, there is little incentive to maintain on-farm conservation practices. This could eventually precipitate the destruction of local food systems and general biodiversity loss. To enhance both agrobiodiversity conservation and income generation through market-based instruments, ways of valuing agrobiodiversity need to be developed that account for its true production cost and contributions to genetic resource usage. Jäger et al. propose that payments for agrobiodiversity conservation schemes and niche market development (e.g., differential marketing, labels, certification schemes, agrotourism) should happen in tandem: stronger activities in agrobiodiversity conservation need to emerge from private sector investment and government funds. These measures offer potential for the successful marketing of agrobiodiversity and its niche products through collective action. However, constraints and possible unintended consequences of market-based approaches to agrobiodiversity conservation must be taken into account.

Key messages that emerged from this group include the following:

- Multiple systems to govern agrobiodiversity coexist and lessons can be derived from this multiplicity. Still, divergent systems are largely incompatible in terms of what is being governed and for whose benefit. Conflicting approaches based on different value systems and their rationales can be distinguished (e.g., stewardship versus ownership approaches). Access, control, and use of agrobiodiversity constitute major expressions of governance (Chapter 14). Future research is needed to understand the interplay of multiple parallel governance systems that vary in scale, object, actors, and purpose.

- Power dynamics involved in governance are crucial in the context of rapidly changing farming and food systems. Multiple power asymmetries affect agrobiodiversity governance as well as the control, access, use, and benefits to diverse actors. As intellectual property systems are costly institutions, the capacity of developing countries to develop and effectively use such systems can be limited. Indigenous Peoples and smallholders manage most of the world's *in situ* agrobiodiversity yet are rarely empowered on equal terms in negotiations. The analysis and self-reflection of power asymmetries, including scientific initiatives, is crucially needed to inform current and future agrobiodiversity governance avenues (Chapters 12–15).

- There are significant differences among modern and traditional peoples, or those outside the mainstream, that have important implications on how people experience, relate to, and seek governance options in agrobiodiversity. The multiple ways in which people relate to agrobiodiversity reveal myriad lifestyles, visions, cultures, and beliefs as well as social systems that determine how resources are owned, exchanged, and distributed. This does not merely translate into different views and experiences, but underlies subtle yet profound associations, unique place-based trajectories, and different ways of living that are vital to understand for the future viability of agrobiodiversity (Chapters 12–15).

- New and innovative models to govern agrobiodiversity have been emerging, often at local to national scales, led both by private and public initiatives to facilitate benefit sharing and farmers' rights. These governance initiatives need to be studied for potential replication and adaptation in different contexts (Chapters 14 and 15). It is important to research the diverse and often less visible initiatives that have emerged. Although a systematic analysis of piloted initiatives has been given priority, many private initiatives need to be evaluated in broader contexts of new corporate responses and potential social responsibility.

Integrating Agrobiodiversity Knowledge: A New Framework

The complexities inherent to agrobiodiversity demand a way to envision and address the contributory elements, linkages, and dynamics that play out on multiple scales (temporal, spatial) and levels (individual, group, regional, global). Such a conceptual framework provides a way to recognize the immensity of the problem and can be used to identify unresolved areas or knowledge gaps as well as research and policy opportunities. It also supports the integration of information that will continue to emerge and can serve to unite wide-ranging actors and institutions to work in concert.

The development of this integrative framework was a motivating force behind the convening of this Forum. The following framing emerged from the discourse and is presented here to promote further work and dialogue within and between research communities, policy makers, and practitioners (see also Zimmerer and de Haan 2017):

- *Evolutionary Ecology and Biocultural Diversity*: This area refers to interactions involving genetics, genetic resources, agroecology, and plant science and addresses how ethnobiology, ethnicity, linguistics, and culture (e.g., use, tradition, practice) impact the ecological system. Emphasis is on comparative diversity, timeline measurements based on genetic markers, and the integration of spatial knowledge systems.
- *Global Change*: This area addresses how climate change, environmental factors (e.g., water, soil erosion, land degradation), and human behavior influence food systems. This includes the impact from socioeconomic drivers (e.g., urbanization, market integration, demographic changes, expansion of global industrial food systems, trade and food policy), processes of valuation and their results (e.g., loss of cultural knowledge systems), as well as socioecological interactions on numerous scales.
- *Food–Nutrition–Health Linkages*: This area focuses on impacts (positive and negative) of food biodiversity on the human diet and links the effects of food and related biota (e.g., microbiome) biodiversity to human health (physical and mental), disease, and well-being. It also addresses the dynamics of food choice (e.g., economic specialization, market-based purchasing power, social movements) and the knowledge needed from the environmental, social, nutritional, and health sciences to create effective policy.
- *Governance*: This area focuses on policy, cultural, and economic practices and involves institutions and legal agreements that span (a) mainstay approaches (e.g., the International Treaty on Plant Genetic Resources for Food and Agriculture, Convention on Biological Diversity, the Nagoya Protocol, and Second Global Plan of Action for Plant Genetic Resources for Food and Agriculture), (b) multiple access and benefit-sharing arrangements (e.g., market-based approaches, cultural movements), as

well as (c) community/grassroots and Indigenous movements, civil society organizations, consumer groups, and the private sector.

Grouping the multiple aspects into these cornerstone areas enables us to visualize broad areas of concern, to identify the requisite expertise that may be needed to address challenges, and to highlight where potential connections and collaborations may be needed. Within each area, multiple relationships are possible on a variety of scales: (a) spatially, from molecule and gene levels to organism, field, community, landscape, region, country, and global systems as well as (b) temporally, from less than 1 year to over 10,000 years (Zimmerer and de Haan 2017:3). Currently, research in all four areas is being framed primarily at global or smaller spatial scales (e.g., organism, field, community). This lack of within-country, regional-scale research demonstrates an important gap that must be filled in all four areas.

Between areas, numerous relationships and interactions are also possible. This framework provides a way to recognize these connections, to access requisite knowledge sources that will help us grasp their significance, and to identify competing interests or systemic tensions that may emerge. It also offers a way to conceptualize the incredibly dynamic nature of agrobiodiversity and appreciate the multiple ways in which linkages form. Understanding the inherent intricacies of agrobiodiversity and integrating resultant knowledge is needed if we are to enhance our ability to develop research, policy, and practical strategies aimed at achieving sustainable food systems.

Looking Ahead

Nutritional and food security, the provision of ecosystem services, and the protection of cultural values are essential components for achieving sustainable food systems. Securing any one of these requires action on institutional and individual levels, as well as a hybrid information base derived from traditional and modern knowledge sources. All of this, of course, is influenced by the highly dynamic processes of global change, which reorganize and modify the conditions under which agrobiodiversity unfolds. Given these dynamics, humanity cannot afford to wait in its response. Collective action is required immediately, molded by priorities and valuation systems, to address a multitude of issues:

- How should agrobiodiversity ideally be used and conserved, by whom, and under what conditions?
- Can on-farm conservation be taken more seriously as a basis of future conservation and a linked, parallel system to gene banks?
- What is likely, or acceptable, to be lost? How can we track this? What might be added?
- What is the role of industrial agriculture versus family farming in future food systems and agrobiodiversity usage?

- Can local market networks that incorporate agrobiodiversity remain or become viable businesses, accessible to different consumer segments?
- Will concentrated wholesale and hypermarket chains become the exclusive global model and what does this imply for agrobiodiversity?
- Is there room for increased emphasis of agricultural research on quality, nutritional, and sustainability traits that build on agrobiodiversity?

Profound choices and creative approaches will be required if production environments are to be generated that reflect economic and sociocultural values. Cross-country and societal comparisons of trade-offs offer important lessons to inform government-level reviews of food production (e.g., Fitzpatrick et al. 2017; GOS 2017; NAFRI 2016). Such studies can also provide insight to civil society organizations, the private sector, donor agencies, academia, and consumers as they seek to undertake actions to use, conserve, and valorize agrobiodiversity at multiple levels (e.g., grassroots implementation, financial and policy support for science and equitable benefit sharing, development of inclusive value chains, awareness raising, messaging within different food environments).

In addition to the various human dimensions discussed throughout this volume, many others demand attention. Demographic attrition, for example, is currently having a negative effect on the global farming workforce, compounded by the trend among young people to reject agriculture as a viable livelihood option. How can we anticipate and respond to impacts that will certainly follow, in terms of associated production modes, knowledge, and culture? Can farming be made to be more attractive to younger generations, perhaps through youth engagement and awareness programs? One possibility would be to incorporate principles of agrobiodiversity and intercultural approaches to learning into educational systems (e.g., through curricula development, digital technology). This challenge holds great potential for participatory approaches that would build on local expertise (e.g., teachers, parents, community elders) and knowledge sources.

Tailoring our response to the multitude of issues involved in creating sustainable food systems requires a new approach along with the requisite sensitivity to diverse interests and systemic tensions. Drawing from distinct disciplines and fields of expertise, this volume offers a framework which we hope will spur further discussion, guide future actions, and lend understanding to the myriad interactions involved in agrobiodiversity across multiple scales.

Acknowledgments

We wish to thank Conny Almekinders, Stephen B. Brush, Timothy Johns, and Yves Vigouroux for their valuable contributions as members of the Program Advisory Committee. We also extend a special word of thanks to all who participated in and contributed to this Forum.

Evolutionary Ecology, Agroecology, Conservation, and Cultural Interactions

2

Crop Evolutionary Agroecology

Genetic and Functional Dimensions of Agrobiodiversity and Associated Knowledge

Kristin L. Mercer, Yves Vigouroux,
Nora P. Castañeda-Álvarez, Stef de Haan,
Robert J. Hijmans, Christian Leclerc,
Doyle McKey, and Steven J. Vanek

Abstract

Agrobiodiversity supports agriculture globally and is used and stewarded worldwide by farming communities that possess traditional knowledge about their crops. This chapter takes an evolutionary ecological perspective on the ecology, use, and conservation of crops and proposes research objectives to advance the study of agrobiodiversity globally. In particular, research agendas are outlined (a) to determine the current state of agrobiodiversity globally and how it is changing through the collection of baseline data; (b) to improve understanding of functions of existing agrobiodiversity and how the historical, evolutionary, and ecological factors have led to that diversity; (c) to increase understanding of the interactions and factors that drive change between crops and their associated agrobiodiversity (i.e., the multitude of organisms that interact with the primary crops); (d) to clarify the role of *in situ* conservation of agrobiodiversity in farmers' fields and how better to link *ex situ* collections to *in situ* use of agrobiodiversity; and (e) to generate a theoretical framework for agrobiodiversity to help us better understand past and future dynamic change. Pursuing such lines of research will enhance humanity's ability to face uncertainty, such as that expected with climate change.

Group photos (top left to bottom right)   Kristin Mercer, Yves Vigouroux, Stef de Haan, Nora Castañeda-Álvarez, Christian Leclerc, Robert Hijmans, Doyle McKey, Steven Vanek, Christian Leclerc, Stef de Haan, Doyle McKey, Kristin Mercer, Nora Castañeda-Álvarez, Steven Vanek, Robert Hijmans, Yves Vigouroux, Nora Castañeda-Álvarez, Stef de Haan, Steven Vanek, Yves Vigouroux, Kristin Mercer

Introduction

Since the beginning of agriculture, humans have increasingly shaped biological diversity. Agrobiodiversity, defined as the biodiversity associated with agricultural production, includes the crops and livestock of prime interest to farmers as well as their wild relatives and other associated organisms (i.e., associated agrobiodiversity) that interact with these domesticates in agricultural ecosystems (see also Chapter 1). In this chapter, we emphasize dynamics of crops and their associated agrobiodiversity, but many of the concepts could apply equally well to livestock. To consider the livestock portion of agrobiodiversity, Hall (2004) and Barker (1999) provide a good starting point for livestock breed biodiversity and surrounding issues. Other more recent studies have also focused on particular livestock species and breeds in different areas of the world.

Crop diversity has been shaped over millennia by interactions between crops and their human guardians as well as by important local biotic and abiotic factors. The diversity of crops and associated agrobiodiversity is intricately linked to local systems of knowledge in agricultural societies (see Chapters 12 and 13). Thus, to increase our understanding of agrobiodiversity in general, we must better understand not only the evolutionary history of crops, but also the societies in which they occur (see Chapter 3).

Since domesticated plants interact with other organisms (e.g., pollinators, soil fauna; see Chapter 4) as well as with their wild relatives, understanding interspecific interactions among components of agrobiodiversity and the resulting evolution is particularly important to assess the functional role of agrobiodiversity. Rapid environmental and management changes that are occurring in agricultural systems affect agrobiodiversity (Chapter 8). Concern about the subsequent loss of crop genetic diversity in farming systems has led to the development of *ex situ* collections (gene banks) (Chapter 14).

Through these efforts, an important slice of worldwide crop diversity has been conserved, but since the collections consist mainly of cultivated seed crops, they are far from complete (Chapter 5). Moreover, the presence of agrobiodiversity used *in situ* is an important complementary asset due to the ongoing adaptation of populations to current agroecological conditions. Assessing the diversity conserved *ex situ* and used *in situ* can serve as a basis for a better understanding of the state of agrobiodiversity and the processes that shape it.

We identified five objectives to prioritize for future work in crop evolutionary agroecology:

1.   Develop a better understanding of the current state of agrobiodiversity globally and how it is changing. Key to this is good baseline data. What types of data are necessary to answer the most salient questions related to the evolutionary trajectory of agrobiodiversity?

2. Improve understanding of the historical, evolutionary, and ecological factors that have led to current agrobiodiversity. Here, it is important to focus on the functional aspects of agrobiodiversity and to understand how it has shaped, and been shaped by, the changes in a variety of agricultural systems (e.g., intensive industrial agriculture, intensive and extensive variants of smallholder systems, agropastoralism).

3. Increase understanding of the interactions and factors that drive change between crops and their associated agrobiodiversity (i.e., the multitude of organisms that interact with the primary crops). Key to this is to understand how interactions with crop wild relatives and other components of the ecological community have shaped, and continue to shape, agrobiodiversity.

4. Clarify the role of *in situ* conservation of agrobiodiversity in farmers' fields and how *ex situ* collections can be better linked to *in situ* use of agrobiodiversity.

5. Generate a theoretical framework for agrobiodiversity to help us better understand past and future dynamic change. In the context of ongoing global change, such insight might help us develop scenarios for understanding the future evolution of agrobiodiversity.

For each of these objectives, the most important points that emerged from our discussions are outlined below to contextualize the objectives and guide future research agendas (basic and applied).

Quantifying Crop Diversity and Farmer Knowledge to Discern a Baseline and Assess Future Change

Our first objective focuses on the need to evaluate the current state of agrobiodiversity that resides *in situ* on the landscape. Only through a systematic collection of baseline data on *in situ* agrobiodiversity will it be possible to understand current patterns and future changes (Chapter 14). To achieve this, however, will require a major, global effort to collect, collate, and store the different types of agrobiodiversity-related data at different spatial and temporal scales and resolutions.

Diversity Metrics and Ethnobotanical Knowledge

Baseline data should be gathered for different levels of organization of diversity: among species (number of species, number of functional groups), within species (number of landraces, molecular genetic diversity, functional diversity) as well as the ethnobotanical diversity for both levels. This would require the collection of crops grown and analyses of genetics, morphology, physiology, and nutrient content as well as a significant investment in surveying the people that maintain and use the diversity. Data collection would need to be standardized to derive the greatest value; it would also need to be comparable across locations and possibly across crops. Data generated would need to be thoroughly documented and kept in stable repositories. The best indicators of diversity are those that are comparable across species as well as across scales: from the field, to the region, to the continent. Genetic diversity indicators should not be specific to a particular molecular marker system, but rather suitable across methods.

Documenting these patterns of diversity offers an opportunity to understand what drives them. Species-level diversity in crops (Hijmans et al. 2016) or crop wild relatives (e.g., Hijmans and Spooner 2001; Jarvis et al. 2003) is fairly straightforward to assess. However, describing crop intraspecific diversity over large areas is much more challenging, and to date there are only a few examples of such efforts (e.g., Mercer et al. 2008; Orozco-Ramirez et al. 2017; Perales and Golicher 2014). New large-scale data collection is needed to investigate these patterns and their human and environmental drivers. For instance, by combining spatial and genomic data we could examine both functional and neutral genetic diversity. Insights into these different aspects of diversity in turn could improve understanding of the role that processes play in structuring diversity (e.g., gene flow by seed sharing or selection by the environment).

Samples and available data may create different forms of bias. For example, collection trips may introduce geographical bias if accessions are collected near well-traveled roads (Hijmans et al. 2000). Gene bank management can also contribute to bias in the following ways:

- Removal of clonal duplicates can mask true geographic variation.
- Evolutionary change may occur due to storage and seed multiplication.
- Human error (e.g., assigning the wrong identifier to an accession after grow out, errors in collection location data) may reduce the value of the metadata associated with collections.
- Loss may occur as a result of economic downturns, war, or civil unrest.
- Some samples may be extremely local, thus causing under- or over-representation of some types.

These biases in gene bank samples may depend on the system of reproduction (clonal, self-pollinating, or cross-pollinating crops).

If diversity is quantified based on the prevalence of named landraces, the methods used to collect these data can be biased and thus need to be adjusted according to different recording efforts, synonymy, and errors in reporting of names (Brush et al. 2015). Despite these issues with gene bank accessions, countrywide data can often provide useful information, and bias affecting our ability to assess local diversity becomes less of an issue with a greater number of samples. Thus, gene bank collections and associated data can allow us to identify regional areas of high diversity.

A critical concern that needs to be addressed is the issue of whether there is correspondence between ethnobotanical and biological data. The unit of local farmer management is often a landrace, yet the genetic bases and ethnobotanical names for landraces may not always correspond, thus making the term "landrace" problematic at times. The level of biological variation within a landrace may be greater than the distinctions between them, which is the pattern that characterizes the case of Andean potatoes (Zimmerer and Douches 1991). Landraces can be defined in two different but complementary ways: from external/scientific and internal/farmer points of view. External observers (e.g., scientists) define landraces as morphotypes and use a classification system and naming convention to compare morphological variation across regions or countries. This approach may not, however, take into account how farmers manage crop diversity. The different ways that farmers name, select, categorize, and use landraces constitute an integral part of agrobiodiversity that cannot be neglected. Since farmer classification can affect farm management, and thereby genetic patterns (e.g., obscuring or enhancing differentiation), these two forms of classification are dependent on one another. We must also recognize that different groups of farmers may refer to the same crop differently. Rare types, referred to using many different names or synonyms, may be identifiable only through interactions with focus groups. Common types, by contrast, may share the same or similar names across large areas and groups of farmers.

How can we access the knowledge that farmers have about their crop diversity? The number of ways in which farmers refer to varieties per area or per amount of genetic variation reveals something about this knowledge, but other knowledge must be considered as well (e.g., folk ecology). Traditional ecological knowledge can include where to plant a given crop variety on the landscape, the length of season for an individual variety, how a variety behaves under stress, or the importance of seed coat color. Knowledge systems are dynamic: they need to take in knowledge continuously for learning to continue, and the loss of knowledge can have a negative effect on agricultural systems (Stone 2007).

Spatial Extent, Strategy, and Databases

A number of strategies exist at different spatial extents to determine a baseline for current patterns in crop agrobiodiversity. Each spatial extent—local (village), subnational (regional), national, and global (Figure 2.1)—may require or use different kinds of collection strategies, such as intensive sample collection in regions of interest, gene bank collections at the national level, or methods for crowdsourcing data globally (van Etten 2011). Subnational data can be coalesced at the national level, and national level data could be used to build international databases. However, it is not clear how or whether understanding

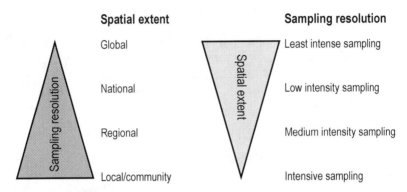

Figure 2.1   Diagram of possible relations between sampling resolution and spatial extent for agrobiodiversity studies.

at the local or regional level can accurately be used to predict patterns and processes at the national and international level.

Sampling strategies may need to be enhanced, since local sampling is known to produce surprisingly high levels of diversity when done strategically (e.g., MINAGRI 2017). In addition to representative sampling, sampling from farmers who are known to be custodians of crop diversity can be helpful, since knowledge of agrobiodiversity is unevenly distributed and the diversity held by farmers with high levels diversity is most dynamic. Similarly, areas with steep environmental gradients or unusual combinations of environments or cultures can provide great diversity (Perales et al. 2005).

Federated databases containing collections of crop and crop wild relatives (e.g., Genesys), traits of the plants collected, and observational data (e.g., landrace names, data from interviews) form the basis for the type of agrobiodiversity studies that we envision. Progress is needed in the standardization and use of unique identifiers to allow for research that integrates data from different databases (e.g., Permanent Unique Identifiers, PUIDs, in the form of Digital Object Identifiers, DOIs). It is also important to enhance access to data to make it less atomized (and invisible in some cases). Data sharing is frequently hindered by issues around data ownership. A wider adoption of open-access policies and licenses, such as creative commons licenses and data journals, are among the options that help overcome such issues.

Ideas for Baseline Construction and Monitoring at Local (High-Resolution) and National or International (Low-Resolution) Spatial Extents

For most of the plants, animals, and microbes that we depend upon for food (i.e., agrobiodiversity), we have only a scant understanding of the patterns of diversity, how these are changing, and how this is affected by local farm management practices. To overcome this limitation, sites where agrobiodiversity is monitored need to be established.

Baseline at Local Scale

Monitoring of agrobiodiversity in key representative areas of high diversity, thereby forming a network of observatories, can produce novel intelligence about the conservation dynamics of landraces. It is important to use semi-standardized procedures and minimal key indicators across all sites to make comparisons in time and space feasible. Such a network of sites can be linked to initiatives to understand patterns across the entire region of interest, through a combination of surveys and predictive modeling. Ideally, one would collect data on (and continue to monitor) the abundance of the elements of diversity studied (e.g., landraces) as well as their spatial and temporal distribution in the landscape and linked farmer and community knowledge. Abundance data allow for the computation of diversity indices, such as Shannon's diversity index (Jost 2006), as well as simpler metrics, such as species richness and evenness.

The establishment of observatories requires solid, long-term partnerships, ideally involving communities, grassroots organizations, and the national agricultural research institute where the observatory is established. It is also crucial to sustain an enabling environment that satisfies key development needs of farmers. This does not necessarily need to be directly related to agrobiodiversity, but may, for example, involve youth, education, or health.

Example of a Local-Scale Observatory: Chirapaq Ñan Initiative

The International Potato Center initiated a potato diversity documentation and monitoring effort called the "Chirapaq Ñan Initiative," involving partners in central and southern Peru, Bolivia, and Chile (de Haan et al. 2016). Over the course of four years, they collected baseline data on the total number of landraces, the inventory of rare to common landraces (creating a "red list"), the spatial distribution of landraces (through participatory mapping), and local knowledge about potatoes linked to primary and secondary education in rural schools. This effort connected NGOs, farmer organizations, universities, and the International Potato Center. Observatories can be combined with research on *in situ* dynamics and citizen science. To motivate communities and farmers, it is important to provide incentives, as in training on pest and disease management, farmer exchange visits, or the development of catalogues (Scurrah et al. 2013).

Baseline at National or Global Scale

Estimates of variation in aspects of agrobiodiversity (e.g., number of species or varieties or amounts of genetic variation) are also needed over large areas. Such estimates are necessary if we wish to develop a broader understanding of the forces that shape and maintain agrobiodiversity. They could also provide a framework to interpret new observations and to make predictions, for instance, about responses to global change (the fifth objective).

Since crop diversity can be observed and exhaustively measured only in relatively small areas, a different sampling approach is needed over larger areas. Typically this involves very sparse samples and the use of modeling with combinations of data sources, including data from gene banks and collected via crowdsourcing, to build spatial predictive models. For instance, smartphone technology is available that can identify wild plants by photographs and register a sighting as georeferenced data. This has been tried on grape varieties that have distinct leaf shapes, so perhaps in the future the same could be used to identify landraces within crop species. Using such occurrence data, models can be built to predict entire geographic distributions. One approach would be to model observations as a function of a set of predictor variables for which we have detailed spatial data, allowing us to predict to all locations. This method is sometimes referred to as species distribution or ecological niche modeling (Elith and Leathwick 2009). In the context of agrobiodiversity, it is relevant to consider using not only abiotic (climatic, soil) predictor variables, but also variables related to variation in human behavior, since humans manage crucial life history stages that affect the ecological niche of a landrace.

An alternative, more theoretical approach could be to create a predictive model based on "first principle" drivers. Such models would increase general understanding and, where data is sparse, could include important processes, such as isolation by distance. Results could also be combined with data from ecological niche models (Kraft et al. 2014).

Example of a Global Baseline Project

Currently, national and global baseline projects are scarce. An example at the national-level is the very comprehensive survey for intraspecific maize diversity in Mexico by the National Commission on the Environment and Biodiversity (CONABIO 2011, 2013). In several countries, existing farmer networks could be used to collect data which could, in turn, lead to new insights while also being used to provide advice to farmers. Different types of data from diverse sources could also be combined. There are three distinct communities that have a lot of data:

- NGOs interested in high diversity, farmers' rights and livelihoods, and creating a baseline for the areas where they work.
- Gene bank managers or collectors interested in conserving diversity.
- Individuals from the scientific community interested in understanding patterns and the processes that create them.

Data from these groups could be made more compatible and form coherent databases, but currently they do not tend to align.

One way to provide sufficient information on the methods and indicators used to monitor agrobiodiversity is to compare and assess data derived from sampling-intensive studies (which usually produce high-resolution data) with data from less exhaustive sources. This should help identify similarities and differences between outcomes and reveal key indicators.

The collection of baseline diversity data will also need to be converted into a format that allows us to visualize diversity across landscapes. This is analogous to other mapping problems, such as soil mapping, where investigators take many samples and interpolate between points. To do this requires significant data, but since data from different sources (e.g., gene banks and interviews) can be combined, this restriction should be able to be manageable.

Thus, we recommend that a massive attempt be undertaken to quantify and document *in situ* diversity and associated knowledge. Establishing a monitoring network or group with interest in this baseline is an important first step. Launching such an initiative could inspire new relationships with funding agencies that value and benefit from this knowledge, while providing feedback to the agrobiodiversity community on the success of different strategies.

Conclusions

To discern a baseline of information about the diversity of agrobiodiversity and associated traditional ecological knowledge, we offer the following research agenda to address existing gaps in knowledge.

1. Further develop community-wide databases for agrobiodiversity:
   - Utilize standards for metrics to ensure data interoperability and dataset aggregation.
   - Design systems for presenting data on different forms of agrobiodiversity.
   - Consider intra- and interspecific crop diversity—genetic, epigenetic, phenotypic, and functional dimensions—as well as knowledge associated with those forms of diversity (traditional ecological knowledge).
2. Develop country-level surveys of crop diversity that can be repeated over time and in other countries. Surveys should collect tissue samples (for genetic analysis, time-tagged DNA banks), local knowledge (names, adaptation, uses), and, if possible, seed (for experimental work). Attempts to estimate relative abundance of different landraces and modern varieties should also be made.
3. Determine where crop genetic diversity is being lost and gained in the field by tracking agrobiodiversity over time.
4. Discern how many named types and how much genetic and functional diversity arise from different sources (environment, knowledge, and culture) in crops and their wild relatives:

- Use comparative techniques to study patterns of diversity across distance, environments, societies (i.e., languages, cultures), and time to increase understanding of the important drivers of diversity.
- Determine the underlying processes by which the above sources of variation affect diversity.

5. Identify thresholds of diversity which, if exceeded, might constitute a crisis of diversity:
   - What parameters are important to monitor?
   - Are there "early warning signs"?
6. Discover the relationships between genomic diversity and diversity of phenotypic and functional traits of landraces.
7. Develop a methodology to identify biological entities of cultivated agrobiodiversity (i.e., landraces):
   - Determine if genetics can be used after initial baseline documentation, rather than having to identify landraces repeatedly.
   - Discern whether landraces might be best identified with a mix of genetics, functional traits, and traditional ecological knowledge.

These data would allow us to (a) understand the status of agrobiodiversity at particular periods of time (whether it is managed *in situ* or *ex situ*), (b) discover patterns and drivers of agrobiodiversity, (c) assess changes and their impacts, and (d) model and project likely changes in the composition of agrobiodiversity and the effects of these changes (the fourth objective). Ultimately, these data are needed to shape strategies for management of agrobiodiversity, such as setting up an overarching network of observatories, development of *in situ* conservation areas, and creating lists of crops and places to prioritize for conservation due to threats to agrobiodiversity.

The Past Evolution of Functional Agrobiodiversity:
Ecologically, Nutritionally, and Climatically Relevant Traits

Agricultural systems evolve in response to natural (biotic and abiotic) factors and human management. Variation between crops and their wild relatives can help us understand the difference in selection pressures at work under wild and cultivated conditions. Increasing our understanding of the phenotypic traits that crops express will help elucidate the process of crop evolution and domestication. Milla et al. (2015) argue that this can be accomplished by studying the changes in plants' phenotypes and ecological interactions. The question thus arises: How might using an ecological lens help us understand how domestication has affected important functional traits of plants, and thereby ecosystem functionality? Such an approach can also inform studies of future evolutionary change (the fourth objective).

Yet social factors can play an important role in creating selection pressures that affect plant traits, so cultural and social norms become a potent evolutionary force. In many regions, there has likely been a coevolution of cultural preferences and crop traits. In the Andes, for instance, it appears that cultural preference has resulted in people selecting similarly for colors of potatoes and colors of threads for weaving. Thus, although crop domestication and diversification are usually described in a biological or genetic sense, it would be of interest to study crop domestication also as a social process.

Indeed, crop domestication has led to fundamental social change beyond the genetic and phenotypic changes that characterize it. The origin of agriculture, while revolutionary, has been a slow, continuous process (e.g., Purugganan and Fuller 2009) that remains observable in hunting and gathering societies that adopt agriculture. Farming gets embedded into preexistent socioeconomic systems that favor continuity, not rupture (Leclerc 2012), so farming adapts to other activities, as with the Pygmy foraging peoples in Central Africa.

Diversity Provides Ecosystem Functionality and Ecosystem Services

Within this historical and present-day perspective on domestication and the emergence of crop assemblages within agroecosystems, we seek to understand the myriad functional roles that species-level and varietal-level diversity play in the functioning of agroecosystems and the wider ecosystems that surround them (see also Chapter 4). Functional roles and functional agrobiodiversity begin at the level of diversity in phenotypic traits, which have a genetic basis (Figure 2.2 below). These traits include

- plant morphology, which contributes to ordering the architecture of plant canopies and rooted soil zones;
- physiologically mediated traits, such as pollen and nectar provision and root exudates that maintain soil bacteria and fungi;
- taste and nutrition components that contribute to food provision and human nutrition; and
- particular functional traits, such as the ability to host nitrogen-fixing bacteria.

In addition, as analyzed elsewhere (see Chapter 4, Figure 4.2), assemblages of phenotypic traits may become functionally important to interactions among agroecosystem biota. For instance, there may be variability in disease resistance or in antagonism to pathogens that interrupts disease cycles and confers resistance to the species and varietal assemblage.

Phenotypic traits may be differentially functional to different actors (e.g., pollinators, ruminant grazers, rhizosphere bacteria; Figure 2.2). In addition, the use of "function" in two-way interactions between crops and associated agrobiodiversity, for example, is different from the complex food webs that drive ecosystem processes (e.g., primary production or cycling of nutrients)

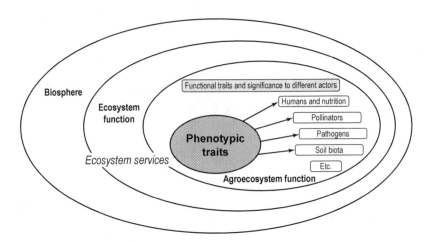

Figure 2.2 Summary of the relationship between functional phenotypic traits, agroecosystem function, broader ecosystem functioning, and ecosystem services. As depicted, functional phenotypic traits impact different actors (e.g., humans, pollinators, pathogens, soil biota) in various ways producing functions which convey services.

that are better conceived at a community level than at a species-to-species level. When taken at this ecosystem level, functional traits become important in providing ecosystem services. For example, the use of varietal multilines or diverse wheat cultivars in one field can impart functional diversity to provide disease suppression (a more binary crop–microbe interaction; Garrett and Mundt 1999) or control pest populations (Tooker and Frank 2012), which contributes to the wider service of enhanced provisioning for human communities.

In a similar way, a distinction can be drawn between *agroecosystem* function (i.e., fields and farms specific to human management for agriculture and food or fiber provisioning) and *ecosystem* function (i.e., agroecosystem components and landscape-level natural ecosystem components that are less influenced by human action) (Hooper et al. 2005) which operate at different scales (Figure 2.2). Pollination of crops by pollinators, for instance, may provide seed set in the agroecosystem, while also provisioning the same service to wild species.

At the largest scales, the linkage to ecosystem function is especially clear: agroecosystems in aggregate along with their natural landscape matrix contribute to regional ecosystems and biosphere functioning (e.g., effects on natural forests, waterways, and the global climate system; Figure 2.2). In reference to the concept of agrobiodiversity observatories discussed earlier, it should be stressed that observatories and related efforts to understand drivers of agrobiodiversity at a fine-grained level could include the opportunity to study the interactions of agrobiodiversity with agroecosystem and ecosystem functioning, which is a key part of the evidence base around the historical and present development of agrobiodiversity.

Past Evolution

There are a number of examples of functional changes in crops with domestication, some of which are enumerated in Milla et al. (2015). Root system characteristics, strength of plant–microbe interactions, phenology of harvest, and nutrient content have all been shown to have shifted with domestication. All this intraspecific variation can be affected by selection, drift, or gene flow (naturally and human-mediated), which can affect the frequencies of particular alleles controlling functional traits across time, space, and culture. The impetus for selection can be social, as in the selection for fruit shape that has more use value or exchange value (Jardón-Barbolla 2015), or natural, as in the selection for drought tolerance. Thus, changes in phenotypes with domestication or other periods of crop evolution will be reflected in the genome. We might expect genetic change that would distinguish crops from their wild relatives (e.g., seed size), affect nutritional traits (e.g., nutrient content), and control environmental tolerances (e.g., heat tolerance) that have evolved through the process of spread.

Crop Wild Relatives

The process of domestication is considered to be inherently about evolving crop wild relatives into cultivated species. While the process is likely gradual in most crops (Purugganan and Fuller 2009), related taxa can include various ancestors of the crop, the direct wild progenitor of the crop as well as other distinct related species. The phenotypic or genetic dividing line between the crop and its wild relatives can be fuzzy (Barnaud et al. 2009). Volunteer crops (i.e., crops which, although harvested, are not planted, and can act as weeds or be left standing) and gene flow between crops and wild relatives can play important roles in agroecosystems. In the Central Peruvian Andes, "k'ita" crops are those that have "escaped" or become feral. With potato, one type falls between wild and cultivated and is classified by taxonomists as cultivated; though weedy, it can be harvested and cooked and is sometimes planted (de Haan et al. 2007, 2012b). All over the world, wild plants that grow in and around fields are harvested as vegetables. In parts of Kenya, for example, traditional vegetables may grow wild in the forest, but are left standing when they appear in a field plot. Interestingly, seed systems are developing around these vegetables, indicating a move from wild to cultivated status. In Mexico, there are wild amaranths and chenopods; some of these "wild" plants can be relic cultigens that were once domesticated (Williams 1993; Williams and Hernández-Xolocotzi 1996).

Wild species can also be thrust into cultivation (i.e., contemporary domestication), often as a result of shifting market demands. Increased demand for novel or fashionable crops has led to expanded cultivation and intensification of species taken from the wild, as in sacha-inchi (*Plukenetia volubilis*) and camu-camu (*Myrciaria dubia*). In the United States, some biofuel or bioproduct crops are actually cultivated wild species. There is considerable human cultural knowledge about the use and management of crop wild relatives and wild or weedy collected species as well as other cultivated and noncultivated agrobiodiversity. Still, we do not know much about how that knowledge evolved and what it encompasses.

Nutrient Content

Domestication and breeding tend to change nutrient content in crops. Longitudinal studies generally show that micronutrient density has decreased over time (e.g., Scott et al. 2006). In grains, this is because the emphasis in breeding has been on weight and led to larger seeds (more endosperm), whereas micronutrient concentrations are often highest in seed coat and embryo tissues. There are, however, cases where some nutrients have increased. Barbeau and Hilu (1993) documented lower iron content, but variable differences in amino acid content between finger millet varieties and their wild relatives. There is an increased interest in breeding for nutrition, as seen in efforts to increase beta-carotene, a precursor of vitamin A, in rice and sweet potato (Low et al. 2007). High-glucosinolate broccoli is being bred using a crop wild relative in response to demand for more nutrient-dense foods by affluent consumers (Sarikamis et al. 2006). Given that some plants are used medicinally (e.g., greens reduce anemia), conscious selection may account for higher nutrient content with domestication or upon further improvement.

On the other hand, selection for nutrient content may be a fortuitous by-product of selection on something else (e.g., color or seed size and thus endosperm to embryo ratios). There may have been more direct selection against antinutritive compounds, which ultimately would increase bioavailability. Selecting for yield or sweetness may ultimately select against more complex compounds or other carbon sinks (e.g., exudates that benefit symbionts, protein content in maize grain) that become too costly. Human management of domesticates may have reduced the plant investment in defense compounds. Thus, greater sweetness or total calories selected for by humans may reduce a crop's nutritional values as well as a crop's defense against pests, tolerance of environmental conditions, or mutualistic interactions. Within the same crop, however, farmers may be interested in different parts (e.g., leaves and tubers), which may provide different nutritional components.

Environmental Adaptation

The process of domestication can be followed by, or be contemporaneous with, short- and long-distance dispersal of crops into new regions. Such geographic expansion results in the crop encountering new biotic and abiotic environments (e.g., different day lengths, temperatures, diseases, and insects). These new conditions can confer novel selection pressures that select for adaptation.

The alleles that confer these adaptations may be sourced from standing variation, new mutations, or via gene flow with local crop wild relatives. For instance, maize, once domesticated from teosinte at low elevations in southern Mexico, required novel adaptations to grow at higher elevations. It appears to have acquired these adaptations through gene flow with another teosinte species that grew under these cooler highland conditions (van Heerwaarden et al. 2011). Understanding the degree of differentiation (genetic, phenotypic) among populations needs to improve if we are to assess the adaptive nature of that differentiation (Mercer et al. 2008).

The incorporation of experimental research methods, such as common garden and reciprocal transplant approaches, as well as evolutionary participatory breeding, can be used to understand the differentiation and adaptation of populations (Enjalbert and Johnson 2011; Mercer et al. 2008; Orozco-Ramirez et al. 2017; Zimmerer 1991b). Since adaptive alleles may move around with pollen or as part of seed exchange, research can explore the implications of exchange on crop adaptation across space and time and among societies (e.g., Bellon et al. 2011; Mwongera et al. 2014; Violon et al. 2016).

Diversity at Different Levels of Organization

Although we have primarily highlighted diversity within species, diversity of species assemblages and across the landscape are also important and can provide ecosystem functions and services. One inductive way to analyze agrobiodiversity from a functional perspective is to study how species are combined in cropping systems, farming systems, landscapes, and cuisines. For example, flavor networks study how ingredients (most of them plant-based) are combined in dishes (Ahn et al. 2011). These networks allow us to explore whether ingredients have complementary roles in food traditions. Different cuisines have different networks, and it may be possible to trace these networks over time. In ecology, networks of species co-occurrence are informative about spatial patterns and change over time (e.g., in response to climate change) (Araújo et al. 2011). Economic studies use "product space" networks which map pairs of products produced in the same country (Hidalgo et al. 2007). Such networks can be used to analyze the complexity of economies. Network complexity is highly indicative of a country's economic development because sets of underlying skills, knowledge, and exchange shape these networks and are demanded for their management. Agrobiodiversity could be analyzed through these different lenses with similar methods to analyze functional relations.

Balanced diets are often intuitively arrived at by humans with access to healthy foods, so finding healthy assemblages in national cuisines is common. Finding a species that provides a particular nutrient to combine with others (e.g., vitamin A) is easier than finding (or engineering) higher-nutrient variants within species. Thus, have we evolved with our diet, and has our diet evolved as well with us? Using wheat and milk as examples, humans have evolved mutations that allow some of us to digest these foods after they became potential food sources. Considering changes in dietary diversity that accompanied domestication, can a reduction in food diversity trigger disease (Larsen 2006)? Domestication of a major food source certainly could have initially shifted (Richards et al. 2003) or narrowed local diets as domesticated species displaced gathered species.

Across a landscape, there may be diversity among its components, which can allow for functional crop diversity across space and time. Such associations among landscape and functional diversity may be especially rich in areas with complex environmental gradients and mosaics of environments. We found it useful to imagine a way to compare the types and amounts of functional diversity found in divergent agroecosystems (i.e., in different areas and in different eras), based on their level of organization of diversity. Initial examples comparing divergent systems are listed in Appendix 2.1 and show that some landscapes provide perplexing contradictions. Take, for instance, the case of maize and soybean farming in the United States (Appendix 2.1, example 2): landscapes have a low crop species-level diversity, minimal associations exist with noncultivated agrobiodiversity, and few cultural and ecological services are provided, despite problematic ecological externalities (i.e., hypoxic zone in the Gulf of Mexico).

Over the past 150 years farm size has ballooned (causing large cultural change), but aggregate (state-level) crop diversity has not changed much (Hijmans et al. 2016). The system has maintained a very high level of productivity, although more diverse systems could also be highly productive and profitable (Davis et al. 2012). Perhaps genetic variation provided by frequent varietal change over time (instead of space), along with high inputs, have made this possible. Swift et al. (2004) provide an analysis of some of these issues, in particular the way that management elements can substitute for agrobiodiversity services at a field level, while not denying the importance of agrobiodiversity at landscape scales. González-Esquivel et al. (2015) frame an analysis of agrobiodiversity services in smallholder-managed landscapes in terms of trade-offs between livelihood and ecosystem benefits. It would be interesting to model how combinations of changes in agrobiodiversity, inputs, and landscape structure affect agroecosystem functions. Finally, some relatively uniform agricultural landscapes are dotted with islands of agrobiodiversity in the form of home gardens. In these areas, there is a great increase in the number of crops per area and not all species present are typical cultivated crops; some may be medicinal or spice plants collected from the wild or left standing as adventitious plants in fields (e.g., among the Mapuche in Chile).

Conclusions

To extend research into the evolution of functional diversity, we recommend the following research agenda:

1. The effects of domestication, spread, and genetic improvement on crop ecology need to be explored to
   - determine whether and how these processes have affected the crop's potential to interact with associated agrobiodiversity;
   - explore the implications for other ecologically, nutritionally, or culturally related traits;
   - discern the roles genetics, environment, management, and culture play in the evolution of functionally diverse traits in crop plants; and
   - examine how the emphasis on the social nature of domestication may change the questions posed.
2. We need to determine how functions can change across situations (space, time, genetic variation, society, and environment) by
   - exploring how one agrobiodiversity system may transform into another over time, or diffuse and adapt spatially into a new region;
   - discerning the roles of traits in affecting different types of services: provisioning (e.g., food), regulating (e.g., evolutionary adaptation), functional (pest and disease control), and cultural services (e.g., cuisine, identity);
   - clarifying different characteristics of the selection or adaptation process;
   - exploring how the selection pressures that farmers initially imposed may have facilitated sociotechnical flexibility and/or preservation of other activities;
   - investigating, from an evolutionary perspective, the degree to which agrobiodiversity historically facilitated adaptation; and
   - identifying typical patterns of crop adaptation and the breadth of adaptation (narrow, broad).
3. Particular relationships between agrobiodiversity and agroecosystem functioning need to be analyzed
   - to identify metrics that best describe or quantify this functioning and the services it provides, especially those that contribute to human well-being;
   - to determine whether increasing diversity affects functions;
   - to explore which ecosystem services are affected by variation within and among varieties as well as within and among species; and
   - to identify the point at which function declines to the point of crisis.
4. Functions are perceived to be valuable when they provide a service that benefits us. While some elements are valued by themselves (i.e., one

can value biodiversity as such), a better understanding of how particular traits provide ecosystem services would be helpful.

5. We need to determine whether adaptation of crops and their wild relatives occurs even without gene flow between them. This requires
    - measuring the capacity of crops and their wild relatives to adapt, and
    - identifying the stressor factors to which they adapt.
6. The roles crop wild relatives can play need to be discerned, not only in plant breeding, but in the *in situ* functioning of agroecosystems. This requires
    - measuring the effects of ongoing gene flow between wild and cultivated types (in both directions),
    - understanding adaptive introgression (in light of global change) as well as the problems such gene flow can pose to farmers, and
    - exploring how local people differentiate between wild or cultivated (i.e., investigate the local knowledge that mediates cocultivation of, and gene flow between, crops and their wild relatives).
7. The process of crop and wild evolution needs to be studied from both biological and social perspectives to determine how continued gene flow results in benefits or detriments to crop populations.
8. We need to understand how the functional diversity found in our crops has evolved and how that diversity interacts with distinct environments and components of associated agrobiodiversity.

Unfortunately, functional diversity and the services provided are often not valued or may be invisible. Nevertheless, a better understanding of this diversity should improve our ability to utilize or enhance functional diversity through breeding or to deploy diversity to increase function, and thereby services, within agroecosystems and more broadly.

Drivers and Effects of Interactions with Associated Agrobiodiversity

Here we expand our discussion of agrobiodiversity beyond crops to encompass weeds and wild vegetation, pollinators, and soil biota (for surveys of these different aspects of associated agrobiodiversity, see Bretagnolle and Gaba 2015; Brussaard et al. 2007; Klein et al. 2007). The interactions of crops with other organisms are strongly influenced by a spatial mosaic of varying natural and human drivers. Understanding these drivers and their relationships is an essential prerequisite to understanding how crop diversity or function might change with agroecosystem change and leveraging their interaction for increasing sustainability. Drivers of associated agrobiodiversity may be broadly categorized into (a) biogeographical patterns of vegetation, pollinators and other arthropods, and soil biota or (b) human management elements, including patterns of land use, agricultural input use, and crop and livestock choice. Within these two categories, the role of interspecific and intraspecific (varietal) diversity and their effects on associated agrobiodiversity as well as the idea that low agrobiodiversity-intensified systems may be replacing the functional role of associated agrobiodiversity (e.g., crop and pasture residues, manure, beneficial insects, N fixation, and phosphorus solubilization by microbes and plant roots) with exogenous inputs (fertilizer, chemically based pest management) are particularly important.

In our discussions, we were unable to adequately consider livestock breed diversity as a component of agrobiodiversity. Nevertheless, it is evident that the domestication and keeping of livestock and fish in many types of mixed farming systems (crop, livestock, and fish) is a strong driver of crop choice across pasture–crop mosaic landscapes. Livestock may have an important influence in motivating the growing of crops and crop varieties that feature leafy and stem biomass as a forage source, versus accentuating only the quality or quantity of seeds, tubers, or fruit product. In addition, perennial and long-season species and varieties with forage uses tend to contribute most organic residues to soils, such as perennial pasture grasses and legumes as well as shrubby grain legumes (Snapp et al. 2010). Species-diverse herds may play a role in promoting stable, diverse plant species mixtures in pastures and field margins (Rook et al. 2004). In Southeast Asia, the use of fish ponds (frequently with local species) are commonly integrated with pig and duck rearing (above the pond) and nutrition-enriched irrigation to rice crops (Little and Edwards 2003).

Kinds of Associated Diversity

Associated agrobiodiversity includes many of the noncultivated species that play roles in agroecosystems through interactions with the cultivated species (see also Chapter 4). Some may promote functions and services. Many associated species are wild, such as pollinators, microbes, pathogens, insects, crop wild relatives, and weeds, whereas others can be somewhat managed. For instance, pollinators can be wild and unmanaged, wild and semimanaged (i.e., if sources of nectar are managed to encourage them), or fully managed (e.g., honeybees whose hives are moved from field to field).

Some unexpected forms of associated agrobiodiversity found in diverse systems can play important roles in ecosystem function. In seasonally flooded savannas in Zambia, for instance, termite mounds are converted by farmers into raised fields around which fish drop feces, creating beneficial growth environments. Another example involves shifts in banana-associated beneficial bacterial communities in legume-based agroforestry systems (Köberl et al. 2015). Agroecosystems, generally low in wildlife compared to neighboring

K. L. Mercer et al.

natural areas, can be managed to create relatively more favorable environments for wildlife. For instance, in California, flooded rice fields can provide a desirable environment for waterfowl in winter and for young salmon during the spring, and in the Netherlands, the timing of mowing is regulated to protect ground-nesting birds.

Social Knowledge and Documentation

Associated agrobiodiversity is an important component of a functioning agroecosystem, and farmers possess a large amount of knowledge (or traditional

	Plant 1	Plant 2	Plant 3	Plant 4	Plant 5	Animal 1	Animal 2	Animal 3	Animal 4	Animal 5
Plant 1		X	X	–	–	X	–	–	–	X
Plant 2										
Plant 3										
Plant 4										
Plant 5										
Animal 1										
Animal 2										
Animal 3										
Animal 4										
Animal 5										

Figure 2.3 Schematic showing how local knowledge of interactions between crop and associated agrobiodiversity can be identified. For a single, cohesive agrobiodiversity setting (e.g., region, cultural group, major farming system), the interaction between a pair of species can be evaluated. For a given pair of organisms (e.g., Plant 1 and Animal 5, see dotted circle), X indicates an interaction, while – denotes no interaction. If the two species interact, details and the direction of the interaction (positive or negative) could also be included. Interactions above and below the diagonal line for the same pair of organisms denote opposite directionality: Plant 1 affecting Animal 5 versus Animal 5 affecting Plant 1. Here we assess interactions involving "plants" and "animals," but any agrobiodiversity components (e.g., soil biota, major pests and pathogens, and arthropods) could be included.

ecological knowledge) about these components (e.g., Cerdán et al. 2012; Pestalozzi 2000; Sileshi et al. 2009; Sissoko et al. 2008). It is possible to represent the presence of interactions between components of agrobiodiversity or knowledge about them (Figure 2.3), clarifying pairs of species most frequently linked by farmers or across the landscape. Information from individual farms, farming systems, environments, or social groups can be of interest, as can comparisons between data generated in different areas or among different groups of farmers or cultural groups (as done by Atran et al. 1999 in the Maya lowlands). Network analysis can also be used to assess these patterns.

These studies are important because it is not always clear how much knowledge farmers have of biodiversity near or outside their field. Some associations may be more obvious or important, while others remain more obscure. In addition, knowledge bases may be affected by different drivers, which would produce differences across farmers' contexts and geographies for amounts of knowledge. Some of the same drivers of diversity uncovered in the first objective and modeled in the fifth may be important here in driving losses or gains in knowledge about diversity. For instance, deagrarianization (i.e., processes by which society moves away from an agrarian mode, e.g., through aging or attrition) and part-time farming may reduce knowledge over time if the density of networks and the usefulness of agrobiodiversity-related knowledge decrease. Even in smallholder systems that are currently maintaining or increasing management intensity, changing patterns of coupling to regional versus global markets can alter the state of agrobiodiversity-related knowledge among smallholders (Zimmerer and Vanek 2016). Some types of associated diversity, like that of microorganisms or insects, may only be appreciated at the level of local knowledge of outcomes (e.g., more stable production) rather than the underlying processes (e.g., symbiotic relationship with nitrogen-fixing bacteria).

Effects on Interactions of Genetics, Environment, Management, and Culture

Many factors can affect the strength and type of associations among agrobiodiversity and knowledge about those associations. Genetics of the crops or associated agrobiodiversity (G), the environmental conditions (biotic and abiotic factors; E), farmer management (M), and culture (C) can all affect agrobiodiversity associations as well as the attending knowledge. This understanding of the interacting effects of genetics, environment, and management/culture (G × E × M/C) could produce value for various types of farming within and outside crop centers of diversity. Some varieties of millet and maize, for example, produce different root exudates (quantity, quality), which can affect associated mutualisms (Li et al. 2016) or soil communities. Since particular genes are associated with exudate production, the crop may manipulate its associations, depending on the conditions (stress, acidity) (Haichar et al. 2008).

It is possible that interactions between cultivated and associated biota may shift over the processes of domestication and spread of a particular crop; crop wild relatives may also have their own associations with other organisms. Genetic and environmental changes that accompanied domestication and improvement may affect interactions in terms of the actors, strength, and type of relationships as well as functional effect. Many pest-resistant genes utilized in breeding come from centers of origin where the pest coevolved with the crop or the crop wild relatives; biocontrol candidate identification uses a similar logic. The invasive plant literature may be similarly useful to increase understanding of how ecological interactions shift with the spread of species to new areas. The directionality of change can either benefit (e.g., by pathogen release) or disadvantage (e.g., by less effective mutualisms) the plant species. In an example of the former, survival and biomass production of Chinese tallow (*Triadica sebifera*, Euphorbiaceae) were affected negatively, or not at all, by rhizosphere biota from their native range; however, rhizosphere biota from areas in North America, where this species is an exotic invasive, affected it positively (Coats and Rumpho 2014; Yang et al. 2013). Do crops similarly respond differentially to soil biota of their native and introduced ranges?

Changes in interaction between crops (compared to their wild relatives) and associated diversity are driven by diverse factors. Genetic changes may affect the plant's ability to interact or the intensity of its participation in an interaction, for example, by reducing energy donated to a mutualist. Genetic changes in interaction strength may have been caused by domestication or subsequent diversification. For instance, we might expect landraces to have greater positive interactions and fewer negative interactions with associated species in the agroecosystem where they originated, compared to improved varieties. Changes in environment due to domestication (i.e., going from wild context to cultivated field), with subsequent spread outside of its area of origin, or when going from low- to high-input systems, could all affect interactions. Thus, interactions may not have the same potential everywhere: genetics, environment, and history matter.

Some interactions may be more likely to be maintained during a crop's human-mediated range expansion than others. Legume-associated nitrogen-fixing rhizobia strains can spread and become established in far-away places, unless environmental conditions (e.g., pH) are prohibitive. In Europe, rhizobia strains introduced 30 years ago can still be found, so perhaps rhizobial associations are relatively easy to maintain with legume crop spread.

Pollinators, by contrast, are harder to maintain with spread, and this affects crop success (Garibaldi et al. 2011). In Malaysian plantations, beetles from the oil palm's native range in Africa needed to be introduced to ensure natural pollination (Dhileepan 1994). Similarly, when *Vanilla* is cultivated outside its region of origin, native pollination does not occur and crop success depends on manual pollination (Lubinsky et al. 2006). In both cases, the introduction of crops to new areas created the need for suitable pollinators: new pollinator communities (sometimes involving humans) were required to ensure production. Yet it is important to note that movement of pollinators can affect local ecosystems and communities. For instance, pollinator communities for cucurbits tend to be diverse in the United States and include species that have expanded their range with the crop, sometimes to the detriment of other local pollinator species.

Associations between crops and their wild relatives (e.g., gene flow) have often been shown to be maintained with domestication; that is, many crop–wild systems experience gene flow in the areas of crop origin: rice in Asia, sunflower in the United States, and maize in Mexico (Ellstrand 2003). That association, however, cannot be maintained in areas where there are no wild relatives to assist in reproduction, unless a wild relative spreads together with the crop.

Wild relatives may spread with crops if their seeds are difficult to distinguish and are often planted together, which can also make them a difficult to manage weed. In other cases, a wild relative may itself be planted for a different use. For example, the tree *Manihot glaziovii*, a wild relative of cassava (*M. esculenta*) and interfertile with that crop, was widely planted throughout Africa in an attempt to produce rubber. Although these attempts failed, the tree became naturalized and is planted widely as an ornamental. Hybrids between *M. glaziovii* and cassava are frequent and were used to breed cassava varieties resistant to important viral pathogens (Beeching et al. 1993; Legg et al. 2014). Alternatively, the crop and the wild relative may arrive in new ranges by different means. For instance, sunflower (*Helianthus annuus*) in Argentina is grown as a crop. Recently, wild sunflower (also *H. annuus*) has become an invasive and was likely introduced with forages (Ureta et al. 2008). In areas where gene flow between wild relatives and the crop is lost, an important source of alleles for evolution in the crop may also disappear.

Some crops may have mating systems (or may evolve traits during or subsequent to domestication) that negate or diminish the fitness consequences of decoupling of pollinators and their crops with expansion into new ranges. Wind-pollinated, self-pollinating, apomictic crops, or vegetatively propagated crops do not require pollinators, and this may have facilitated their spread (Garibaldi et al. 2011). Banana, cassava, yam, and potato are examples of clonal species that have spread, but many have had disease epidemics associated with genetic uniformity in their introduced ranges. Clonally propagated crops can benefit from the surges of diversity that come with outcrossing (McKey et al. 2010b, 2012). For many crops (e.g., almonds, tomatoes), managed pollinators are moved around to facilitate pollination, compensating for variation in the composition and size of local pollinator communities.

How these interactions play out on the landscape and provide ecological and social services has become an important area of study in agroecology. Thus, the ways that landscape-scale diversity affects ecosystem functions are important. There are different examples of this. One is the management of shade coffee plantations to provide overwintering sites for birds (i.e., bird-friendly coffee). Pest- and disease-suppressive landscapes provide another example (Bianchi et al. 2006). The uniform timing of rice planting in Bali and Vietnam, the use of sectoral fallowing in the Central Andes (Parsa 2010; Parsa et al. 2011) or of sorghum in Africa to protect seedlings from bird predation amounts to group pest management. Pollination rates by wild insects can be higher in more diverse landscapes (Kremen et al. 2002).

Clearly there could be value for communities in maintaining a diverse landscape. Great variation may be found in knowledge of crop associations as well as in how to manage or encourage such associations. Some may see advantages or disadvantages in emphasizing the associations and may find replacement services (e.g., replacing natural pollination with managed pollination) to be a more cost-effective practice. This would likely depend on the particular ecosystem function, social and political pressures, and various externalities. Diverse landscapes may also be appreciated for their cultural value, such as the provisioning of fishing and hunting spots (another form of biodiversity within agricultural landscapes).

However, changes in associations can impact crop performance to varying degrees. Farmer management may be able to affect interactions, thereby turning knowledge of associations (Figure 2.3) into a management tool. Examples of this include the application of rhizobia to increase nitrogen fixation of legume crops or the management of landscape diversity to stimulate the pollinator community. At the field scale, management of ecosystem functions related to associations may prove easier than acting at the landscape scale (e.g., with pest-suppressive landscapes), because farmers can manage their fields individually. In the future, we need to consider other factors that affect the strengths of associations or the responsiveness of crops to changes in associations, such as environmental conditions, natural systems versus cultivated ones, amount of time in cultivation, and the type and intensity of management.

Conclusions

Given our limited understanding of what drives associations across evolutionary time and ecological space, we suggest the following research agenda:

1. Assess how interactions between crops and associated agrobiodiversity impact the productivity, function, and capacity of agroecosystems to provide services (including health benefits). This requires that we

- identify the associations that have the greatest positive effect on production,
- explore the degree to which farmers possess knowledge about different aspects of the associated agrobiodiversity they employ,
- discern the cultural services provided by associated agrobiodiversity,
- determine the degree to which particular crop-associated agrobiodiversity might be important for overall modeling of diversity of agrobiodiversity, and
- clarify how data for research in this area should be gathered and structured.

2. Understand the interacting effects that genetics, environment, and management/culture (G × E × M/C) have on agrobiodiversity. We need to
   - determine the degree and processes by which domestication, breeding, management, and geography affect how crops are associated with their wild relatives, other plants, pollinators, and microbial communities;
   - identify examples of farmer management that affect agrobiodiversity interactions (e.g., between the crop and the microbial community), positively and negatively;
   - determine how different crop–agrobiodiversity associations, including microbiomes, are within crop centers of origin and have (or not) spread to distant locations;
   - discern the ways that crop spread influences the spread of associated diversity and effects on local agrobiodiversity and wild biodiversity in the extended range.

3. Determine the degree to which the amount of diversity that farmers use reflects the amount of diversity in their environment.

4. Discern the degree to which contemporary, ongoing gene flow and introgression between crops and crop wild relatives affect adaptation of crops (but also of crop wild relatives) to biotic and abiotic factors.

5. Investigate, more fully, inter- and intraspecific crop diversity and its relation with soil diversity.

With this knowledge, we can come to better understand the complex web of interactions at work among components of agrobiodiversity (and associated traditional ecological knowledge), how that web is affected by various factors, and the ecosystem services these interactions provide. With this additional knowledge, we can promote the importance of associated agrobiodiversity and its services to different actors (farmers, NGOs, breeders, communities, gene banks, funders) that are, or could be, engaged in considering associated agrobiodiversity. Eventually, it may be possible to exploit interactions by managing or enhancing them, but it must be undertaken with care, due to the potential harm done by introducing species (following the precautionary principle).[1]

Redefine *In Situ*, Its Connections to *Ex Situ*, and Its Role in Maintaining or Increasing Agrobiodiversity

Many have argued that *in situ* and *ex situ* agrobiodiversity management systems are complementary and should benefit from such a linkage. Yet the end users of, rationale for, and actors involved in or affected by *in situ* and *ex situ* management systems are quite different: for example, farmers versus breeders, direct versus long-term benefits, evolution versus static preservation, and civil society versus state-linked networks. While some biological entities or sources of knowledge can only be conserved *ex situ* (e.g., DNA banks, sequence databases), the same is true for *in situ* (e.g., place-based traditional ecological knowledge[2]). *In situ* management is predominantly shaped by informal processes such as farmer seed systems and informal markets. In contrast to *ex situ*, conservation of diversity is not a prime objective for *in situ* management by farmers; but it can be an emergent property. The term *in situ* conservation can therefore be a bit of a misnomer and has been perceived by formal and state-linked actors to be less reliable and accessible compared to *ex situ* conservation, which can occur in highly formalized institutional contexts with registers and documentation systems for collections.

The crop genetic resources community has raised the question several times if, and how, the *in situ* and *ex situ* systems could be more connected. The underlying rationale for connecting the two systems commonly includes: intelligence (Elzinga et al. 2001), coverage (Castañeda-Álvarez et al. 2016), and pressure to show use of accessions (Fowler and Hodgkin 2004), among other factors.

Use of intelligence has resulted in important opportunities for crop conservation. Intelligence refers to the capacity of *in situ* monitoring initiatives to "take the pulse" of on-farm conserved landrace stocks and observe shifts in relative abundance, conservation status, or spatial distribution. Intelligence can determine which landraces are apparently extinct in *in situ* contexts, or present *in situ* and conserved in *ex situ* collections, or present in farmers' fields yet not represented in gene banks. This latter information can be used for targeted additions to be made to *in situ* collections. Conversely, when *in situ* diversity is lost and local stakeholders such as farmers or village authorities request

---

[1] Care should be taken in deploying associations about which little is known and in assuming that intervention or disruption of agroecosystems (e.g., by introducing new genetic material or mutualists) is better than letting them function as is. Similarly, no- or low-cost improvements may be better than complex ones. We can only know what effects we are having once we know more.

[2] There are many exceptions: through its documentation, traditional ecological knowledge can be partially catalogued *ex situ* as outlined above for objective 1.

the reintroduction of landraces, then repatriation may be an effective linkage mechanism (Huaman et al. 2000). Unfortunately, repatriation is often carried out blindly without solid knowledge about past diversity, farmer demand, or evidence of loss. Repatriation also assumes that on-farm crop diversity is static rather than dynamic. While connections between *in situ* and *ex situ* management of agrobiodiversity exist in theory, the practice of building systematic linkages is rarely realized. Conservation of crop wild relatives can also benefit from these *in situ–ex situ* linkages, as collection location data from gene banks can be used to prioritize *in situ* conservation and collection concerns (Castañeda-Álvarez et al. 2015).

Understanding the intraspecific classification and nomenclatural system of farmers is essential to interpret on-farm management and thereby to enhance *in situ* conservation of knowledge together with genetic resources. If farmer crop selection is based on prerequisite morphological "mental images" with a particular place in the classification system, as suggested by Boster (1985), farmer categories may exist in farmers' minds before they exist in their fields, though farmers may also create a mental image once a variant occurs in their field. It supposes that there is a cultural consensus on how crops are classified and named in such a way that both the transmission of knowledge over generations and the communication between farmers are ensured. There is thus an immaterial knowledge component that cannot be stored in an *ex situ* gene bank and needs to be considered in *in situ* projects.

What Is Currently Encompassed Within *In Situ* Systems?

In general, two types of *in situ* management of crop diversity can be distinguished (Brush 2000). First, in its most natural form, it entails the continued use and evolution of landraces in farmers' hands ("*in situ* use"). In this case, there is no outside intervention—just farmers farming. This is the case in most farmer communities that are not obtaining seeds from large seed companies through the formal market. This recognizes that on-farm management is both an historical and ongoing process and largely autonomously driven by farmers. In this case associated scientific research may monitor diversity without actively intervening in the gene pool. In essence such programs aim to understand conservation dynamics (loss, conservation, enrichment) and their drivers.

A second type of *in situ* management involves interventions to actively pursue the conservation of landraces by supporting farmers and their communities ("*in situ* conservation"). A whole range of interventions exists and these may have trade-offs and also involve different visions of development (Table 2.1). Ideally, before any intervention takes place, a project will critically examine the initial agrobiodiversity and knowledge present. Such baseline documentation is required to determine priorities and measure outcomes and impacts of

Table 2.1 Types of *in situ* management.

	Autonomous Use	Outside Intervention for Conservation
Role of Researchers	Passive: To observe farmers' own capacity to use diversity, conserve, and adapt	Active: To encourage farmers to conserve and adapt
Justification	Diversity is perceived as present and dynamic	Diversity is perceived as lost or threatened
Activities	Farmers farming with active seed networks (Coomes et al. 2015; Thiele 1999) Systematic monitoring (de Haan et al. 2016; Hunter and Heywood 2010) Cataloguing (Scurrah et al. 2013)	Community seed banks (Vernooy et al. 2015) Biodiversity seed fairs (Scurrah et al. 1999; Tapia and Rosas 1993) Payment for environmental services (Midler et al. 2015; Narloch et al. 2011b) Cultural reaffirmation (Apffel-Marglin 2002) Conservation education programs (Guitart et al. 2012) Value chains and markets (Keleman and Hellin 2009; Ordinola et al. 2007) Participatory breeding (Camacho-Henriquez et al. 2015; Ceccarelli 2009) Rewards for custodians (Gruberg et al. 2013; Sthapit et al. 2015) Park system (Argumedo 2008, 2012) Repatriation or introduction (Huaman et al. 2000)

interventions. Yet, commonly, *in situ* projects do not collect traceable metrics at the onset of interventions.

Active intervention programs differ widely in their portfolios. Some combine multiple interventions while others are predominantly based on a single type of intervention. Furthermore, some of the interventions seem mutually exclusive. For example, programs working on cultural reaffirmation rarely promote value chains and *vice versa*. Unfortunately, there is a lack of studies that document outcomes and trade-offs, or assess the impacts of interventions and how they are perceived by local stakeholders. How *in situ* conservation is promoted depends on different, sometimes conflicting development philosophies and on the perception of conservation itself: purist and frozen in time versus dynamic and changeable.

We propose a third type of *in situ* management, "*in situ* diversification." Emphasis on using crop diversity in farmer fields has been given to areas where those crops are, or historically were, diverse. For example, most *in situ* research on potatoes takes place in the Andes. However, if we believe that crop diversity is an important way to sustain farming systems, perhaps *in situ* diversification could be prioritized elsewhere. For instance, for many years only one potato variety ("Mira" originating from an East-German breeding program) was grown on a million hectares in southwest China. Varietal diversity has recently been expanded by the introduction of late blight resistant varieties and processing varieties. Expanding *in situ* diversity into such areas of low diversity could be beneficial. Yet such cases have not been recognized or studied.

Links between *In Situ* and *Ex Situ*: Intelligence to Inform Feedback Loops (Case 1)

Farmers throughout the world are managing diverse landrace populations, whether it concerns maize in Bolivia (Zimmerer 2013) or Mexico (Perales and Golicher 2014), sweet potato in Papua New Guinea (Roullier et al. 2013b) or rice in Laos (Schiller et al. 2006). Despite historical and continuous change and factors that may be perceived as having a negative effect on agrobiodiversity, smallholder farmers around the world find sufficient incentives to continue growing diverse sets of landraces (Brush 2004). Trust in farmers' ability to manage and adapt diversity, therefore, could arguably be greater than it generally is. Assuming that all diversity will inevitably be lost is a dated paradigm and there is a renewed scholarly interest to understand how diversity changes, adapts, and evolves under contemporary smallholder management in an ever changing world (de Haan et al. 2016; Dyer et al. 2014; Montesano et al. 2012).

An enhanced understanding of the conservation dynamics of crop genetic resources in the field, whether at the spatial, reproductive, genetic, or population level, holds considerable potential for continued feedback between *in situ* and *ex situ* conservation. This potential has not been realized and few examples exist, yet we argue that such studies provide relevant intelligence. An obvious example concerns the use of *ex situ* collection data to define *in situ* diversity hotspots, gaps, and collection priorities (Khoury et al. 2015b). The use of areas of high diversity as *in situ* observatories to document conservation dynamics and make regular comparisons with *ex situ* collections or to past reference data from the same area would allow for spotting new and lost diversity, or shifts in habitats as well as conducting red listing (Cadima Fuentes et al. 2017). Intelligence about the conservation status of species and landraces *in situ* could then in turn provide an evidence base for the actual potential to repatriate diversity from *ex situ* collections to original collection sites or climate analogue sites.

An additional function of modern gene banks and associated science communities could thus be one whereby the *ex situ* collection serves as a reference population for systematic comparison and is linked to a network of key *in situ* reference sites where temporal and spatial change processes are regularly documented to provide intelligence about unique uncovered genotypes, shifts in abundance, and eventually loss or enrichment of genetic diversity. Such a model links naturally with the first objective, as described above.

Links between *In Situ* and *Ex Situ*: Intervening via Repatriation and Introductions (Case 2)

Ex situ collections from national or international gene banks can be reintroduced into farming communities to replenish diversity used in the field. An example of this repatriation strategy was used in France, where sales of seeds between farmers is illegal but farmer-to-farmer exchange networks exist. Wheat seeds that had been kept in a seed bank at the French National Institute for Agricultural Research (INRA) since the 1940s were reintroduced and incorporated into farmers' portfolios of varieties (Thomas et al. 2012). Similarly, in a crisis environment, where drought in northern Mexico wiped out seed lots, farmers combined seed from existing fragmented populations and *ex situ* national gene bank collections to compensate for seed loss and genetic bottlenecks.

A temporal effect is introduced into the system with reintroduction of *ex situ* collections, since their use implies use of diversity acquired at a previous time (with past genetic structure), some of which may not be as useful upon repatriation as it was upon collection. However, the opposite may also be true. In the case of clonally reproducing crops, the presence (*in situ*) or absence (*ex situ*) of viruses can affect phenotypes to the point that genetically identical clones may look completely different depending on their disease status. Finally, although repatriation programs can be high-profile projects for gene banks, implementation should really be demand driven and focused on areas where loss is evident rather than due to pressure to show that the gene bank distributes its accessions.

Gene banks can also support *in situ* diversification (or "assisted migration") in areas where diversity may not have previously existed or that may require novel diversity to face novel challenges. Introduction of varieties other than that circulating on the landscape among farmers may provide new opportunities to better adapt to novel conditions. As the climate changes, for example, repatriation of past diversity may not be enough to provide the evolutionary potential that a crop will require to produce well as temperatures increase. A useful approach may be to add *ex situ* collections from different environments to current *in situ* diversity from an area. This mix of diversity may then recombine on the landscape and reassemble (through the processes of selection, drift, and gene flow mediated by the environment and farmer management) in ways that may prove adaptive (Mercer and Perales 2010).

While repatriation or novel introductions can allow diversity to rearrange on the landscape over time, these initiatives can also dovetail nicely with participatory plant breeding efforts, especially ones using evolutionary breeding approaches (Ceccarelli 2009; Ceccarelli et al. 2001; Dawson et al. 2008; Suneson 1969). Such approaches potentially allow for the functional incorporation of diversity into potentially genetically impoverished regions. With climate change, participatory plant breeding and varietal selection may be especially useful as it can introduce diversity from materials that are adapted to what may be considered "future environments" for a given location. For outcrossing species in areas with ample crop genetic diversity, such adaptation may be autonomous (Vigouroux et al. 2011b).

Key to the process of using *ex situ* collections in an *in situ* system is knowing something about the *ex situ* collections. Many collections are well documented, but, depending on the species, *ex situ* collections may vary in the quality of their documentation of location of origin and of relevant phenotypic information. For U.S. Department of Agriculture collections of crop wild relatives, 1/6 to 1/4 have no geographic coordinates. The same can be true for crop accessions. *Ex situ* descriptions also are often missing phenotypic (e.g., flowering time) and genotypic information—a lack that has been long bemoaned by breeders. However, some programs of reintroduction have overcome some of these challenges by testing for adaptation prior to repatriation (e.g., vegetables in Cambodia).

Therefore, systems that improve *ex situ* information for *in situ* programs may become increasingly beneficial, as climate change and other environmental changes can be partially mitigated by the introduction of new variation. Likewise essential to the process of repatriation is to know something about the conservation status of the *in situ* populations to be replenished and to avoid mixing-up already diverse gene pools or introducing materials in a social context where demands are absent.

Is *In Situ* Diversity in Crisis or Fit for Adaptation?

In situ conservation systems are potentially threatened by a number of socioeconomic, technological, and climate change factors, among others. Such change has been a constant throughout history and may or may not affect agrobiodiversity (e.g., Brush et al. 1992; Deu et al. 2010). In many areas, smallholder agroecosystems have experienced great outmigration and other demographic changes in recent years. The question is whether value will remain ascribed to agrobiodiversity to such a degree that the system can be resilient to other trends and shocks (e.g., caused by climate change). Nevertheless, there are also areas where, with the growing global interest in healthy and attractive food, there is an increasing market demand for "heirloom" varieties that can spur a renewed interest in crop diversity. Climate change has also generated a renewed interest in participatory and evolutionary breeding approaches that make enhanced use of genetic diversity (e.g., Camacho-Henriquez et al. 2015).

With climate change projections, crop wild relatives distributions are projected to move considerably (Jarvis et al. 2008a). When average species migration rates are set, optimistically, at 1 km/year between now and 2050, crop wild relatives would not move far enough to keep up with the velocity of climate change (Castañeda-Álvarez et al. 2016). Like other wild plants, crop wild relatives may in part be able to survive climate change through phenotypic plasticity and adaptive evolution (Jump and Penuelas 2005), though it depends on their degree of plasticity and their speed and degree of adaptation. The same concerns and stipulations are relevant to forest species and crops adapting to climate change (Mercer and Perales 2010).

With crops, however, seed networks and farmer choices regarding the use of diversity and management also play important roles. For instance, in Europe there has been guided redistribution or assisted migration of forest genetic resources. In Mexico, Bellon et al. (2011) calculated that most communities have some seed exchange with environments that resemble their "future" environment under climate change, so adaptive alleles should be circulating. In some crop species, in bad years, the distance at which farmers exchange seed goes up (Violon et al. 2016), possibly sampling from a broader range of functional diversity or adaptation. Still, pessimism about the future of *in situ* agrobiodiversity exists, even though crop genetic diversity under pressure has frequently shown to be highly resilient (Richards and Ruivenkamp 1997; Sperling 2001).

A better mapping of projected environmental niches for crops and crop wild relatives can inform us as to where and in what species we may see extreme limits and where we see the biggest issues. Then, systematic monitoring approaches can help to group truth and provide intelligence on the adaptive capacity and conservation dynamics in key hotspots. In landscapes with elevational gradients, the distance that species need to move to catch their envelope compresses (although the top of the mountain is a stringent limitation to movement). Variation in land use can cause impermeable landscapes and fragmentation, which will hinder movement, and different species will have different optimal dispersal distances. Additionally, it is not clear if associated species necessary for the crop's performance will move with the crop or not. In natural systems, some have seen this uncoupling of species associations with climate change adaptation. Crop wild relatives may also find themselves in plant communities of altered composition as climate changes, producing new competitive environments.

Conclusions

Many have indicated that there are potential functional links between *in situ* and *ex situ* management of crop diversity that can benefit both systems (Maxted et al. 1997; Nevo 1998). Yet, the distinct objectives of the two systems and their disparate users and actors also imply that there are many independent components that do not logically or naturally link. The relevance of *in situ–ex situ* linkages, who actually benefits, and how, can be highly case and context specific. Continued scholarship in this area could help us clearly understand how they can be co-leveraged and to what ends. To this end, we propose the following research agenda:

1. Redefine *in situ* conservation to better represent the range of situations (autonomous or farmer-driven versus conservation interventions), strategies employed, and the dynamism of the system. This requires that we
    *   consider what *in situ* encompasses under different circumstances and discern what it should be called,
    *   identify the autonomous drivers of farmer-driven conservation under global change and how these can be strengthened, and
    *   determine the circumstances under which monitoring of farmer-driven conservation or the promotion of intervention approaches provide best-bet options.

2. Explore in depth whether *in situ* and *ex situ* can really be linked. This requires that we
    *   integrate *in situ* and *ex situ* efforts to conserve and track landrace and crop wild relatives populations where linkages make sense and are demanded,
    *   explore whether it is necessary for the *in* and *ex situ* systems to be connected and linked, and under which circumstances, or consider whether these systems (as a whole or in part) are too different to achieve true complementarity, and
    *   discern whether diversity observatories can play a linking role by testing empirically in the field for key crops and geographies.

3. Identify and characterize key strategies for reintroducing and repatriation of genetic materials and explore their relative success for different scenarios (e.g., reintroduction of the same diversity from the past versus future climate homologous gene pools, distribution of original landraces versus evolutionary breeding populations derived from landraces). In doing so, we would
    *   determine the best management practices for reintroduction of diversity into *in situ* systems (e.g., repatriation or expansion of working diversity),
    *   identify how to choose what to introduce and the ways by which introduction occurs,
    *   quantify how well the introductions augment diversity and production and whether there are any trade-offs or impacts,
    *   determine through monitoring whether one can see changes in genetic diversity with these reintroductions immediately and over time,
    *   identify the functional diversity that is introduced and selected,

- clarify subsequent strategies that may be used to further integrate diversity (e.g., evolutionary breeding and participatory varietal selection or increasing adaptive capacity),
- discern whether assisted migration across the landscape of *in situ* populations may be effective in the face of climate change, and
- clarify how such projects could augment existing seed networks and farmer-driven conservation.

4. Understand the impact of different *in situ* conservation interventions (i.e., seed banks, seed fairs, market linkages, park systems) on conservation outcomes systematically. To do this, we need to

- determine the degree to which the population structure and functional diversity are affected,
- discern whether it matters whether the intervention was performed in high- versus low-agrobiodiversity areas or promoted in a top-down or bottom-up manner,
- conduct quantitative and qualitative impact assessments documenting rates of adoption and user perceptions, and
- research the cost-benefit ratios of different interventions and their scalability.

Improved understanding of the opportunities available to *in situ* and *ex situ* diversity management, as well as linkages between the two, will move the field forward. This may ultimately allow us to apply a common framework (including methods, metrics, and tools) for enhanced intelligence between *ex situ* collections and *in situ* populations, which could ultimately result in a red listing strategy for highlighting the conservation status of important crop diversity (e.g., landraces). Such work may also allow for evolutionary breeding to augment diversity and the promotion of networks of *ex situ* and *in situ* linked observatories for key crops and centers of diversity.

Generate a Theory of Agrobiodiversity and Project Trajectories of Agrobiodiversity in Response to Social and Environmental Change

There are basic processes that govern the generation and maintenance of agrobiodiversity. We have ideas about many of these basic factors (e.g., genetic bottlenecks, dispersal, isolation by distance, environmental adaptation, mating system) influencing patterns of diversity, so it may be possible to build a theory of agrobiodiversity, or at least a theoretical framework, that can be used to interpret specific cases and to build theoretical models. Such models, whether simple conceptual models or more complex mathematical models and computational simulations, would describe how these processes have worked up to the present and could help us predict future changes in agrobiodiversity.

The general theory would acknowledge both natural and human-mediated processes that have generated diversity. Theoretical models based on fundamental social and biological processes should help improve understanding of the spatiotemporal patterns we observe. They could allow for the unification of empirical observations that would otherwise seem disparate, and produce predictions that can be verified by empirical data. Relevant questions include: what are the ways by which basic processes and geography affect diversity, what are the other main drivers of changes in agrobiodiversity, what are the best strategies for model building, which processes are most important to consider for projection of the future state of agrobiodiversity, how can we assess system tipping points, and how can we determine whether agrobiodiversity is at a desired state, insufficient, or in crisis?

Types of Diversity to Predict and Drivers

Predictions of past and future patterns in diversity could be made for the diversity metrics we have discussed (see the first objective). The primary ones are interspecific diversity (i.e., number of species or groups of species); intraspecific diversity, such as the number of named landraces; genetic diversity (i.e., variation determined from genetic markers or DNA sequences); or functional diversity (e.g., variation in ecophysiological traits). Perhaps diversity of associated knowledge could also be predicted.

There are basic evolutionary drivers of inter- and intraspecific diversity that can be considered. These include *selection* (of all kinds: diversifying, directional, and stabilizing selection), *gene flow* (propagule or pollen-mediated), *genetic drift*, and *mutation*. Similarly, *isolation by distance* (from lack of gene flow) and *genetic bottlenecks* (due to selection and sampling effects) are important factors that can shape diversity. Thus the *distance to the area of crop origin* might be used to predict patterns of diversity (van Etten and Hijmans 2010), and vice versa, the area of crop origin from observed patterns (Kraft et al. 2014). Each of these drivers can be naturally occurring, affected by social processes, or both. For instance, environmental conditions may select for particular ecophysiological traits and farmer preference may select for seed color.

Aside from those evolutionary processes noted above, there were other biological, environmental, socioeconomic, and geographical factors that may need to be considered. *Mating systems* (self-pollinated, wind-pollinated, outcrossing, clonal reproduction) have been shown to be important for determining diversity patterns across the landscape. *Environmental variation* (e.g., diversity higher on mountains) also affects diversity. *Global commodity trade* has led to the expansion of intensive agrobiodiversity systems where modern varieties replace landraces. *Land use change* can be an important force driving change in agrobiodiversity: it can include urbanization, where the footprint of urban areas increases, as does that of peri-urban areas, and it can also go along with *dietary changes*, which can drive changes in what is grown on the landscape. *Climate change* may be a driver of change in agrobiodiversity by influencing yields or affecting selection for ecophysiological traits, but it is interconnected with drivers throughout the rest of the system.

There are important demographic drivers that change agricultural landscapes. *Deagrarianization*, along with the transition to part-time farming and outmigration from rural to urban areas, can affect the density of farmers and reduce time spent farming, thereby affecting farmer choice. The resulting lower population density in rural areas can have effects on seed networks. A farmer's *access to land* can change the crop's population size and his or her *access to water* affects the environment the crop experiences. *Cultural variation* (Perales et al. 2005) and changes in *farmers' knowledge base or preferences* can affect diversity. *Farmers' innovation* can have its own effect on agrobiodiversity, as can *innovation and human adaptation* to change or other drivers. *Compensation* for use of agrobiodiversity or *farmer insurance* can influence agrobiodiversity use and levels. The *speeding up of social time* can make events happen more quickly. This may have effects on agrobiodiversity and affect stability of choices. *Agricultural policies* can incentivize or deincentivize use of diversity for conservation or sustainability outcomes.

There are thus many drivers and processes to consider. Modeling efforts could consider general drivers and then move to more concrete scenarios. The particular directionality and strength of impact on diversity of some drivers may need empirical data to understand, but the effect of other drivers may be more difficult to determine. By studying processes through the lens of evolution, we can better understand the effects of particular drivers on diversity. We should also note that the importance of drivers can depend on the degree of knowledge of the farmer and how that knowledge directly influences diversity.

Models

There are a number of approaches to modeling agrobiodiversity. One approach is to create a comprehensive and complex model. It would be unlikely to be fully complete and could get very complicated. Another approach is to create either a null model (producing patterns that would be expected given some rules) or a very basic model to which you can add or remove factors to understand their roles. Thus it is possible to test particular effects by introducing them individually or in groups, and by including stochastic factors (e.g., genetic drift). The results can then be compared and verified against reality. Differences could reveal the presence of important additional factors influencing patterns of diversity. Null models have some benefits in that we can test whether we have seen loss or gain of diversity, both of which might be expected. If we always assume that diversity is being lost (and do not test for differences compared to null expectations), it may seem that conserving diversity in gene banks would be the only viable option. This less deterministic approach may help us understand how and why diversity is maintained or increases in some regions. Additionally, individual drivers can be included in the model as individual factors unto themselves or modify basic evolutionary processes (e.g., deagrarianization reduces seed networks and thus propagule-mediated gene flow) and tested for their ability to explain current patterns.

There are other approaches that could be used to explore variation in agrobiodiversity and drivers thereof. One is to focus less on the question of what influences levels of diversity since that change is contingent. In other words, exploring cases where diversity is declining or increasing can give insight into factors that drive the dynamics of agrobiodiversity. Another modeling approach would use individual choice and fuzzy logic to discern patterns of diversity. Models that can discern tipping points (i.e., points at which systems flip to another state) can be helpful. However, we mostly see opportunities for retrospective study when there is a loss of a particular crop from an area (e.g., changes from growing maize to growing sorghum in Africa with climate change) or a major decrease in diversity due to environmental conditions (e.g., loss of landraces and increased use of maize hybrids in Belize after a multiyear drought). Better understanding of the conditions that lead to state changes can be informative. Finally, one can use heterogeneity studies, which are powerful where heterogeneity exists, and can be followed up with ethnography to explain patterns.

Different models would need to be developed to understand patterns in intraspecific and interspecific diversity. However, there might be important similarities in terms of relevant drivers and whether those drivers have similar or different effects.

Future Trajectories and Scenarios

Some modeling or analytical approaches may provide basic information about factors at play in creating current patterns in agrobiodiversity, which are the culmination of past processes (short and long term). However, there can be value in modeling or exploring the future trajectories of diversity and considering the impacts of different scenarios. For instance, one could modify the strength of different drivers and discern the implications for diversity.

One trajectory that is often discussed is the possibility for crisis in particular systems. In other words, precipitous declines in agrobiodiversity can cause loss of agroecosystem function. While agrobiodiversity is argued to provide flexibility, fluidity, and resilience to systems, extreme loss can constitute a crisis. However, it may be hard to decide what states would qualify as a crisis. A crisis could relate to social metrics, production, or environmental impacts. Most work on agrobiodiversity focuses on traditional agriculture. That is, small-scale family farming often in mountainous areas of developing countries. It is important to connect what we have learned from these systems to what happens in industrialized agriculture since the *in situ* dynamics in those systems merits more study.

A dominant narrative is that most diversity in such intensive systems has been lost and replaced by a single, or very few, varieties. While this may be true in some cases, it does not have to be true in all cases. A particularly interesting aspect of industrialized agriculture is that it seems to have shifted from diversity in space to diversity in time. The lack of crop diversity in the maize and soybean system of the U.S. Midwest might suggest crisis, yet the system is very productive (even if there are problematic externalities and loss of social benefits). Perhaps diversity over time (i.e., through changes in varieties by farmers) may provide benefits. An example of the ability of temporal diversity to support sustained productivity in this system was the rapid and successful response to the 1970 southern corn blight epidemic in the United States. This was a crisis caused by a lack of diversity in maize hybrids. That is, the temporal turnover of varieties may have increased at the expense of the spatial turnover. This is perhaps similar to the role of crop rotations. Crops are generally not grown in mixtures, but a number of different crops can be grown on a plot across seasons. An important theoretical question that needs to be looked into is whether, where, or how much, diversity in time can be equivalent to diversity in space (Denison 2012).

Conclusions

Given that many current projections of future agricultural change do not incorporate a diversity perspective and rarely use data from landraces, a model based on landraces and diversity would be novel. It would help us clear up the confusion about the degree to which diversity is being lost and patterns thereof, and how diversity can increase resilience to agriculture in a rapidly changing world. A modeling approach necessarily simplifies what happens in the real world, but such simplifications can still be useful. Our proposed esearch agenda follows:

1. Generate models of various sorts, which could include the following:
   - Model 1: Origin, spread, and diversification using most basic drivers.
   - Model 2: Roles of additional drivers currently affecting agrobiodiversity, including those which maintain agrobiodiversity.
   - Model 3: Projections into the future based on changes in various drivers and their interactions.
2. Determine which factors have generated patterns of diversity in the past (first objective) and how best to predict those patterns from these first principles.
3. Look across different systems and different areas to identify similarities and differences in the patterns and drivers.

4. Investigate interactions between drivers (may require empirical work to parameterize models). To do so we need to determine the strength and direction of effects of social and environmental drivers on diversity and clarify how drivers interact.

5. Determine how future scenarios will change genetic diversity. Specifically, we could discern the relative importance of different factors in future change and project diversity into the future with increased force of particular drivers (e.g., outmigration or climate change or the interactions between them).

6. Determine what combinations of factors could get systems to crisis points in the future.

7. Perform longitudinal studies to recount varieties or discern functional or genetic diversity in order to explore population and landscape structure of diversity and document change in "observatories" (fourth objective) or "working laboratories" and highlight pathways of change to make change visible and researchable.

A thorough investigation of the factors and processes shaping crop diversity and predictive work to understand the future trajectories of crop diversity could propel our understanding of existing crop diversity. It may also help promote conditions where we might expect positive changes in diversity, but also enhance the quality of information about factors most important to understand and predict losses of diversity.

Discussion

Humanity relies on crop agrobiodiversity for food, fiber, medicine, and fuel production. Yet, we have an incomplete understanding of how agrobiodiversity is affected by human activity and environmental variation under global change scenarios (Chapter 6). Crop agrobiodiversity is a product of a complex combination of historical factors (e.g., location of crop origin), interactions with human society (e.g., markets) and associated biodiversity (e.g., pollinators), and local knowledge. As these factors continue to be affected by societal and global change (Chapter 8), agrobiodiversity will as a result continue to change. Understanding the factors shaping agrobiodiversity and any future changes in agrobiodiversity will augment its local use as well as strategies to cope with any shift or loss of use. This further understanding will also allow acknowledgment of the value of agrobiodiversity to global society, enabling its conservation for future generations.

Appendix 2.1

To illustrate the importance of different scales of diversity, as well as connectivity, in providing ecosystem functions and services, two examples are provided on the following pages. Example 1 covers high-diversity, moderate-intensity systems that grow maize, potatoes, cereals and pulses, as in the Andes or West Africa. This situation may not be typical for all smallholder agroecosystems, where higher-intensity and lower-diversity examples are also seen. Examples of connectivity include links to regional or global markets and other food systems. Example 2 represents a low-diversity maize–soy system with potential for cover crop use in the U.S. Eastern Seaboard and Midwest.

Example 1 Key: functional diversity impact is minimal or does not apply (—), doubt exists as to its importance or meaning and may represent a research gap (??).

Ecosystem Functions and Services	Scale of Diversity			Connectivity
	Genotypic and Varietal	Species	Land-Use Composition	
Supplying nutrients to crops	Varietal differences in uptake and residues	Greatest impact on functional diversity of residues (e.g., N fixation)	Depends on fallow (short, pasture, swidden forest)	Low importance, some purchase of fertilizer
Yield and primary production	Moderate to high importance	Moderate importance	Moderate importance	
Yield stability	Moderate to high importance	Moderate to high importance	Forest and wild species may stabilize yield	Low to moderate substitution by outside crops to buffer risk
Pest and disease protection	Frequently important, variable resistance	High importance (e.g., rotation vs. pests)	Moderate to high importance	Variable importance, seedborne diseases, some purchase of pesticide
Pollination	??	Important in food source diversity	Important in food source diversity	low
Hydrologic buffering	—	??	Important if forest or grassland matrix exists (or is threatened)	—
Food provision and health	Important in taste, nutrition, and seasonality of varieties	Important in dietary diversity	Forest food provisioning	Some connectivity
Cultural services	Diet importance, prestige, identity	Diet importance	Cultural importance (e.g., grazing, ceremonial, local governance of fallows)	Seed exchange, barter, reciprocity, remittances, rural urban linkages
Externalities and disservices	Buildup of pests, potential pesticide effects on soil biota	Implications for soil cover, soil erosion	Effects of forest loss	Soil erosion (can be service to outside systems downstream)
Human security	Redundancy of varieties in multiple communities, use of feral types during lean periods (??)	Redundancy of species	Hiding places for conflict; signifies livestock presence for cash conversion (??)	Can assume extreme importance: protection, relief, or assault from external actors

Example 2 Key: functional diversity impact is minimal or does not apply (—); doubt exists as to its importance or meaning and may represent a research gap (??).

Ecosystem Functions and Services	Scale of Diversity			Connectivity
	Genotypic and Varietal	Species	Land-Use Composition	
Supplying nutrients to crops	Low	Some importance (e.g., variable need for chemical fertility among crops, N fixation, cover cropping)	Low	High importance of purchased fertilizer
Yield and primary production	Moderate	Moderate	Low if noncrop uses are low % cover of landscape	—
Yield stability	Low to moderate: year to year diversity to evade pests (??)	Moderate	Low	Crop insurance is a buffer against yield variation
Pest and disease protection	Low	Some importance (e.g., sequential year-to-year biodiversity)	Low to moderate, depending on field and forest matrix	High importance, purchased pesticide/herbicide
Pollination	Low	Low	Low to moderate, depending on field and forest matrix	Low
Hydrologic buffering	—	Low	Low to moderate, depending on field and forest matrix	
Food provision and health	Low	Determined by external markets	Determined by external markets	High connectivity: assemblages of regions comprise food system diversity
Cultural services	—	Agritainment and agrotourism uses	Fishing/hunting, other recreation, agrotourism	??
Externalities and disservices	Pesticide effects on soil biota (??)	Implications for soil cover, soil erosion	Water pollution versus retention of nutrients	Water pollution/greenhouse gas emissions (N_2O)
Human security	—	—	??	??

62

3

How Does Population Genetics Contribute to an Understanding of the Evolution of Agrobiodiversity?

Yves Vigouroux, Christian Leclerc, and Stef de Haan

Abstract

Agrobiodiversity results from the domestication and continued selection of crop and livestock species. Understanding the evolution and population dynamics of agrobiodiversity in terms of its genetic, reproductive, ecological, or anthropogenic dimensions, requires both long-term and contemporary perspectives. Population genetics can supply valuable information about the short- and long-term dynamics of agrobiodiversity by describing the trajectory of the frequency of an allele (a genetic variant) within and among given populations. The resultant information makes it possible to understand the relationship between populations and individuals within populations. It is also particularly pertinent to an understanding of how agrobiodiversity has evolved on different timescales. Advances in modeling population genetics enable hypotheses to be tested on the different drivers that shape crop and animal diversity. With the increasing availability of genomic markers, population genetics offers new opportunities to assess, test, and understand agrobiodiversity dynamics.

Introduction

Agrobiodiversity is the result of the domestication and continued selection or diversification of crop and livestock species as well as cultivars (or breeds), including the wild relatives of domesticates. Selection occurs in response to production, cultural and societal demands, and environmental variability. Understanding the evolution and population dynamics of agrobiodiversity, whether it concerns the genetic, reproductive, ecological, or anthropogenic dimensions, requires both long-term and contemporary perspectives. Early

domestication events and the resulting genetic bottlenecks, geographic isolation, cultural divergence, environmental selection, and historical (extreme)
events have all influenced the long-term evolution of agrobiodiversity (Gepts
et al. 2012). Contemporary phenomena, including formal crop improvement
by breeders and the continued selection of diversity by farmers, continue to
drive evolution today (Ceccarelli et al. 2009; Parra et al. 2010). Apart from
the purely biological dimensions (i.e., the populations themselves), it is just as
important to understand what drives evolution (i.e., environmental and sociocultural changes) at multiple scales, from the local to the global (see Chapters 6
and 8). Analyzing agrobiodiversity through both a historical and contemporary
lens is appropriate because these different timescales generally imply distinct
research approaches.

The crop and livestock histories as well as contemporary evolutionary processes that shape the population genetics of domesticated species are often
complex (Meyer et al. 2012). They are partially determined by the crop's reproductive biology, geographical distribution range, and cultural exposure as
well as by global investment in genetic improvement and breeding (see Chapter
6). All these processes shape the diversity of the varieties at the molecular
level. Population genetics provides a specific framework that enables comparison of diversity among and between individuals, populations, varieties, and
breeds. Population genetics describes the trajectory of allele (genetic variant)
frequency within and among given populations. Consequently, this framework
is particularly suited to understanding the evolution of agrobiodiversity.

In this chapter we explore some of the overarching questions related to the
population genetics of agrobiodiversity by taking into account such dimensions as time, space, drivers of change, and conservation systems:

- What are the complex genetic, evolutionary, cultural, and ecological interactions of diversity based on continued cultivation in
  agroecosystems?
- How does the ongoing evolution of diversity function as an emergent
  adaptive mechanism in response to environmental change?
- What are the potential complementarities of *ex situ* and *in situ* approaches to genetic resources and how can these be strengthened?
- How do we meet the challenges of fuller integration of the geospatial
  and temporal scales that characterize agrobiodiversity?

We begin with a review of the basic concepts of population genetics and will
illustrate how population genetics is used to study the origin of domestication as well as the diffusion and selection that results. Thereafter we discuss
how population genetics can be used to understand how agrobiodiversity is
shaped by social factors and present research that can help us understand
recent changes in and dynamics of agrobiodiversity. Finally, we offer our
view on the future outlook for population genetics approaches to the study of
agrobiodiversity.

Population Genetics and Its Use in Agrobiodiversity Research

Conceptual Overview

Population genetics describes gene allele frequencies in individuals and populations and addresses how they are shaped by four major forces: drift, mutation, selection, and migration. *Drift* involves how allele frequency changes from one generation to the next, simply as a result of chance (i.e., random draw). This random draw is the probability of picking an allele from a previous generation and passing it onto the next generation through reproduction; it is directly linked to the size of the population. Genetic drift occurs in all populations of limited size, but effects are strongest in small populations. A large population size will slow down changes in allele frequency. *Mutation* is the process by which a new genetic variant emerges. *Selection* is the process that leads individuals bearing a given allele to have more descendants. Mutation and selection can occur naturally or be purposefully induced. Finally, *migration* is the movement of an allele from one population to another through gene flow, seed exchange, or other migratory processes. Population genetics is the study of how these different forces shape the diversity of a population and the inferences that could be drawn from this diversity with respect to the different forces at play. It can help us address multiple issues important for agrobiodiversity research (Table 3.1) by questioning, for example, how local farmers impose selection on specific genes or genotypes.

The level at which populations show similarity—in terms of allele frequency, because they recently diverged or drifted from an ancestral population, or how they are related, with or without recent gene flow—can be tested by comparing allele frequencies. Proximity of wild and cultivated populations can provide insights into the origin of domestication. We expect a cultivated population to resemble more closely the wild population from which it was derived rather than a secondary or tertiary gene pool (see Harlan and De Wet 1971). Based on this simple theoretical framework and the specific research questions we pose about drift, mutations, migration, or selection, inferences can be made by comparing allele frequency between populations. Several statistical methods have been developed to make such inferences about population histories from allele frequencies. Below, we explore sample studies that have used population genetics to pose questions and look for answers.

Studying the Origin, Domestication, and Diffusion of Crops

While archaeology and the social sciences, more broadly, provide invaluable information about crop history, major insights have been gained over the last thirty years using a population genetics framework. The basic questions addressed were: When and where does domestication occur? How many

Table 3.1 Population-level evolutionary forces: impact and importance to agrobiodiversity.

Evolutionary Force	Parameter	Impact and Importance	Sample Studies
Drift	Effective size N	History of the crop (e.g., domestication bottleneck, growth of the population with diffusion) Human choice of the quantity of seed used in the next generation	Hufford et al. (2012); Matsuoka et al. (2002)
Migration	Migration rate m	How the history of a crop is shaped by gene flow from wild relatives	van Heerwaarden et al. (2011)
		How agriculture migrates from its original setting to a larger scale	Roullier et al. (2013a)
		Diffusion of seed and pollen within one crop at the farmer scale, from field to field	Celis et al. (2004)
		How much seed or how many animals are transmitted between farmers	Leclerc and d'Eeckenbrugge (2012)
		How seeds are exchanged between farmers who share (or not) the same set of values (e.g., language)	Labeyrie et al. (2016)
		How wild relatives or hybridization is used in a cultivated setting	McKey et al. (2010b); Scarcelli et al. (2006)
		Introduction of new or modern varieties	Thomas et al. (2012)
Selection	Selection coefficients	Early selection during domestication	Hufford et al. (2012); Wright et al. (2005)
		Environmentally imposed selection (biotic and abiotic) both ancient and very recent	Vigouroux et al. (2011b)
		Selection imposed on specific genes or genotypes by local farmers Breeder selection	Mariac et al. (2016)

independent domestication events are observed? How and when do crops diffuse after domestication? What specific genes were selected by early farmers?

Single or Multiple Origin: Simple Hypothesis, Complex Answer

Studies of the origin of crops frequently use a population genetics framework. Such studies are based on genetic markers using amplify fragment length polymorphism, simple sequence repeats, or single nucleotide polymorphism. Inferences based on such approaches have shed light on the origin of einkorn wheat (Heun et al. 1997), barley (Badr et al. 2000), emmer and hard wheat (Ozkan et al. 2002), maize (Matsuoka et al. 2002), rice (Huang et al. 2012), potato (Spooner et al. 2005), and pearl millet (Oumar et al. 2008). These studies were mainly based on a population genetics framework. One of the most important queries several of these studies sought to answer was whether the domestication process was a single event or involved multiple events. A multilocus phylogenetic analysis (generally using a neighbor joining algorithm) was classically performed to assess whether cultivated plants grouped together before they grouped with wild relatives. The grouping together of cultivated plants (monophyly) was then interpreted as evidence for a single domestication event. Multilocus phylogenetic analysis is widely used, even though it is not a standard approach for inferring intraspecific evolutionary history. A major concern associated with intraspecific phylogenetic frameworks for multilocus genomic markers is the complex nature of the crop varieties or populations studied, which are not isolated. Interbreeding between populations or varieties undermines the assumptions that underlie the phylogeny framework. One of the emerging pictures in plant domestication is that we have largely neglected the impact of gene flow on our understanding of crop evolutionary history. Two recent examples can be used to illustrate this point.

The first concerns maize domestication. Maize was domesticated from lowland teosinte, specifically the *Zea mays* ssp. *parviglumis* from the Balsa river in Mexico (Matsuoka et al. 2002). In terms of diversity, lowland maize should be proximally closer to this lowland ancestor, but Matsuoka et al. found that highland maize was genetically closer to this wild lowland teosinte. This discordance was explained by a confounding factor: the evidence of gene flow from the high-altitude wild relatives *Zea mays* ssp. *Mexicana* into cultivated maize (van Heerwaarden et al. 2011). This gene flow is thought to have allowed maize that was originally cultivated in lowlands to adapt to high altitudes by acquiring specific alleles already available in these high-altitude wild relatives.

The second example involves Asian rice. The evolutionary history of Asian rice was initially postulated to be associated with two distinct domestication events (Londo et al. 2006): one in China (*japonica* form) and one in India (*indica* form). Using genomics data from one of the largest-scale studies, Londo et al. (2006) concluded that a single domestication event of *Oryza sativa japonica* occurred in the Pearl River area in southern China (Huang et al. 2012). Thus, *Oryza sativa indica* was the result of recurrent crossings between these initial cultivated forms with wild rice in Southeast Asia and South Asia (Huang et al. 2012). This understanding of rice evolutionary history was possible because (a) the study of domestication was based not only on "neutral" alleles but also on identified "domestication alleles" and (b) discordant phylogenetic signals existed between neutrally behaving genes and domestication genes. Indian rice looked like its wild relatives found in India, and Chinese rice looked like its wild relatives found in China. However, the gene that was selected early on during domestication, irrespective of the origin of the rice, was traced to China. The question is still open regarding the number of domesticated alleles or genes and if the gene flow in India does not qualify as separate domestication (Civáň et al. 2015). This is an important question of broader interest, yet condensing the question into a single or multiple origin hypothesis does not capture the whole story. For this, we need to study the complex history of the crop.

These studies were able to pinpoint the origin of the major crops, but again, this is not the whole story. After initial domestication, wild relatives continued to play a significant role in shaping overall crop diversity in maize (van Heerwaarden et al. 2011), rice (Civáň et al. 2015), potato (Hardigan et al. 2017), and pearl millet (Oumar et al. 2008). The simplicity of the hypothesis to be tested—single versus multiple origins—should not mask the fact that crop history is not that simple, as evidenced by a recent study highlighting the very complex structure of crop and animal diversity (Frantz et al. 2015).

Diffusing Agricultural Society and Reshaping Agriculture

After domestication, crops spread through human migration and exchange. Using human skeletal remains, genetic analysis of relationships between modern humans, and linguistic analyses, it has been possible to trace the spatial and temporal dispersion of human populations (Diamond and Bellwood 2003). Traditionally, the dispersion of domesticated plants and animals has been analyzed separately or by juxtaposing archaeological evidence (farmer or crop archaeological remains) and genetic data. For example, using mitochondrial DNA of modern and ancient samples, Larson et al. (2007) showed that pigs were introduced into Europe during the Neolithic from the Near East. In addition, the presence of starch, pollen, or other vestiges of crops at archeological sites enables the presence of domesticated crops in specific parts of the world to be dated. There are limitations to both approaches: archaeological analyses often have limited data and genetic analyses can be compromised when inaccurate geographical models are used to simulate migration. Some studies have tried to combine both archaeological and genetic data to improve insight into domestication and dispersal of domesticated crops and animals (Hanotte et al. 2002; Larson et al. 2007; Perrier et al. 2011).

More recently, geographically explicit analysis has used spatial data to improve the modeling of dispersal and has been extensively performed to understand how our own species moves and colonizes (Handley et al. 2007). One of its core ideas holds that after the dispersal of a population from its original location, successive colonization leads to changes in allele frequencies during dispersal. This pattern leads to a correlation between the genetic distance between populations and the distance of dispersal between populations, and a decrease in diversity (e.g., heterozygosity) along the dispersal route. Initially, geographically explicit analysis takes the geographical distance between samples (individuals, group of individuals, or populations). This distance could be the geographical distance between samples (a simple benchmark model) or the distance between samples based on geographical barriers (mountains) or preferential paths (e.g., rivers). Some have calculated a regression between pairwise genetic and geographic distances of samples (Handley et al. 2007; Manica et al. 2005) or evolution of genetic diversity along a dispersal path (Prugnolle et al. 2005). This approach could also use genetic and archaeological data sets (van Etten and Hijmans 2010).

To understand crop diversity dynamics and diffusion, more elaborate tools need to be employed (Gerbault et al. 2014). One promising approach is approximate Bayesian computation (ABC) (Beaumont et al. 2002), a class of methods in Bayesian inference that provides the means to evaluate posterior distributions when the likelihood function is not analytically available. ABC algorithms enable inferences to be made on data considered intractable only a few years ago. The use of ABC methods has found an increasing number of applications in molecular genetics, ecology, epidemiology, and evolutionary biology. One of the most interesting approaches, in terms of domestication and diffusion, is to directly use spatially explicit simulations. ABC provides an efficient way to assess phylogeographic scenarios and to address questions about the origins and diffusion of plants and other organisms (François et al. 2008). It also makes it possible to assess different scenarios on the basis of coalescent simulations (Clotault et al. 2012).

The above models have been used to infer the expansion of *Arabidopsis* in Europe (François et al. 2008) and the spread of the domesticated horse (Warmuth et al. 2012). Other model-based tools have been developed to infer evolutionary history and variations in the effective size of

- a single population, such as the pairwise sequentially Markovian coalescent or the multiple sequentially Markovian coalescent (Li and Durbin 2011; Schiffels and Durbin 2014);
- a few populations using a likelihood approach, like $\partial a \partial i$ (Gutenkunst et al. 2009); and
- several populations using the composite likelihood approach implemented in fastsimcoal (Excoffier et al. 2013).

In addition, they are increasingly being used in evolutionary studies of the domestication of crops and animals.

Clearly, the field could move forward with better and more efficient testing of hypotheses using such model-based approaches. The complex interplay between early origins, secondary contact with wild relatives, and the impact of ancient human migration could be statistically assessed in the context of model-based inferences (Gerbault et al. 2014). Population genetics offers the opportunity to move from a description of diversity to statistical assessment of the main drivers of past dynamics. Finally, one of the most exciting approaches now being developed is the use of ancient DNA to infer crop history and selection (da Fonseca et al. 2015). Coupling archaeological data and model-based inference of diffusion is certainly one of the most promising approaches to understand past dynamics.

Uncovering the Genetic Basis of Domestication and Crop Adaptation

Population genetics offers the opportunity to identify the genes that were selected early on during domestication. Selection led to a very specific genetic signature at the molecular level, and this signature can be identified relatively easily. Several studies have succeeded in this using whole genome resequencing in rice (Civáň et al. 2015; Huang et al. 2012) or maize (Hufford et al. 2012). Dehiscence loss is a trait directly associated with cereal domestication considered to be the signature of domestication (Li and Olsen 2016). Harvesting seed is greatly facilitated if grains stay on the cob or panicle. In rice, three major genes associated with dehiscence were found based on the selection signature and further validation with genotype or phenotype association: *Sh4*, *qSH1*, *OsSh1* (Li and Olsen 2016). Further it is possible to identify not only the genes but also the timeframe under which the selection occurred (Nakagome et al. 2016). These new approaches enable a glimpse of selection dynamics from the point of domestication until the present day.

In addition to selection during domestication, another interesting issue involves how crops adapt to a different climate or environment (see Chapters 2 and 4). Analysis of crop adaptation relies on correlation with environmental variables and association between genotype and phenotype (for a review, see De Mita et al. 2013), such as approaches based on genome-wide association mapping or quantitative trait loci (Bazakos et al. 2017). These techniques build primarily on population genetics and should certainly be used in future studies of agrobiodiversity.

Population Genetics: Understanding How Agrobiodiversity Is Shaped by Social Factors

Together with biological and environmental factors, social factors (including cultural processes and other human activities such as cognition and classification) shape crop genetic diversity. Two complementary approaches can be distinguished (Leclerc and d'Eeckenbrugge 2012):

1. A farmer's individual decisions, motivations, and actions (individual-based approach), where the social component can be interpreted as the sum of individual choices.
2. The characteristics of social groups (e.g., language, marriage, residence rules, and preferential exchange systems), where individual

actions are partially determined by social norms and identity (group-based approach).

A crucial point in population genetics is that the populations and factors to be tested are intrinsically related: together, they determine the sampling and testing procedures. Social factors, for example, are usually associated with the individual-based approach to agrobiodiversity, whereas populations are associated with farmers' fields or villages, which were rarely characterized sociologically. This demonstrates a fundamental challenge for social anthropology and the group-based approach to agrobiodiversity studies (Leclerc and d'Eeckenbrugge 2012).

Individual-Based Approach

The study of root and tuber crops offers some very interesting insights into the influence of farmers' choices on diversity dynamics and the creation of plant varieties (McKey et al. 2010b). Many root and tuber crops are propagated vegetatively. Consequently, they are clones propagated from one generation to the next, either by a root, a tuber (e.g., yam, potato), or stem cutting (e.g., cassava, sweet potato). This allows varieties to be propagated with an identical phenotype (same color, taste, flowering behavior, morphology) and well identified by farmers through many distinct and stable categories. In their broad comparative studies across species, cultures, and countries, Jarvis et al. (2008b) showed that farmers use more detailed intraspecific names for vegetatively propagated crops than for sexually propagated crops.

Such propagation, however, does not allow new varieties to be created nor does it favor adaptation to a changing environment (climatic variation, new devastating pest or viruses). In the case of yam, farmers collect wild relatives or hybrids between cultivated and wild relatives and introduce them into their cultivated fields. The resulting "new" plants are the result of sexual reproduction, representing genotypes that are a completely new combination of alleles from the previous generation. Some of these plants are adopted, vegetatively propagated, and lead to new varieties (Scarcelli et al. 2006). In cassava, a similar mixed reproductive model is implemented by using both the vegetatively propagated stem, while also allowing outcrossed natural seed to be incorporated as a future vegetatively propagated stem (McKey et al. 2010b). This mixed reproductive system is rare yet quite widespread in what we often call "clonally propagated crops" (e.g., Bonnave et al. 2014; Scurrah et al. 2008). This practice allows farmers to test new genotypes to see if they produce a higher yield in a changing environment. Is sociocultural transmission of knowledge associated with such practices? How can other farmers appropriate a new variety if an isolated farmer initially selected the variety in question? How could this creative process shed light on adoption and diffusion processes, which are known to be constrained by social barriers and norms?

Social Group and Network-Based Approach

Cultivated populations have been studied using wild plants as a theoretical framework, with emphasis on "natural" diversification factors (e.g., geographic distances and environmental variations with attendant natural selection). Thus far, geographic and social factors have not been analyzed separately; most authors followed a classical genotype by the environment approach. To consider the close interdependence between crops and societies, we need to consider a triple interaction: $G \times E \times S$, where "G" represents genotype, "E" environmental variables, and "S" social factors (Leclerc and d'Eeckenbrugge 2012).

Codistribution between farmers' cultural and crop genetic diversity has been described at different scales: for cassava in Gabon (Delêtre et al. 2011), for banana in New Guinea to West Africa (Perrier et al. 2011), and for sweet potato in South America to Polynesia (Roullier et al. 2013a). Sociolinguistic barriers were included (Westengen et al. 2014) as factors that play a key role in shaping the genetic structure of sorghum in Africa. Here, Westengen et al. showed, at the continental scale, that the three major sorghum populations identified with molecular markers (central, southern, and northern) were associated with the distribution of ethnolinguistic groups. These patterns suggest that seed exchanges occurred primarily within, as opposed to between, linguistic or social units.

In studying social factors, individual-based approaches emphasize selection as an evolutionary force, whereas group and social network-based approaches highlight the role of migration (see Chapter 8). Still, only a few good ethnographic descriptions enable analysis of the underlying social processes; for example, ethnicity and residential rule in the Mount Kenya region (Labeyrie et al. 2014, 2016). Crop diversity dynamics is complex and requires integrative, interdisciplinary studies in the future.

Population Genetics as a Tool to Monitor Current Agrobiodiversity Changes

Ongoing global changes associated with climate, land use, globalized trade, human migration, and growing urban populations are increasingly impacting agrobiodiversity (see Chapter 6). How a local farmer responds to these changes could directly or indirectly impact agrobiodiversity. Here we focus on adaptation associated with climate change and discuss the trajectory of agrobiodiversity through monitoring as well as the main drivers of such observed changes or resilience (see also Chapters 2 and 14).

Analysis of Selection Associated with Recent Environment Changes

The continuing impact from climate change constitutes new threats to agriculture; for example, the increase in temperature tends to reduce wheat yields

(Asseng et al. 2015). How agrobiodiversity will adapt to climate change is an issue that can be addressed using population genetics. Not many studies have focused on this specific issue, although one study found a relationship between adaptation of pearl millet to climate variation and available function variation (Vigouroux et al. 2011b). More specifically, an early flowering allele of the *PHYC* gene was positively selected and is associated with earlier flowering varieties in a context of a shorter season. Earlier flowering plants produce seed before the end of the wet season. Such functional alleles might undergo environmental selection with spatial and temporal variation (Gerbault et al. 2014).

Functional alleles already present in landraces could thus be easily selected from standing variation. Maintaining high diversity in landraces allows such selection and adaptation. We wish to emphasize, however, that human and environmental selection also plays a role and that divergent selection pressures may not necessarily be heading in the same direction. The same two alleles of this gene impact the shape of the spike, and farmers who select the shape of their spike from one generation to the next might favor what they consider to be the true varieties—those with a slightly longer spike. This spike phenotypic selection will increase the longer spike allele, an allele that allows later flowering. Human and environmental selection may thus not be going in the same direction. How are landraces selected and shared by farmers, and how have they evolved over time? Understanding the interactions between environmental and human selection requires studying all the different aspects of selection, both at the level of the farmer as well as environmental selection. The respective roles of human and environmental selection are still unresolved. Although challenging, it would be interesting, for example, to know whether some cultural feature enables or counteracts these adaptations.

Population Genetics to Monitor Agrobiodiversity

A comprehensive understanding of spatial and temporal changes within and between crop–livestock populations requires robust monitoring frameworks. *In situ* populations are highly dynamic. Comparisons of time series focused on crop or livestock populations at representative benchmark sites or diversity hotspots, however, are notably scarce in agrobiodiversity (see Chapter 2). Exceptions do exist (Dyer et al. 2014; Salick 2012), but these are typically nonsystematic in the sense that they do not cover multiple benchmark sites or countries at the center of origin of a particular crop–livestock species: they apply standard procedures that are concerted and easily accessible or replicable across species, and foresee regular time intervals with local partnerships for sustained monitoring. While monitoring is arguably a basic requirement to make inferences about changes in crop–livestock populations (i.e., genotype or allelic loss, enrichment, or shifts in abundance), the institutional and methodological platforms to facilitate monitoring are as yet rare and dispersed (see Chapter 2). On the other hand, researchers who study the conservation status of wild flora and fauna have adopted systematic monitoring frameworks, as in the Global Observation Research Initiative in Alpine Environments or the IUCN Red List of Threatened Species.

Methods that utilize population genetics and molecular or genomic marker platforms enable agrobiodiversity monitoring. To evaluate the conservation status of a particular landrace population at a given point in time, de Haan et al. (2016) proposed four levels: (a) total diversity, (b) relative diversity, (c) spatial diversity, and (d) associated collective (traditional) knowledge. The first three levels implicitly involve population genetics. Total diversity quantifies the total number of unique genotypes or alleles present in a population, as in catalogues (Scurrah et al. 2013) or population genetic databases (de Haan et al. 2013), whereas relative diversity indicators focus on frequencies (red listing). In addition, spatial diversity metrics can characterize the distribution of populations in terms of latitude, longitude, altitude, and agro-ecological range.

Applying systematic monitoring approaches to agrobiodiversity in landraces, wild crop relatives, or animal breeds involves several challenges. Ideally, methods across sites and time series should to some extent be standardized. This requires agreement on the tools and metrics to be used (Maxted and Kell 2009). Molecular and genomic tools available for population genetics change fast, yet such change can be circumstantiated by establishing time-tagged DNA banks as reference populations. Another challenge concerns monitoring efforts that focus on wild flora and fauna, and involves the system level at which monitoring is conducted. Monitoring intraspecific diversity is particularly important for crops and livestock, and this requires appropriate methods for monitoring landraces, breeds, genes, and alleles. Such methods are more tedious and expensive than species-level monitoring. Importantly, good baseline information is essential but frequently absent or biased.

Ideally, future monitoring of crop–livestock populations would establish a network of representative benchmark sites for key species and geographies. A network of long-term observatories could include many different scientific disciplines to provide a solid evidence base for documenting the evolution of agrobiodiversity. On this basis, comparisons would be possible among crop–livestock species, geographies, farming systems, and ethnic groups. Furthermore, active linkages among conservation approaches (*in situ* and *ex situ* conservation), and thus potential intelligence concerning gaps in global gene bank collections, could be actively pursued (see Chapters 2 and 14).

Discussion and Future Outlook

There are many dimensions to agrobiodiversity that must be assessed. Here, we provide a summary and discuss how population genetics can contribute some answers.

Time

From a historical perspective, population genetics enables us to reconstruct the evolution of key events and stages for a given species. Usually, however, it is challenging to obtain a full picture of changes that have occurred throughout history. Evolutionary stages typically involve the onset of domestication (through either single or multiple events), an *in situ* increase in desirable alleles, continued formation of cultivated populations adapted to new environments and cultural preferences, and deliberate breeding and selection (Meyer and Purugganan 2013). Based on genomics as well as linguistic and archeological studies, it is often possible to reconstruct time-tagged evolutionary events, such as the proximate date of the original domestication, geographical spread, and crop wild relative ancestors involved at different points in time. Recent studies have identified genes associated with the initial domestication and subsequent diversification of agrobiodiversity. Using quantitative trait locus mapping, genome-wide association studies, and whole genome resequencing, new studies reveal the functions of genes involved in the evolution of crops under domestication, types of mutations as well as the drivers that trigger these mutations in response to geographical adaptation or exposure to stressors (Huang et al. 2012; Wright et al. 2005). Although the new tools available for population genetics studies have an increasingly high level of resolution, the lack of high-resolution baseline data limits temporal comparisons at the level of the gene pool. From a contemporary research perspective, these new tools provide a unique opportunity to develop robust baselines for future time series research.

Spatial Scales

Population genetics can be addressed at different geographical scales, where genetic diversity is commonly the richest in centers of origin. Interestingly, however, the genetic diversity of some crops is higher, or as high, outside their primary centers of diversity. Although the level of resolution of population genetic inquiry at the global, regional, or hotspot level is often different (Choudhury et al. 2013; Ríos et al. 2007; van Etten et al. 2008; Vigouroux et al. 2008; Zimmerer and Douches 1991), the integration of scales requires particular attention and is rarely addressed in either population genetics or spatial studies involving agrobiodiversity. Spatial dimensions with a clear population genetic link include agroecological patterning, land use, and seed flows.

Drivers of Change

Population genetics studies can enable a finer-grained understanding of shifts within and among crop–livestock and wild relative populations, both within or outside centers of origin. Change is intrinsically linked to time and space; it is also widely accepted that change has been a constant in shaping the total and relative diversity, population genetic structure, and distribution pattern of agro-biodiversity. Typical drivers of change assumed to affect population genetics (and hence the conservation status) of agrobiodiversity include war and conflict (Sperling 2001), technological innovation (Brush et al. 1992), and climate change, among other factors (for further discussion, see Chapters 7 and 8). A dominant paradigm for agrobiodiversity research has been the notion of diversity loss or genetic erosion. However, for over half a century, discussions on crop genetic erosion have remained largely anecdotal: the presence of modern varieties in a farming system was (and still frequently is) taken as *prima facie* evidence of diversity loss (Dyer et al. 2014). Loss of *in situ* diversity has typically been considered a consequence of global crop improvement and landrace replacement. Studies on several key crops, including maize in Mexico and potato in the Andes, however, have shown that contemporary species and intraspecific *in situ* diversity is still high and evidence for genetic erosion limited (de Haan et al. 2010; Dyer et al. 2014; Monteros 2011). While the dynamics underlying the population genetics of *in situ* populations is not static, and thus includes allelic loss and enrichment in confined spaces, it rarely results in a full-fledged wipeout of an entire gene pool (Bretting and Duvick 1997; Huang et al. 2007; van de Wouw et al. 2009). In-depth research on drivers of change and the interplay with crop–livestock population genetics (including temporal–spatial dimensions) is needed to enable a more fine-grained understanding of contemporary evolution (see Chapter 6).

The Conservation System

In situ and *ex situ* approaches to conservation have both different and complementary needs from a population genetics perspective. For gene bank applications, it is useful to obtain knowledge of the structure of the *ex situ* conserved population, define core collections, maintain population integrity, prevent genetic drift, and to develop regeneration protocols and robust sampling procedures (e.g., Camadro 2012). For *in situ* conservation, population genetics has applications to measure the temporal, spatial, and change dimensions outlined above. Highly complementary applications for filling gaps and obtaining intelligence about the representativeness of particular populations can potentially benefit both conservation systems (see Chapter 2). Indeed, it is now increasingly recognized that the long-term creation and maintenance of genetic diversity for plant breeding will depend on better ties to on-farm and *in situ* conservation and selection (FAO 2010b; Jansky et al. 2015).

Biosystematics

Population genetics addresses diversity at the intraspecific level where subspecies, cultivar groups, cultivars (landraces, races, bred varieties), and allelic diversity are common taxa. Cultivar groups and cultivars, in particular, are commonly recognized in folk taxonomic systems. The formal or informal structure of populations at the intraspecific level themselves can be subject to genetic analysis, for example, comparing folk taxa (Moscoe et al. 2017) or subspecies (Garris et al. 2005). Another approach is to consider only the lowest ranks (i.e., allelic diversity) and focus population genetics inquiry at this level. The choice of single or multiple intraspecific ranks has implications for linkages to the other dimensions outlined above. One example involves races of maize in Mexico and Peru that are typically separated geospatially (Grobman et al. 1961; Wellhausen et al. 1952). Another involves conservation systems and the conservation of vegetatively propagated crops as either genetically fixed landraces or as botanical seed. What do we conserve in this case: gene pools or landrace stocks? While population genetics offers a "lens" through which we can study agrobiodiversity, the biosystematic scale has implications for the inferences that can be made.

Conclusions

Population genetics has the ability to address crucial issues inherent to agrobiodiversity research with a high level of resolution and precision. Furthermore, it provides the means to compare crop–livestock populations systematically in space and time to inform diverse actors (e.g., researchers, conservation practitioners, decision makers) as to the conservation status of agrobiodiversity amidst global environmental change. Population genetics is not the only "lens" available, but it can be an integral part of a comprehensive monitoring approach that also considers morphological, cultural, or other factors. Clearly, a multidisciplinary approach to the study of agrobiodiversity is needed: population genomics offers both the opportunity and the tools needed to answer questions about how agrobiodiversity accumulates and is shaped by short- and long-term environmental and social changes. One major challenge is the need for coordinated, practical, and replicable frameworks that are standardized, robust, and accessible.

4

Crop and Varietal Diversity Impacts on Agroecosystem Function and Resilience

Steven J. Vanek

Abstract

This chapter identifies major areas of impact from agrobiodiversity at crop inter- and intraspecific levels on agroecosystem functioning relevant to management and resilience of cropping systems. Impacts from biodiversity on agroecosystem function are summarized as are impacts from pollination services, pest and disease impacts and resistance, soil biota and soil nutrient cycling, and abiotic stress resistance. The distinction between production characteristics related to plant phenotypes (provisioning services of ecosystems) and functional traits that support ecosystem services (supporting services of ecosystems) is highlighted, including the tendency for there to be trade-offs between these two and the need to harmonize them to a greater degree for agroecosystem resilience. Discussion follows on how these production and supporting services are linked also to wider social and economic contexts and ecosystem resilience. Important questions, challenges, and research areas are raised that may be productive in the scientific framework for sustainability proposed in this volume.

Introduction

The relationship between biodiversity and ecosystem function (Cardinale et al. 2012) is a broad and important question with far-reaching implications for human management and policy decisions (Hooper et al. 2005). A more narrow version of this question, and one that evokes a great deal of practical significance as well as a socioecological analysis and relation to both global change and food systems studies, involves the relationship between functioning of agroecosystems and the inter- and intraspecific (or crop varietal) agrobiodiversity created by human land managers. This chapter reviews what is currently known about biodiversity and ecosystem function in agroecosystems, addressing both the categories of impacts that diversity creates within agroecosystems (e.g., pollinators, pests, soils) as well as the wider socioecological context that drives the choices of land managers and receives the productive and supporting services created by agroecosystems. Where appropriate, I highlight important areas that would benefit from further discussion, future study, and practical action, especially in regards to local knowledge and management strategies of farmers, basic science, plant breeding, and promotion of resilience to climate change.

Conceptual Framework: Types of Agrobiodiversity, Axes of Trait Variation, and Agroecosystem/Ecosystem Distinctions

Genotypic and Functional Diversity

It is useful to create a heuristic framework for the types and functions of agrobiodiversity discussed in this chapter. This need arises from the fundamental difference between species or genotypic diversity related to numbers and evenness in the distribution of species or varieties, and the concept of functional diversity which prioritizes key functions and functional groups as well as the redundancy of organisms performing these functions (e.g., primary producers carrying out C fixation, plant disease pathogens and their microbial antagonists, nitrogen fixers, and plants which host or facilitate pollinators).

Distinguishing genotypic from functional diversity is not trivial given that the imperative to maintain a critical mass of germplasm resources may cause scientists and practitioners to think expressly about species and varietal numbers as an estimate of each crop having a full complement of possible traits (including those perhaps not yet thought of as important or functional). By contrast, consideration of the functional role of agrobiodiversity and its relation to associated biodiversity (e.g., weeds, arthropods, and soil biota), ecosystem services, and human nutrition causes us to think more about the need for functional traits and functional redundancy to be retained in systems (Chapter 2).

Social and Ecological "Axes" of Mediation of Functional Traits for Agrobiodiversity

In this chapter, I will focus primarily on the latter concept of functional diversity, in which the phenotype of a species or variety *in relation to* a particular function, process, or other mediating condition within an agroecosystem is the central building block of agrobiodiversity within agroecosystems (vertical axis in Figure 4.1). However, these functional traits in agroecosystems also interface with local socioecological and knowledge systems that create and sustain agrobiodiversity. Cultural preferences, socioeconomic influences, and knowledge systems may value a number of key genotypes or landraces for cultural or taste reasons that complement but also go beyond their

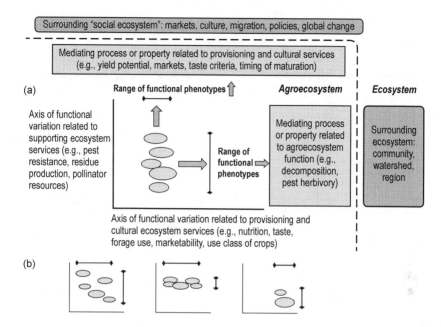

Figure 4.1   Conceptual diagram of diversity of crops and crop varieties (ovals) within agroecosystems: (a) Different variation in production characteristics important to yield, taste, market, and overall livelihood provisioning (horizontal axis), contrasted with functional characteristics related to agroecosystem function (vertical axis). Each functional trait for an ecosystem service needs to be referred to a mediating process where the phenotypic variation of the trait can have an impact (e.g., residue amount and quality for decomposition, pollinator attraction). Functional traits also are driven by and affect both a surrounding natural ecosystem and "social ecosystem" beyond the boundaries of the agroecosystem. Multiple vertical axes would be needed to represent multiple functional traits for these ecosystem services. (b) Depiction of different scenarios that combine production characteristics and functional traits for agroecosystem functioning. From left to right: wide range of both; wide range of production characteristics and narrow range of functional traits; collapse of production characteristics and range of functional traits for agroecosystems (e.g., industrial monocultures).

functional traits for agroecosystems (e.g., Brush and Perales 2007; Jarvis et al. 2008b; Swift et al. 2004; Zimmerer 2014). These taste, cultural, and market considerations can be thought of as a separate social "axis" and set of mediating processes governing the management of crop and landscape use and varietal diversity. Note that both of these axes are not immutable or objectively determined. They are projections from different disciplinary interpretations of underlying genotypic diversity and ecosystem processes, and are thus subject to discussion and recombination, and might well contrast with local knowledge systems regarding agrobiodiversity and functional roles of species.

Within this interpretive framework, human crop domestication, varietal proliferation, and the suites of agroecological management (agrodiversity;

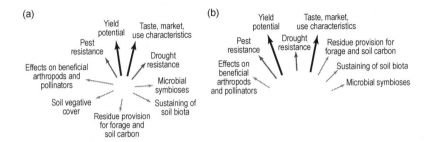

Figure 4.2   Two (potentially extreme) scenarios: (a) Provisioning services and yield characteristics dominate while supporting services are secondary; production and agroecosystem sustainability are not aligned and subject to trade-offs. (b) Provisioning services and yield characteristics still dominate but supporting services are aligned through changes in perception and investment, increased abiotic and biotic stress risk, and innovations that capitalize on synergies, breeding, soil management, and investment.

Brookfield et al. 2002) accompanying crop diversity represent different instances of the need to secure

- the biomass or harvest appropriated by human management, and the characteristics of that appropriated biomass (e.g., nutritional diversity, food vs. forage uses of crops), sometimes referred to as provisioning ecosystem services and aligned to a greater degree with the horizontal axis in Figure 4.1; and
- the opportunity presented for the suite of species and associated agrobiodiversity to support agroecosystem functioning (or regulating and supporting ecosystem services) by expressing a set of functional traits aligned along the vertical axis (e.g., inclusion of a nitrogen fixer, use of cereals with high yields of accessory forage biomass to feed animals and recycle manure to soils).

As noted by Power (2010) and González-Esquivel et al. (2015), these two goals frequently involve trade-offs for land managers. Aligning these goals to the greatest extent possible is a central and laudable goal for land managers, researchers, and policy makers (see Figure 4.2). In addition, farmers' knowledge includes, and sometimes ignores, the ecological aspects of agrobiodiversity (e.g., Cerdán et al. 2012; Sileshi et al. 2009) such that systematization and further research regarding local knowledge of agroecosystem function relating to crop and varietal diversity is a priority within the heuristic framework represented in Figure 4.1.

Agroecosystems versus Ecosystems

In considering agrobiodiversity management by humans, it is useful to distinguish between impacts that occur *within* the agroecosystem (e.g., provision of food and forage to livestock, pest resistance, positive interactions among crops within a rotation) and impacts and interactions *between* the agroecosystem and the larger ecosystem (e.g., natural, peri-urban, watershed, regional ecosystems at different scales). Humans are the overall "keystone species" within agroecosystems (e.g., Stahl 2015) through the management of crop and livestock types, which are in some sense deployed to be joint keystone species as principal biomass providers (crops) and primary consumers (livestock). Human management, especially in annual cropping systems, therefore imposes a plant community with a few dominant species intended for biomass, seed, and fruit production as food, animal forage, and saleable products.

Pastures, unmanaged fallows, as well as managed forests and hedgerows within agroecosystems exhibit this tendency to a lesser degree, being more diverse and "wild" and highlighting the important roles of associated agrobiodiversity. Nevertheless, they form part of a managed whole where it is usually easy to pick out influential "keystone" crop and livestock species that dominate an agroecosystem.

Meanwhile, agroecosystems interact with wider ecosystems (e.g., facilitating and benefiting from pollinators of noncrop species, watershed-level impacts on hydrology from field and hedgerow matrices, and carbon sequestration impacts). Often the relationship between agroecosystem diversity and larger ecosystem function creates and sustains supporting ecosystem services that are public goods because they exist within a shared regional or community ecosystem, partially explaining the economic trade-offs between the locally perceived benefit within a managed agroecosystem and the wider benefit to the larger ecosystem (Power 2010). In this chapter, I will occasionally use the term "ecosystem" in a generic sense to refer, for example, to "the soil ecosystem," and "ecosystem services."

Agrobiodiversity, farmer knowledge, and seed systems (whether farmer-managed or largely exogenous) exist, therefore, within agroecosystems, farms, and communities, and they also link to larger natural and social environments. In accordance with the framework of functional axes presented in Figure 4.1, the first of these along the vertical or ecological axis is the traditional definition of a landscape-level environment or ecosystem; the second represents the wider cultural, market, knowledge, and otherwise "social" ecosystem of regional and global factors that codetermine the development and fate of agrobiodiversity and their impacts on local agroecosystems and ecosystems (Keleman et al. 2009; for further exploration of these historical and global frameworks for modeling agrobiodiversity, see Chapter 2 and Zimmerer and Vanek 2016). In addition, the consideration of environmental and social contexts along with potential trade-offs between production and other services from agrobiodiversity prompts the question of where the impacts and incentives to maintain agrobiodiversity are adequate and where they need to be supported, which are addressed at the end of this chapter.

Overview of Mechanisms That Explain Agrobiodiversity Impacts

Before considering ecological and social mediating processes for the functional importance of biodiversity, it is useful to address the general types of mechanisms that operate to create positive impacts. In an earlier review that addressed biodiversity and the functioning of agroecosystems, Hajjar et al. (2008) allude to fundamental hypotheses about why diverse systems are more productive, self-regulating, and resilient. Borrowing from their framework (Figure 4.3), these types of facilitation include

- improved resource capture via *complementarity in niches*, leading to increased amount or temporal duration of biomass and rooting on fields (e.g., complementary rooting strategies in soils that increase primary production and fuel other ecosystem services);
- improvements in function based on a greater *number of functional traits*, especially with regards to key functional traits such as pest resistance, abiotic stress in the face of perturbations, or a key attractive or habitat function for pollinators; and
- *direct facilitation*, such as reduced spread of a pathogen or insect pest, provision of habitat to antagonists or beneficial insect predators, or the nutrient cycling benefits of biomass that is more nutrient rich or easily decomposable, and effects on subsequent crops (Figure 4.3).

In addition to these important distinctions, this chapter incorporates the insight that Tilman and other authors make with reference to the concept of a "sampling effect" in increasing the number of functional phenotypes: rather than diversity per se being linked to benefits to agroecosystem function, higher biodiversity increases the likelihood of including a particularly influential phenotype or functional trait (e.g., Fargione and Tilman 2005). For example, adding a well-adapted and productive legume to a system without any legumes will tend to dramatically alter nutrient cycling and nutrient availability.

Hajjar et al. (2008) also state the widely appreciated principle that agroecosystems differ in key ways in comparison to natural ecosystems, thus making it difficult to extrapolate biodiversity findings from natural systems to agricultural systems. As noted above, agroecosystems exist at the intersection of social and environmental contextual factors, helping to explain this difference. Along with the initial impression of difficulty and the important suggestion that more should be done to understand agroecosystems on their own terms, seeing agroecosystems as managed elements of socioecological food systems actually helps to focus our thinking and to develop research questions about agrobiodiversity and the functioning of agroecosystems which may be of use in assessing the beneficial effects of species and varietal diversity on agroecosystems. First, the presence of humans as an ordering "keystone species," which recruits codominant plant and livestock species via the development of agrobiodiversity, highlights the key interactions

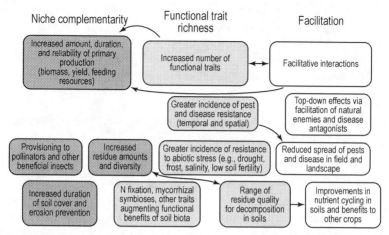

Figure 4.3 General classes of positive impacts from biodiversity on ecosystem function. These impacts fall roughly into categories such as increased resource capture from niche complementarity, increases in functional traits that may include particular functional traits that provide resilience and benefits, and facilitation effects involving sometimes complex relations among varieties, species, and trophic levels. The way that trait richness and facilitation can also positively impact the provision of biomass and other resources for ecosystem function is noted by arrows along the top, while relations between residue amounts, diversity of quality, and facilitative effects among species by improvements in nutrient cycling from residues are also noted. Additional examples of interactions and impacts, and relations between them, may also be present. Adapted from Hajjar et al. (2008).

between crops and associated biota, for example, even if the relationships are as complex as they can be in natural ecosystems. Second, given that we are often interested in varietal diversity as a component of these managed systems, the criteria that Hughes et al. (2008) delineate for intraspecific diversity impacting functional diversity support the notion that varieties can have impacts on ecosystem function:

1. Communities are dominated by a small number of keystone species.
2. Genetic diversity of one species directly affects the abundance, distribution, or function of a keystone species.
3. Genetic diversity is reflected in trait diversity.
4. Genetic diversity is present in changing and variable environments.

These criteria are very often satisfied in one or more ways in agroecosystems. This makes it seem likely that varietal diversity can, in fact, alter ecosystem function, the hypothesis that will be queried using literature for different classes of ecosystem services in the next section. Examining these criteria one by one for agrobiodiversity, it is evident that

- crops and livestock are usually dominant or keystone species imposed by human managers to appropriate harvest and biomass;

S. J. Vanek

- the range of varieties deployed may (or may not) bear relation to the functional role or distribution of crops;
- trait diversity is, of course, the key genotype factor manipulated by farmer selection and formal breeding methods, although it is important to understand the degree to which these traits are of functional relevance for supporting ecosystem services (Figure 4.1); and
- the near certainty that agroecosystems face varying conditions, in terms of normal fluctuations and long-term changes (e.g., climate change), draws attention to resilience and how varietal diversity can affect the performance of crop fields that face drought or disease.

Two additional conceptual contributions deserve mention here that extend the application of the general framework for agrobiodiversity impacts presented above, and counter the risk that these agrobiodiversity impacts are understood in a piecemeal versus a holistic manner. First, the concept of food webs has been underutilized in the assessment of the role of agrobiodiversity (e.g., Tixier et al. 2013). Emphasis on manager choices and production impacts in more agronomic and human-system oriented literature has tended to narrow the focus on particular functional traits and their relation to culture, pests, or markets, rather than on an overall appreciation of crops and livestock as primary producers and consumers within trophic levels and food webs. Considering food webs as an additional level of analysis can lead to productive appraisal of crop and varietal diversity and their impact on soil biota and pests, for example. These food web considerations are present as key parts of Figure 4.3 (e.g., biomass and pollinator resource provisioning) and will be important in formulating new research questions (see below). Second, Wood et al. (2015a) have proposed that agrobiodiversity impacts can best be understood in relation to ecosystem services by measuring functional trait impacts across environmental and management gradients. This knowledge then becomes a management tool for proposing trait assemblages to improve ecosystem services. These authors also highlight the need to understand the spatial nature of functional traits: for example, detailing the way in which landscape patches expressing key functional traits interact with surrounding land uses.

Categories of Impact: Pollination, Pests and Diseases, Soil Ecosystem Services, Abiotic Stress, and Local Knowledge Regarding Impacts

Let us now consider the categories of impacts on ecosystem function: pollination, disease and pest impacts; soil biota and soil ecosystem services; resilience to climate change with an emphasis on abiotic stresses, particularly drought; and the effects of local knowledge on functional diversity. In general, and for practical purposes, emphasis is given to positive interactions, although negative interactions are a constant in agroecosystems (e.g., herbivory, disease transmission, competitive use of resources, and competition between appropriation of biomass and its benefits to agroecosystem function). Finally, local knowledge of these types of impacts is an important area of ongoing research that can interact productively with the formal scientific knowledge presented here.

Agrobiodiversity and Pollination

Several plausible and demonstrated mechanisms link interspecific diversity to increased function of pollinators in agricultural ecosystems, which is potentially of great importance given the large ecosystem service value provided by pollinators globally, estimated at thousands of USD per hectare (for wild pollinators, see, e.g., Gill et al. 2016). The most obvious among the mechanisms linking agrobiodiversity and pollinator services is the fact that a diverse crop field or landscape arrangement of species provides a more varied and temporally stable habitat and feeding options for pollinators, tending to sustain their diversity and function in agroecosystems as well as in wider ecosystems and landscapes. Data analysis from 39 globally representative ecosystems has showed that diverse fields with mixed crop and border as well as organic management had higher wild bee abundances and species richness (Kennedy et al. 2013). This finding is in agreement with other studies, including those on weedy components of associated agrobiodiversity (e.g., Bretagnolle and Gaba 2015; Morandin and Kremen 2013) and is particularly relevant in light of the documented predominance of wild bee species in crop pollination (e.g., Imbach et al. 2017; Mallinger and Gratton 2015; Winfree et al. 2008). Temporal diversity of crop and cover crop sequences has also been shown to influence pollinator communities: the species of cover crops employed and planting date influence the bee communities visiting cover crop flowers (Ellis and Barbercheck 2015). However, other research provides counterexamples by showing that land-use diversity had little impact on pollinator diversity (e.g., Wood et al. 2015b) or emphasizing the importance of wild habitat in sustaining pollinator function in contrast to the type of agricultural landscape implemented (Forrest et al. 2015). In addition, while confirming hedgerow impacts on maintenance of pollinator diversity, Sardinas and Kremen (2015) point out that the stability and diversity of pollinator populations, which tend to correspond to habitat diversity, may differ from the pollination services at any given moment, which may not respond to crop and border diversity.

Intraspecific diversity impacts on pollinators are less well documented, although we would expect that these conclusions regarding interspecific diversity would extend to varietal diversity among crops, at least in circumstances where there are clear impacts of phenology on abundance or stability of habitat and feeding resources for pollinators. The staggered availability of pollen and nectar from maize landraces of differing maturation time in Yucatán, Mexico, provides an interesting example (Tuxill 2005; Tuxill et al. 2010), which likely serves to maintain pollinator communities in highly variable wet or dry environments. However, it has not been researched sufficiently how widespread this sort of staggered resource provisioning is in "traditional" smallholder systems, many of which increasingly incorporate modern-bred varieties, which could serve to restrict or expand the range of phenologies. The varietal impacts on resource provisioning to associated agrobiodiversity (pollinators and other components explored below) in smallholder and low-input agroecosystems merits additional research attention, since impacts on pollination from varietal phenology likely parallel impacts on other aspects of ecosystem function considered below: crop residues for forage and soil nutrient cycling, pest and disease resistance of crop assemblages, and overall cropping system resilience to drought and other abiotic stress (as in this example from Mexican smallholder agroecosystems).

Other crop examples show patterns of facilitation and the importance of particular functional traits, as compared to the overall abundance or stability of feeding resources described above. For instance, in almond orchards, pollinator-preferred varieties have been shown to confer "accidental" pollination benefits to adjacent rows of almond trees of a more market-preferred variety (Jackson and Clarke 1991). At a more mechanistic level, interesting interactions of crop intraspecific and pollinator diversity have been demonstrated by analyses of mixed varietal plantings of sunflowers visited by nectar- and pollen-specializing species of pollinators plus generalist species. This created greater pollinator diversity, "chasing" specialized pollinators to more widely disperse their foraging patterns, and thus fostered greater cross-pollination in a diverse sunflower field (Greenleaf and Kremen 2006). At the level of crop traits, varietal differences in nectar and pollen resources for pollinators have been demonstrated in crops such as rapeseed (*Brassica napus* var. *oleifera*; Bertazzini and Forlani 2016), and could affect pollinators at a field or landscape level if these varietal differences were a focus of mixed planting schemes or breeding approaches.

In addition to the effects of functional and phenotypic diversity per se in fostering greater resource abundance, stability, and diversity for pollinators, we need to consider the impact that management of crop types or crop varieties within rotations (*agrodiversity* associated with agrobiodiversity) can exert on pollinators. Here, the use of particular harmful pesticides implicated in bee decline is often associated with the use of modern, relatively nondiverse crop assemblages in industrial farming systems. For a discussion on neonicotinoids, see Henry et al. (2015), Krupke et al. (2012), and Pettis et al. (2013); for methodological issues surrounding contradictory findings on neonicotinoid contributions to bee decline, see Schaafsma et al. (2016).

Agrobiodiversity and Biotic Stressors: Pests, Diseases, and Weeds

The principles of agrobiodiversity and risk limitation from pests and diseases via interspecific diversity is well known from the principle of crop rotation, with particular sequences and pairs of crops known to limit incidence of pests and diseases (e.g., Wright et al. 2015) as well as the pest and disease benefits of diverse landscape assemblages (reviewed in detail for low-input small-holder systems by Ratnadass et al. 2012). Rotational diversity is an established principle, and crop species-level and land-use diversity are productive ongoing research priorities for the development of ecologically based and more sustainable management. It is also true that there are particular cases where species diversity per se does not foster benefits in ecosystem functioning. A classic example is the emergence of suppressive soils under wheat and barley rotational monocultures, which tend to suppress the take-all disease caused by *Gaeumannomyces graminis* var. *tritici*, in comparison to more diverse rotations with multiple crop species (Kwak and Weller 2013). Schroth et al. (2000) also point out that in agroforestry systems, if diversified perennial components include alternate hosts for crop diseases or other pests, greater species diversity can increase pests and diseases.

Beyond rotational diversity in time, there are myriad examples of the use of intraspecific diversity and facilitative interactions in mixed cropping to manage pest and disease incidence. In a review of intraspecific diversity and its deployment to manage diseases and arthropod pests, Tooker and Frank (2012) list many examples that show an overall positive response of pest and disease resistance to intraspecific diversity, including effects on plant biomass and other productive indices (see Table 1 in Tooker and Frank 2012). They note that facilitation effects among crop varieties to reduce pests and diseases can be either *bottom up* or *top down* with respect to crops, pests, predators, and antagonists within trophic levels. *Bottom-up* effects are exemplified by direct varietal resistance to a disease or arthropod pest that hampers its primary consumer role to spread through a mixed varietal population. *Top-down* effects are manifested through the limitation of pests via facilitation of antagonist organisms to the pest or disease (e.g., through beneficial predators of arthropod pests). Practical examples of bottom-up effects are mixed plantings of wheat and barley used to control rust, powdery mildew, and other diseases, which currently comprise significant proportions of land areas (e.g., 7–50%) planted in Europe, Asia, and the western area of the United States (also Creissen et al. 2016; Mundt 2002). Shoffner and Tooker (2013) also demonstrated the ability of mixed plantings of wheat to deter aphid populations in comparison to monoculture, perhaps related to greater volatile emissions by the varietal mixtures. Mulumba et al. (2012) found that in smallholder cropping systems of Uganda, greater varietal richness in bean and banana crops was associated with reductions in both the overall damage and the variance of impacts from a number of common bean and banana diseases. Top-down effects of diversity that facilitate the effects of beneficial arthropods and disease antagonist microbes include the observations in barley that mixtures of cultivars can increase the attraction for parasitoids and ladybird beetles that feed on aphid barley pests via some airborne means of plant–plant signaling (Glinwood et al. 2009). This observation parallels the interspecies communication found to limit aphid herbivory in potato due to intercropping with onion (Ninkovic et al. 2013). In addition, Jones et al. (2011) demonstrated that increases in parasitoid diversity (which could reduce pest incidence) in more diverse mixes of ryegrass cultivars occurs through complementarity among cultivars in habitat characteristics.

These interesting mechanisms suggest that positive influences of varietal mixtures to reduce pest and disease pressure are possible and may even be a dominant effect (Johnson et al. 2006). We lack, however, extensive testing of the intentional deployment of intraspecific diversity, or the retention of mixed cultivar plantings in intensifying smallholder systems. These include mixed-variety potato plantings in native Andean potato fields and current evolutionary breeding efforts that express a large genotypic diversity within the same field (e.g., Brush et al. 1981; Phillips and Wolfe 2005). As in the other categories of ecosystem function that we explore here, the management associated with crop assemblages of low or high varietal diversity may be as important as the interactions among varieties themselves. An obvious example is the chemically and transgenic intensive management and effects on pests, weeds, and beneficial insects at multiple trophic levels in industrial monocultures. Smallholder management can also provide counterexamples to the principle of improved function with diversity: Parsa et al. (2011) found a pest dilution effect of growing potatoes together in a single landscape block in adjacent farm fields of an Andean sectoral fallow scheme.

Weeds pose an important challenge within low-input farming systems as well as part of the associated biodiversity that these agroecosystems help to maintain. Crop varietal diversity is a part of strategies for managing weeds. For example, Midega et al. (2016) found that maize landraces were less susceptible to the parasitic weed striga than were improved varieties, whereas some wheat landraces were found to have better competitive ability against weeds related to plant height (Costanzo and Barberi 2016; Murphy et al. 2008). Rotational diversity among different crops and other land uses is an integral part of managing many weeds for farmers (Kremen and Miles 2012). In addition, and partly because they parallel the role of plants as primary producers in agroecosystems, weeds have beneficial functions as crop wild relatives that may even be a source of gene flow to crops (Beebe et al. 1997; Felix et al. 2014; Scurrah et al. 2008) and as hosts to beneficial soil microbes that may serve to maintain or augment populations of these microbes in soil (Sturz et al. 2001).

Soil Biota and Soil Ecosystem Services of Crop and Associated Diversity

Effects of crop biodiversity on soil biological communities and ecosystem services are potentially of great interest and importance given the tremendous importance of soils in nutrient retention, transformation, and supply to crops; water cycling and drought resilience of systems; and as a reservoir of carbon storage and biological diversity in Earth's biosphere (Barrios 2007). A previous review of the relationship between above- and belowground biodiversity (Hooper et al. 2000) suggests caution in assuming that any relation (positive, negative, none) is predominant between plant and soil biodiversity, due to the complexity, patchiness, and functional redundancy of soil ecosystems. Therefore, as in any probing of functional relations between agroecosystem diversity and function, it is important to posit and test plausible functional mechanisms.

In this vein, a meta-analysis of 50 studies focused on land-use and crop diversity in agroecosystems and livelihood spaces of smallholder agriculture (Zimmerer and Vanek 2016) found that land-use and crop diversity were associated predominantly with (a) changes in the community structure of soil biota (92% of cases), (b) increases in soil biological diversity (55% of cases), and (c) ecosystem functioning of soil biota (78% of cases). Findings of increased soil biodiversity were less frequent when changes in biodiversity were related to less dramatic changes of crop diversity within rotations versus those cases where diversity was altered by wholesale changes in land use. In addition, soil whole community structure, including in particular the structure of enormously diverse bacterial and archaeal communities, was impacted by longer-term (i.e., decadal) shifts in land use, whereas the diversity of soil macrofauna (e.g., earthworms, arthropods > 2 mm) as a functional indicator group of organisms as well as plant symbionts (e.g., rhizobia, mycorrhizas) responded more quickly and definitively to crop and land-use diversity. These results bear out the caution stated by Hooper et al. (2000), while allowing us to conceptualize direct associations between agricultural and soil biological complexes with associated management strategies into a single assemblage (see Figure 4.4) called an "AGSOBIO" by Zimmerer et al. (2015). There are at least two direct mechanisms that operate via important functions to link crop and soil biodiversity and soil ecosystem services. Crop residue (shoots and roots) properties—such as quantity and quality (i.e., decomposition rate and other functional attributes) as well as timing of residue and root exudate introduction to soils—constitute the first mechanism and is based on plants' role as the dominant terrestrial primary producers for both aboveground and belowground soil ecosystems. The second is the relationship of crop species and variety to their ability to host plant symbionts and associative rhizosphere microbes: some are more generalized across plant species (e.g., *Pseudomonas* spp. as a common type of rhizosphere bacterium) while others are specific to crop species (e.g., rhizobial strains and specificity to groups of legume

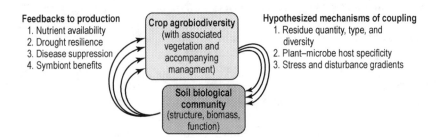

Figure 4.4   Conceptualization of the direct associations between agricultural and soil biological complexes and associated management strategies, along with mediating linkages and effects on production. After Zimmerer and Vanek (2016).

species). In addition to these direct mechanisms, there are likely differences that derive from the associated management and inputs of fertilizer and manure that accompany land uses and assemblages of crops. These are not completely separate from the assemblage of crops in many cases, since manure sources in the system, for example, are usually based on forage resources that are partly or wholly locally sourced.

In comparison to the impact of land use and crop rotation, the impacts of intraspecific diversity on soil biotic diversity and function have not been well documented, especially their functional importance to agrobiodiverse systems managed by smallholders. One example which demonstrates the plausibility of intraspecific differences in effects on soil biota is the finding that maize genotype contributes an appreciable amount of variation to differences in the microbiome of the maize rhizosphere (Peiffer et al. 2013) when analyzed at a whole community level. In other research, chickpea genotypes were also shown to establish different rhizosphere microbiomes; however, these differences disappeared when both plants and microbes experienced drought stress (Ellouze et al. 2013).

In more general terms, Johnson et al. (2012) advanced the importance of understanding the connection between intraspecific variation and mycorrhizal colonization as a highly functional component of agroecosystems. Martinez and Johnson (2010) found that maize landraces of the southwestern United States were more responsive to arbuscular mycorrhizas, in terms of infection rate than modern varieties, and were able to carry out more mutualistic symbioses in soils managed for lower fertility. A similar variation in mycorrhizal responsiveness among native landraces and one hybrid variety was found in Central Mexico (Sangabriel-Conde et al. 2014). Relations between arbuscular mycorrhizas and crop genotypes may be especially important because of the role of mycorrhizas in accessing nutrients under drought conditions that may confer resilience to climate change (see below). Nevertheless, the relationship between crop varietal diversity and the functional role of soil biota is an area where we may expect the null counterhypothesis of "no significant impact" to be especially strong, given large effects of land use and species, and large amounts of noise from agroecosystem variability.

The cited work on arbuscular mycorrhizas suggests that to derive actionable knowledge on cultivar diversity and soil biota, it may be useful to look for particular beneficial impacts in the context of participatory breeding and evaluation work with germplasm from widely varying management contexts (e.g., landraces vs. modern hybrids, high fertility or chemically intensive "infields" vs. marginal "outfields" in smallholder systems). A final note is that arbuscular mycorrhizas have been particularly studied, and for longer, because (a) they are thought to have large functional impacts on crops as key symbionts and (b) they are relatively easy to study. Similar important conclusions regarding other more elusive rhizosphere-associated microbes are already emerging from research and will continue to be elucidated in coming years.

For highly redundant soil biotic communities (e.g., bacteria, archaea, soil fungi beyond mycorrhizas and endophytes), where species richness is bewilderingly large and does not relate explicitly to function, it is important to understand the functionality of the community composition rather than species identities. Research that investigates the impacts of different management paradigms (e.g., conventional, organic) rather than crop diversity per se (including residue inputs related to crop species and landscape-level diversity) suggests hypotheses and methods that could be applied to the impact of crop varieties, varietal assemblages, and related management on microbial community function. For instance, differences in key functional enzymes and genes linked to particular soil nutrient cycling functions can indicate greater potential of soil biota to carry out nutrient cycling functions resulting from different crop diversity and management cases (Acosta-Martínez et al. 2010). Assessing the impact of management in promoting microbes adapted to higher resource availability in soils, including nitrogen availability (Fierer et al. 2012) or high-organic matter versus low-organic matter input systems (Berthrong et al. 2013), may also be seen as promising models for testing different crop and crop varietal assemblages.

In more general terms, we need to view hypothesized plant–soil assemblages from the standpoint of how each component (crops, soil biota) responds and influences the other. Once again, mycorrhizas serve as a well-studied "model system": arbuscular mycorrhizas' adaptation to soil environment, which includes both nutrient levels and other soil microbes, and the effect of arbuscular mycorrhizas on the same species of plant across multiple sites (Johnson et al. 2010) indicate the existence of mycorrhizal ecotypes which mirror the aboveground varietal effects on crop–soil biotic relations. This is a more "mycocentric" approach to the AGSOBIO hypothesis cited above. As in the other ecosystem services, this necessarily involves soil fertility, water, and soil organic matter management and their effects on the coemergent assemblage of crop varieties, microbes, and soil fauna (e.g., nematodes, collembola, macrofauna) in a system. The coupling effect of crop residues may be particularly important in this regard. For example, Mexican maize landraces with different phenologies, cited above for their impacts on pollinators and drought resilience (Tuxill et al. 2010), create plausible effects on the timing of root exudate release as well as root and leaf senescence in soils. This, in turn, would create different dynamics in the soil microbiome and maintenance of rhizosphere habitat across the growing season (albeit in spatially separated soils, unless these maize varieties were associated).

Another example is the differing varietal assemblages of maize in Mexico deployed either to maximize earnings in fertile soils (modern varieties) or to minimize risk and protect fragile and infertile soils (rustic landraces; Bellon and Taylor 1993), which presumably are associated with differing soil biota via differing management and varietal characteristics. More research is, however, needed to understand the drivers of such assemblages, since varietal assemblages responding to different cultural and market demands that exhibit this mixing of modern and landrace varieties are relatively common in smallholder systems with maize, potato, and millet main crops, among others.

Varietal diversity may also affect soil biota and ecosystem functioning at a systemic level by affecting the spatial architecture of the soil ecosystem. As pointed out by Hooper et al. (2000) and Lavelle et al. (2004), the soil ecosystem is exceptionally complex and ordered with different functions corresponding to different spatial scales: nutrient transformations at the scale of single soil and organic matter particles; root–microbe interactions at the scale of root microcosms composed of many particles and aggregates; macroarchitecture and organic matter redistribution produced by macrofaunal "ecosystem engineers" at the scale of centimeters (earthworms, termites, and other soil arthropods; Lavelle et al. 2004). Varietal differences and associated management could support these functions, either by contributing different amounts of residues to soil ecosystems that foster greater macrofaunal "engineering" activities, or directly through the structuring activity of root architecture, which has been incorporated as a breeding goal related to crop phenotype for water and nutrient access (Bishopp and Lynch 2015; Szoboszlay et al. 2015).

Abiotic Stress and Cropping System Resilience to Climate Change

Agrobiodiversity—especially the intraspecific variation of tolerance to drought and other stresses—is considered one of the major resources for crop breeding in response to climate change (e.g., Ortiz 2011). Diversity of landraces and interspecific crop diversity has long been employed within agroecosystems to stabilize yields in the face of drought and other abiotic (e.g., frost, low soil fertility) stresses (Condori et al. 2014; Tuxill et al. 2010). The use of intraspecific diversity in resilience to perturbations embodies the principle that genotypic and phenotypic diversity within a species can improve agroecosystem function when conditions are temporally unstable. The use of varietal mixtures of wheat in fields to increase drought resistance has been proposed in a detailed way by Adu-Gyamfi et al. (2015).

Similarly, the designed use of the local diversity of germplasm by participatory plant breeding has led to increases in drought (and disease) tolerance in wheat (Gyawali and Sthapit 2006). Analogous efforts to maintain or breed frost-tolerant varieties seem plausible and are ongoing in the case of Andean potato landraces for growing areas in the Central Andes (Condori et al. 2014). A household-level study of rice varietal diversity found that in an area dominated by modern rice cultivars (the Nepali Terai), varietal diversity on farm was perceived by farmers to decrease production loss to drought over a number of years (Bhandari 2009). Campbell et al. (2013) document that for monoecious crops like cucurbits, open-pollinated varieties which exhibit a wider range of flowering phenologies may be better adapted to risks of asynchronous male and female flowering and low fruit set under drought conditions.

Other studies of agrobiodiversity and climate change highlight antagonism and trade-offs between different overall strategies of adaptation deploying different types of germplasm. Mercer et al. (2012) contrast a current "transgenic adaptation" strategy for climate change in Mexican maize-based cropping systems (i.e., the deployment of transgenic, drought-adapted maize varieties) with evolutionary approaches to breeding and adaptation, using local varieties that maintain the cultural and economic importance of this local agrobiodiversity (see also Feitosa Vasconcelos et al. 2013). They suggest that farmers use the altitudinal adaptation of maize varieties as an adaptation strategy for climate change, as maize adaptation zones move up in highland Mexico, while noting the threat posed to high-elevation varieties due to warming climate (Mercer et al. 2008). Meanwhile, Mukanga et al. (2011) document the trade-offs in the maize systems of Zambia that are made between drought-tolerant, short-season hybrid varieties and longer-season, established local varieties which have better culinary characteristics and resistance to ear rots.

It is likely that varietal resistance to abiotic stress is not only an expression of plant morphological, physiological, biochemical, and genetic traits but may also be related to the plant symbioses and associations with rhizosphere microbes and endophytes. Varietal differences in relations to mycorrhizas, for example, may confer greater or lesser drought tolerance, since mycorrhizas can access phosphorus and other nutrients at higher levels of drought in soils. Mycorrhizas and other microbes also engage in complex signaling with plants that can enhance drought resistance (Belimov et al. 2015; Lopez-Raez 2016; Wittenmyer and Merbach 2005). These soil biotic and varietal differences in drought stress, as well as biotic factors that can benefit smallholder-managed soils, are reviewed by Fonte et al. (2012), including the role of rhizosphere associative microbes in allowing plants to mount more effective drought responses.

In a broad sense, associated management and the fate of residues from different varietal types (e.g., forage vs. grain varieties of oats or barley; accessory legumes, such as grass pea in India and Pakistan or pigeon pea in southern Africa, that generate abundant forage biomass that is then available for recycling to soil as manure) may also play an important role in mediating drought resilience in systems resulting from different levels or types of agrobiodiversity (see Figure 4.4). Root architecture and root biomass amounts dependent on species and cultivar choices may also have effects on the level of aggregation and thus infiltration and water retention of soils so that soils and crops are buffered from drought stress. Root architecture and root–microbial interaction functional traits are also being targeted for improvement in breeding programs, using traits that form an important part of agrobiodiversity management among smallholders that reflect crops' adaptation to soil biota and organic inputs, and nutrient stress (e.g., Bakker et al. 2012; Lammerts van Bueren et al. 2011; Schmidt et al. 2016).

Local Knowledge of Functional Diversity: Impacts on Agroecosystems

Because social, cultural, and economic considerations play a central role in decisions regarding agrobiodiversity (Figure 4.1), future research needs to consider the degree to which farmers view functional diversity in relation to agroecosystem function and other factors. Previous research has shed some light on this aspect of local knowledge, despite the fact that (perhaps justifiably) many studies on agrobiodiversity management address the socioenvironmental nexus pictured in Figure 4.1 in a wholesale way, rather than trying to section off farmers' conscious attention to goals such as pest management, nutrient availability, or drought adaptation. Farmers do, in fact, manage diversity to contribute to agroecosystem functionality, including the finding that coffee plantation managers ascribe functions such as soil formation or soil water retention to different tree species in shade coffee systems, and link these to functional traits of forest species that comprise associated agrobiodiversity in these systems (Cerdán et al. 2012), or the widespread knowledge of farmers about varietal differences in pest or drought susceptibility, which is likely enacted through crop choices (Mukanga et al. 2011; Okonya et al. 2014; Teshome et al. 1999).

Farmers certainly appreciate the regenerative properties of associated agrobiodiversity and vegetation, of livestock agrobiodiversity in fallowed fields, and the application of livestock manure in their systems (e.g., Pestalozzi 2000). Nevertheless, farmers may exhibit less knowledge about invisible pests and ecological processes such as plant parasitic nematodes in soil (Kagoda et al. 2010). Bentley (1991) argues this tendency to be more generally true for microscopic processes in agroecosystems versus macroscopic or landscape-level processes.

A hypothesis for future work is that farmer knowledge regarding general notions of soil fertility and macroscopic functional traits of agrobiodiversity, including some aspects of insect pest susceptibility, are readily incorporated along with other cultural and market factors that shape crop and varietal mixtures, whereas mechanistic impacts (e.g., microbial nutrient transformations, or disease development at a rhizosphere or leaf scale) are not readily perceived. Future research should consider these links between farmer knowledge and functional aspects of plant traits in agroecosystems, wherever agrobiodiversity is investigated in a detailed way. In this regard, the concept of agrobiodiversity "observatories" (see Chapter 3) may help us understand the trajectories of agrobiodiversity at a local and fine-grained level of study.

Research Issues Defining Future Efforts

As we seek to appreciate the linkages and mechanisms involved in agrobiodiversity, ecosystem function, and resilience, and to improve the functioning of systems that face challenges from intensification, climate change, and the erosion of biodiversity resources, many important issues come to the fore. In this concluding section, I delineate a number of these issues, highlight relevant literature, and propose potential directions that researchers may wish to pursue in the future.

Embracing the Complexity of Agrobiodiversity Assemblages, Agroecosystem Functions, and Social, Economic, and Ecological Drivers

As agroecosystem function and ecosystem contributions that derive from crop and livestock diversity are considered, we need to recognize that many smallholder systems now combine modern and traditional varieties, including many perceived "landraces" that represent crossed modern and traditional germplasm (e.g., van Heerwaarden et al. 2009). Often these assemblages are organized by parallel and overlapping strategies that involve market access for some varieties, cultural and taste importance for others, differential levels of earliness and seasonal food availability, and specific ecological imperatives for system sustainability, productivity, and pest management. Research is needed that appreciates this complex context and understands multiple rationales of crop choice and production. Variants of such approaches are embodied in participatory breeding alongside traditional and hybrid varieties in maize systems of southwest China (Li et al. 2012), the documenting of landrace diversity across elevation zones in confronting climate change in Mexico (Mercer et al. 2008), and the potentials and pitfalls of fostering finger millet agrobiodiversity in Nepal using urban niche markets (Pallante et al. 2016). Meanwhile, at a somewhat larger scale, the concept of "regional agrifood and livelihood diversification spaces" (Zimmerer and Vanek 2016) may be helpful to approach this complexity across multiple contexts, with impacts on soil biota, pests, pollinators, and other aspects of supporting services for agroecosystems.

Building an Evidence Base for Agroecosystem Function and Ecosystem Service Impacts of Agrobiodiversity

Notwithstanding the complexity and frequent nonecological drivers of agrobiodiversity discussed in this chapter, there is still a tremendous lack of evidence of the (a) agroecosystem impacts and services provided to larger ecosystems and (b) species and varietal assemblages *within smallholder and other farmer contexts*. Much of the existing evidence derives from on-station trials and results extrapolated from large-scale industrial agriculture. Where evidence does exist from the modernizing "traditional" systems mentioned in the previous section, it is limited to single research sites. Thus, models and regional data are needed to understand how widespread and ecologically important such findings are. In particular, in the context of smallholder and low-input agriculture, we need increased information on the relations between crop and varietal diversity and impacts on residue return to soils, microbial and arthropod communities as well as the architecture of soil and aboveground ecosystems as structured by plants and macrofauna, and other aspects that contribute to ecosystem function. In addition, it is vital to understand whether and how managers perceive and value these outputs in terms of ecosystem services, and how managers relate the outputs in local knowledge systems to the provisioning of services (e.g., yield, yield stability, taste and market characteristics). In building this evidence base, we should anticipate some null results; that is, places or systems in which some aspects of agrobiodiversity countenanced by social or economic drivers play little or no role in agroecosystem function.

A prescriptive variant of this effort to build an evidence base is suggested by Wood et al. (2015a) who propose gathering data on functional traits across environmental and management gradients with an awareness of spatial interactions among communities and their expressed functional traits. The intent of their proposal is to design and further test agroecosystem innovations with key functional traits to maximize ecosystem services in a hybrid effort at applied research joining the efforts of ecologists and agronomists. Such applied research can provide useful insights, especially if it can address the risks of being overly prescriptive and not sufficiently integrating the social and economic drivers identified as priorities above with the biophysical performance of functional trait assemblages.

Adopting a Food Web Approach to Functional Agrobiodiversity

Returning to the ecological "axis" of functional trait diversity in Figure 4.1, it may be useful to think in terms of how functional trait diversity contributes to a whole food web, rather than how it relates to individual elements of the agroecosystem (e.g., resistance to a pest, palatability for livestock considered separately). In this way, cascading effects of functional diversity throughout

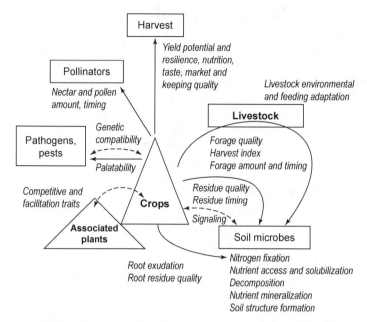

Figure 4.5   Functional trait diversity conceptualized as part of agroecosystem food webs. Triangles represent main agroecosystem producers; rectangles represent consumers and human appropriation of harvest biomass. Elements labeled in bold represent the major managed components of agrobiodiversity. Solid arrows indicate energy flow through trophic levels in the food web; dashed arrows represent nontrophic interactions such as pathogen–host genetic specificity or root–microbe signaling. Italicized labels on arrows highlight functional traits of food web components that affect food web functioning.

the agroecosystem can better be hypothesized and understood. Figure 4.5 attempts to illustrate how functional diversity of crops and livestock could be visualized, with reference to both energy flow and trophic levels in the food web (e.g., producers, consumers, decomposers) as well as nontrophic interactions such as signaling or plant–pathogen genetic compatibility.

Models of When and Where Addition of Agrobiodiversity Can Achieve Impact

One potential result of the evidence base is to identify areas or systems in which bolstering agrobiodiversity can have substantial to maximal impact, either because of recent loss, particularly promising impacts in comparable systems, or resources available to land managers or local policy makers to invest in these agrobiodiversity outcomes. Given some level of trade-off between the production of appropriated biomass by smallholders (provisioning ecological services, which are extremely important for farmers) and the maintenance of soil, water, and production environments in the longer term (supporting ecological services; Figure 4.2), we need to know what investments and opportunities will effect synergy between current production and supporting services to sustain future production (González-Esquivel et al. 2015). Jarvis et al. (2011) stress that capacity building and empowerment of farmers and communities is a crucial part of this process.

In addition, the thinning of seed systems and lack of availability of land-races or functionally diverse assemblages of varieties are significant issues, as in the Zambian maize system, where more pest-tolerant and taste-valued varieties have been displaced by drought-tolerant early hybrids (Mukanga et al. 2011). In some situations, the reintroduction of important functional traits from locally unavailable landraces, or new breeding efforts, may be called for via informal or formal networks (Orindi and Ochieng 2005).

In Situ Conservation and Evolutionary Breeding

In situ conservation of varieties is not static. It has expanded beyond a purely conservation role to become an important way to examine functional diversity within local farming systems and to test traits for future adaptive capacity of these systems to stresses (Chapter 2). Participatory and evolutionary breeding efforts that promote taste and quality characteristics as well as aspects related to drought and pest resistance, microbial symbioses, nutrient access by pro-motion of rhizosphere nutrient cycling processes, and supporting ecosystem services more generally (Figure 4.2), seem a particularly pragmatic way to use agrobiodiversity to address climate change and production challenges in smallholder systems (Ceccarelli et al. 2010; Lammerts van Bueren et al. 2011; Murphy et al. 2016; Schmidt et al. 2016). Such approaches are particularly valuable in situations where local diversity levels are insufficent to allow for adaptation.

An interesting, additional reflection raised by plant breeders is that the func-tional traits of species or varieties as well as of individual crop plants within a varietal population become important as breeding and selection proceeds. Therefore, the functional traits needed for pest or drought resistance or other benefits may not even be apparent at the outset but rather emerge and become the basis for adaptation under changing conditions. The same is true for farmer-managed *in situ* populations in centers of crop origin. All too often, however, the lack of baselines and varietal diversity timelines limits the necessary un-derstanding of how farmers use Darwinist selection to adapt agrobiodiversity to the changing dynamics of environments.

Adaptation to Global Change Should Not Be an Assault on Agrobiodiversity

The development and widespread diffusion of new, "improved" crop varieties is often misprescribed as the primary actionable response to climate change and also the emergence or movement of arthropod pests and plant disease strains (Chapter 7). As pointed out by Mercer et al. (2012) in the concept of *transgenic adaptation*, these new "silver bullet" varieties enter landrace-rich agriculture, and their adoption may place further stress on the genotypic and functional diversity of smallholder-managed cropping systems that form one of the repositories of *in situ* varietal and functional trait diversity. We need to determine whether (and how) a more creative response can be found to leverage local germplasm resources, to think carefully about the full range of production and support involved in ecosystem service functional traits (Figures 4.1 and 4.2) in breeding approaches, or to anticipate what is usually unanticipated about breeding for abiotic stresses and disease.

5

Spatial Analysis for Crop Genetic Resources Conservation

Selected Approaches for *In* and *Ex Situ* Efforts

Nora P. Castañeda-Álvarez, Robert J. Hijmans, and Stef de Haan

Abstract

In and *ex situ* conservation are two approaches that can support each other and benefit from spatial analysis to inform about genetic resources distribution, change dynamics, and conservation priorities. This chapter reviews approaches that integrate spatial analysis to support *in* and *ex situ* conservation and their linkages; specifically, the mapping of crop diversity patterns at various scales, identification of conservation gaps, modeling to reveal unique or uncovered diversity, participatory cartography, and the identification of candidate sites for establishing conservation networks. It highlights areas for future research and action, which include improving data quality and availability, establishing baselines of the spatial distribution of intraspecific diversity, experimenting with remote sensing tools and citizen science, as well as further work needed to explain spatial patterns in crop diversity and develop models to elucidate how crop diversity responds to different drivers of change. Finally, it argues that integrating different spatial scales, such as field, local, regional, and global is a major challenge that requires attention.

Introduction

Crop genetic diversity consists of plants that have sprung from domestication and selection trajectories (Leclerc and d'Eeckenbrugge 2012b) and their wild and weedy cousins (i.e., crop wild relatives). Crop genetic resources are the raw materials for crop breeding that have supported the rapid increase in agricultural productivity over the past 50 years. Breeding programs have produced high-yielding hybrids and varieties that, for many crops and regions, have resulted in the replacement of local traditional varieties (i.e., landraces). This loss of crop genetic diversity, or "genetic erosion" diminishes the options for future crop improvement and may affect resilience of farming systems (Harlan 1975). The concern about genetic erosion has led to the further support and creation of new crop genetic resources conservation programs. Nonetheless, high levels of crop genetic diversity still persist on farms globally, particularly in rainfed, marginal, and complex environments within smallholder systems and family farms (Brush 1995; Brush et al. 2003; Jarvis et al. 2008b). What is unclear, however, is how much genetic diversity actually exists, its past and current spatial distribution, and the degree to which it continues to change.

Two main approaches have been established to address the risk of crop genetic diversity loss. The first, and dominant approach, involves the creation of *ex situ* collections or gene banks. Most countries in the world have gene banks, and a collection of major food crops is maintained in international gene banks (e.g., by the Consultative Group on International Agricultural Research, or CGIAR). The second approach is *in situ* conservation, which consists of conserving landraces on farm and crop wild relatives in their natural habitats.

In this chapter, we begin with a review of these two approaches. We discuss selected examples of how spatial data analysis has been used to support these conservation efforts and highlight linkages between them. We conclude by looking at some of the current challenges and benefits of using spatial analysis as an element to enhance the integrated conservation of crop genetic resources.

In Situ and *Ex Situ* Conservation

Since the late nineteenth century crop genetic resources have been formally collected for introduction into new areas, use in crop improvement programs, and conservation. After the 1960s, the notion of rapid genetic erosion in areas of high crop diversity fueled an increased interest in conserving crop genetic resources in *ex situ* collections (Gepts 2006b; Harlan 1975; Pistorius 1997). In *ex situ* collections, samples of crop wild relatives and crops are maintained and stored alive, be it as plants, seeds, or in tissue or cryopreserved cultures. A stated ideal is that gene banks conserve 95% of the rarest 5% of alleles (Oka 1969), but it is not known how much genetic diversity is actually conserved. More than 7 million samples are conserved globally, but it is estimated that 65–70% might be duplicates (FAO 2010b) as samples are exchanged and a sample may end up in multiple gene banks, or because essentially the same genotypes are collected multiple times. To evaluate the effectiveness of conservation programs, it is important to estimate the difference between the genetic diversity that exists, or used to exist, and what is conserved. Such "gap analysis" can be used to plan additional collection efforts to improve the coverage of *ex situ* collections, or to prioritize *in situ* conservation efforts. *Ex situ* conservation largely occurs through formal institutions, including public organizations and the private sector. Gene banks conserve a snapshot of crop diversity, maintain it through time, and provide access to crop genetic resources. These collections are largely used by researchers, yet may be unknown to potential other users, particularly farmers (Bjørnstad et al. 2013).

In situ conservation of crop landraces occurs at the farm level, where conservation is not the ultimate goal. Instead, conservation may be an emergent property of farming systems in which seed is generated locally and there is no demand for product uniformity. There can also be explicit demand for varietal diversity for a number of reasons, including local cuisine, risk management, identity, and prestige. From a global perspective, there is an interest in the persistence or creation of farming systems that maintain landraces or traditional animal breeds to permit their evolution through anthropogenic and natural selection. The idea is that high levels of crop genetic diversity at farm and community levels can enable adaptive evolution in response to changing pests, disease, climate, and consumption patterns (Bellon et al. 2017).

In situ conservation is thus mostly an outcome of autonomous smallholder farmers' decision making. A number of intervention approaches has been created to promote *in situ* conservation among farmers, such as seed fairs (FAO 2006; Tapia and Rosas 1993a), development of value chains to commercialize specialty landraces (see PapaAndina in Devaux et al. 2016; Gonsalves 2013), community seeds banks (Vernooy et al. 2017), organization of agritourism routes and park systems (e.g., the Potato Park in Peru: Argumedo 2008; Shepherd 2017), adult education programs (e.g., in pest and disease management), and youth engagement plans as well as direct payments to farmers (Wale et al. 2011). Little is known about the effect of such interventions, and more research is needed to evaluate their effectiveness (see also Chapter 2). The following examples of work by the International Potato Center and others on Andean potatoes illustrate the diversity of *in situ* conservation approaches and how they can be linked with *ex situ* conservation:

- The establishment of community gene banks (Huamán 2002)
- A park model linking accumulated repatriated diversity with tourism (Argumedo 2008; Shepherd 2017)
- Development of value chains that connect varietal diversity to urban markets (Ordinola et al. 2007b; Tobin et al. 2016)
- Long-term systematic crop diversity monitoring (de Haan et al. 2016)

The first three of these examples have introduced potato landraces from a gene bank to farms without baseline inventories or monitoring of local landrace populations. The fourth strategy established an *in situ* baseline of farmer-managed landrace diversity and had the goal to explore gaps in gene bank collections. This gap analysis was not successful due to methodological challenges (i.e., the relatively high cost and turnover of genetic marker systems to compare sizeable populations).

Ideally, *ex situ* and *in situ* approaches would act in a complementary fashion, such as establishing sites for *in situ* conservation while collecting the target genetic resources for storage in a gene bank (*ex situ* conservation). Permanent monitoring of the crop diversity targeted for conservation in such a site, as part of the *in situ* conservation strategy, can produce relevant information about shifts (i.e., loss or enrichment) in *in situ* populations. It can also help identify new alleles or genotypes which, in turn, can be conserved in gene banks (for further discussion, see Chapter 3).

In reality, *in situ* and *ex situ* approaches are often not fully integrated or linked, as the objectives, mechanisms, institutions, and actors involved are very different (Chapter 2). Important reasons for this disconnect include:

- The institutional setup underlying each conservation approach differs. *In situ* conservation tends to be decentralized and follow more informal dynamics, whereas *ex situ* conservation is governed by formal structures and institutions. Both domains are governed by different norms and dynamics that need to be reconciled for complementary conservation action.
- Paradigms and objectives of the conservation communities are seldom aligned. While gene banks focus on maintaining a snapshot of genetic diversity that is accessible to users, stakeholders involved in *in situ* conservation stimulate the evolution of gene pools and their activities support primarily local use, benefits, and rights.
- Global and national legal frameworks discourage the connection between *in situ* and *ex situ* conservation: Collecting new crop diversity to conserve in gene banks can result in lengthy bureaucratic processes, depending on the policies established at the national or even regional level, where material is planned to be collected. Conversely, the lack of documented and practical benefit-sharing schemes that should accompany complementary efforts may discourage participation by smallholder and traditional farmers (Chapter 14).

Putting aside the conservation strategy to be implemented, baseline information and analytical tools are necessary to support decisions needed to maintain and increase crop diversity efficiently. Spatial and genetic analyses are increasingly being used to understand patterns of crop diversity at different temporal and spatial scales. Such approaches can help prioritize species, landraces, or gene pools requiring conservation action. In addition, they can set priorities for complementary *in situ* or *ex situ* conservation while adequately allocating the resources available. A main advantage of using spatial data analysis for informing conservation planning of crop genetic resources resides in the possibility of revealing diversity patterns at different scales. These patterns can be combined with other spatial data to study the processes that affect genetic diversity (e.g., isolation, and environmental and cultural variability) and how these may affect the composition and distribution of landraces and crop wild relatives in the future.

Diversity Distribution Patterns

A number of studies focused on the diversity patterns of crop wild relatives have been used to collect local data on herbarium specimens and gene bank accessions for wild potatoes (*Solanum* section *Petota*: Solanaceae; Hijmans and Spooner 2001; Hijmans et al. 2002), wild groundnuts (*Arachis* spp., Jarvis et al. 2003), and wild vignas (*Vigna* spp., Maxted et al. 2004). Generally of a descriptive nature, these studies provide a good starting point to understand the evolutionary and biogeographical history of a group, to assess the environmental conditions in which the group occurs, to identify highly diverse areas where reserves could be established for *in situ* conservation, to prioritize areas favorable for the collection of samples for *ex situ* conservation or for *in situ* monitoring, and to identify areas where crop landraces and their wild relatives overlap, making evolutionary changes in the genetic composition of such populations more likely to occur.

There are only a few examples of studies mapping intraspecific diversity of cultivated species. Perales and Golicher (2014) and Orozco-Ramirez et al. (2017) mapped maize race diversity in Mexico. Their analyses were enabled by the CONABIO database on maize landraces. This unique database includes gene bank records but mostly consists of a nation-wide survey. The maize diversity patterns show clear geographic variation; most diversity is found in the highlands, which are more isolated, less market oriented, and have a large Indigenous population. While there is insufficient historical data to allow for a very rigorous evaluation of changes in race diversity, Perales and Golicher (2014a) concluded that there was no evidence for substantial genetic erosion. Their analysis was done using vernacular nomenclature of maize races as a unit of analysis. In contrast, Dyer et al. (2014) reported loss of maize diversity in Mexico at the farm level based on household surveys between 2002 and 2007.

Genetic data has also been used extensively to increase understanding of the geographic origin of a crop, its dispersal, and the structure that crop diversity may have over a delimited geographic area (e.g., Bradbury et al. 2013; Rabbi et al. 2015; Vigouroux et al. 2008). Combining both genetic data and spatial analyses has proven useful to determine priorities for *in situ* and *ex situ* conservation (van Zonneveld et al. 2012). Producing large data sets of genetic data has become relatively cheap and we expect that the use of such data for mapping crop diversity will become more common.

Over the last decade, an increased number of landrace catalogues have been prepared to record the landrace diversity found in *in situ* hotspots at different geospatial scales. These catalogues, which often contain ploidy, nutritional, and phenotypic information, document diversity from the village to the provincial level. One example is the large number of Andean potato catalogues that has been released over the last two decades (MINAGRI 2017). They contain spatially explicit, time-tagged, and genetic data that offers the possibility to elucidate the spatial patterns of crop diversity in the center of origin. To date, however, no efforts have been made to consolidate the information using spatial modeling tools.

Gap Analysis

Over the past decade, several studies have attempted to assess the degree to which *ex situ* collections represent the diversity within crop wild relatives (Castañeda-Álvarez et al. 2015; Khoury et al. 2015a, b; Ramírez-Villegas et al. 2010; Syfert 2016). The studies cited estimated the entire geographic range of species with species distribution models. Such models use occurrence (locality) data from herbaria and environmental predictor variables to identify geographic areas with similar environmental conditions where the species may be presumed to be present (Elith and Leathwick 2009b). The predicted potential distribution of a species was compared with the observed range of the species from the gene bank samples using the circular area method (Hijmans and Spooner 2001). This comparison can be done in terms of area covered and in terms of environmental diversity covered, and combined with other indicators to determine the extent of representativeness of each taxon in *ex situ* holdings and identify geographic regions where further germplasm collecting missions can be prioritized. The data collected in these studies can also serve to derive information about regions where multiple crop wild relatives are likely to be found, and thus their viability as candidate sites for *in situ* conservation for multiple species (Castañeda-Álvarez et al. 2016; Hijmans and Spooner 2001).

Fielder (2015) conducted a study to identify candidate sites for *in situ* conservation of crop wild relatives in the United Kingdom. She compiled georeferenced records of crop wild relatives and used an iterative method to identify a network of complementary sites for conservation in their natural habitats (Rebelo 1994). The network consists of 15 sites and includes populations for 148 priority crop wild relatives. Further refinements included field visits and molecular analysis performed for a range of crop wild relatives populations. These refinements indicate that the priority species are geographically structured, suggesting that conservation efforts for crop wild relatives in the United Kingdom need to involve a network of conservation sites to conserve wild relatives in their habitats, rather than focus on a single site. Complementary conservation strategies (e.g., *ex situ*) can benefit from this approach, as it reveals patterns of richness and sites where multiple species occur, thus allowing for collections to be assembled more efficiently. Molecular characterization and diversity analysis revealed the structure of populations in their habitats and thus provided intelligence for the gene bank minimum sampling size required for effective collections of seeds and propagules. Future work on gap analysis methodology could investigate

- the coverage of crop wild relatives rich areas within the network of established protected areas,
- the level of coincidence with global biodiverse hotspots to identify regions where there is a potential to conserve crop wild relatives together with other species (Myers et al. 2000),
- the likely vulnerability of crop wild relatives due to land-use change (Watson et al. 2016), and
- the expected impact of climate change on the future spatial distribution or survival of crop wild relatives (Jarvis et al. 2008a).

A weakness of the gap analysis methods employed so far is that there is no distinction made between species; that is, there has been no attempt to adjust for phylogenetic diversity, or the known usability of species. There has also been a strong reliance on the probably weak assumption that a large range implies a large infraspecific diversity. Ideally, infraspecific diversity should be estimated and taken into account to set conservation priorities (e.g., Camadro 2012; Camadro et al. 2012).

The species level is not a very relevant unit of comparison for cultivated species. For crops we need to consider intraspecific diversity (i.e., races in maize, landraces of cassava). One plausible approach is to model premodern patterns of diversity and then to check if it has been sampled appropriately. For the Americas, van Etten and Hijmans (2010) present a spatially explicit model of maize dispersal that can be used to determine the geographic distribution of maize genetic diversity. Thus, this model provides elements for a gap analysis in crops to set priorities for *in situ* and *ex situ* conservation.

Modeling to Find Novel Genetic Resources

The use of geographic analyses as data support tools for the conservation of crop genetic resources is not restricted to the definition of highly diverse hotspots or species-rich areas. Spatial models can also be used as an input for more efficient germplasm-collecting missions to complete *ex situ* holdings. This was done for a rare chili pepper in Paraguay (Jarvis et al. 2005), for lupins that displayed particular adaptations to microclimates along the Iberian Peninsula (Parra-Quijano et al. 2011), and to research forages in Russia (Greene et al. 1999). In addition, spatial models can be used as a tool to discover new species (Särkinen et al. 2013). Of course, the same tools simultaneously draw attention to unique locations that can be considered as *in situ* conservation sites. Other studies have focused on the distribution of traits of interest. For example, Hijmans et al. (2003) mapped the distribution of frost tolerance in potatoes and Bonman et al. (2007) mapped the distribution of stem rust resistance in wheat. Insight into the distribution of particular traits of interest can be of use as a guide for further screening, but also for targeted collection and conservation.

Local-Level Studies

Research on local-level (e.g., community) diversity can capture the spatial distribution of intraspecific diversity at a very high resolution (e.g., field level or within fields). Such studies have associated diversity with social variables (e.g., ethnicity, culture, social organization, language) and other variables associated with on-farm management of intraspecific diversity (Delêtre et al. 2011; Labeyrie et al. 2014, 2016). Local-level data collection can enable communities to participate in the research process and the establishment of observatories for systematic monitoring (see Chapter 2).

Since 2012, the International Potato Center has conducted baseline studies and *in situ* monitoring of potato landrace diversity in the Peruvian and Bolivian Andes at the community level (de Haan et al. 2016; Polreich et al. 2014). This is an effort to compile high-resolution information on the current spatial distribution of landraces managed by local communities. The communities are involved through participatory cartography and this has been instrumental in systematically capturing detailed spatial data at the landrace level (Juarez et al. 2011). This high-resolution information is valuable as a baseline to characterize the genetic and spatial distribution of landraces at sites of high landrace diversity (de Haan et al. 2010). Furthermore, field-level mapping and sampling with farmers allows quantification of the relative abundance of specific landraces, which in turn can be used to construct indicators of risk of extinction (e.g., Red List) of rare landraces. This approach can be replicated to track spatial shifts and changes in the conservation status of landraces for other crops as well, and this is currently being promoted for banana, cassava, and yam by partners of the CGIAR Research Program on Roots, Tubers, and Bananas. Systematic monitoring of agrobiodiversity at benchmark sites allows researchers to better understand change dynamics in time and space. This approach offers potential ways to connect conservation strategies:

- It provides information about resemblances between *in situ* and *ex situ* populations.
- It enables the detection of specific landraces that may be rare and possibly warrant an *ex situ* backup.
- It provides gap analysis at the cultivar level and the possibility to enhance gene bank coverage through the addition of uncovered landraces to the collection.

Satellite Image Analysis and Modeling Approaches

Remote sensed images together with survey data are also used to understand and model the dynamics of agrobiodiversity. Processes of interest include decision making and management choices of farmers and communities regarding the cultivation of crops and cropping systems in the context of global change (Meles 2011; Rostami et al. 2016; Zimmerer 2013). For instance, satellite images have been used to estimate landscape-scale attributes related to the management of both the diverse maize crop and the market-driven expansion of peach growing in Bolivia (Zimmerer 2013). These estimates were incorporated into a model illustrating the trajectories of agrobiodiversity between 2000 and 2010 in the context of global change drivers related to environmental factors (water availability) and socioeconomic conditions (market change). Additional farmer and land-use surveys were used to determine how agrobiodiversity use and *in situ* conservation can be compatible with agricultural intensification under conditions of favorable markets, the continued cultural valuation of agrobiodiversity, and generally adequate availability of water resources. Zimmerer and Rojas Vaca (2016) used joint count statistics to demonstrate high levels of spatial clustering of highly diverse maize fields. Remote sensing can also be used to infer upon the fine-grained, field-level dynamics of agrobiodiversity. One approach focuses on field-level interactions involving the spatial externalities of short-distance spillover processes (e.g., information sharing, coordination of labor, seed exchanges). This approach has been used to evaluate organic and high-agrobiodiversity farming through the incorporation of econometric modeling and farm surveys (Lewis et al. 2008; Zimmerer and Rojas Vaca 2016). Satellite images have also been used to differentiate tree species assemblages and cultivars (Turner et al. 2003), but more attempts are needed to use images to disaggregate landrace assemblages, particularly for annual crops.

Future Challenges and Opportunities

To inform conservation action, we need to understand both the challenges and opportunities of using spatial analysis to anticipate future spatial changes in agrobiodiversity. Global environmental change, including climate change, urbanization, and land-use intensification, are likely to affect the distribution of crop genetic resources (Chapter 8); in particular, the distribution and relative abundance of species and intraspecific diversity at the landscape level. Spatial shifts that result from climate change have been demonstrated for wild flora and fauna (e.g., Bodin 2010b; Morueta-Holme et al. 2015; Seimon et al. 2007), and have been predicted for crops (e.g., Hannah et al. 2013) but have only rarely been documented for agrobiodiversity (cf. Skarbø and VanderMolen 2015). In addition, high spatial resolution and regularly updated distribution maps of agrobiodiversity at intraspecific levels are still rare, but when available, they prove useful for conservation action (Pacicco et al. 2018). Thus, systematic approaches are needed to link scales and cluster interactions among landscapes (including different environments and production zones), species (including subspecies), cultivars (including races, cultivar groups, landraces, and bred varieties), and genetic diversity (including genes and alleles).

Data availability, completeness, and quality are important challenges for spatial analysis of crop diversity. Several of the studies presented in this chapter relied on gene bank collection locality data. There has been enormous progress in the availability of such data over the past twenty years or so, thanks to an expanding culture of open data access for public institutions, such as gene banks, albeit mostly in Europe, the United States, and international institutions; much less so in national-level gene banks in other countries. Thus, despite heroic global efforts to digitize and publish biological-related data, notably through the Global Biodiversity Information Facility, Plants JSTOR, Eurisco, and Genesys, a large share of global crop collection records are not digitally and publicly available (FAO 2010b). The reluctance by some to share locality data, especially when coordinates or detailed geographic descriptions are available, remains a major obstacle to research in this field. While spatial modeling can be used to estimate distribution patterns based on small samples, as has been done in the gap analysis work we described, this method will likely not perform well if there are major geographic gaps in the sample that the model is based on.

Nomenclatural identity (i.e., scientific name) and sampling bias are key elements that determine the quality of crop genetic resources collection data, and thus directly affect the consistency of spatial analyses. Plant taxonomy is dynamic and changes over time. Keeping accessions properly identified and updating taxonomical classification is crucial to knowing what crop wild relatives and cultivated (sub)species are being conserved (see Syfert 2016). Continuous updates require verification (i.e., whether a species name has changed according to the taxonomic consensus of the moment). Of course, lack of consensus can further complicate matters (Oliveira et al. 2012). Herbarium specimens are often subject to updates, yet this rarely happens for crop genetic resources in gene banks. In addition, spatial bias is a result of the patterns followed by germplasm-collecting missions. Most georeferenced, *ex situ* collections targeted fields next to roads, rural markets, and locations where collections could quickly be made (Hijmans et al. 2000); often the frequency of collecting missions spanned numerous years and different lengths of time. In an effort to overcome such biases, several questions have emerged:

- Are areas rich in crop genetic diversity actually properly sampled?
- Is the information we can access enough for studying the dynamics of crop diversity?
- What is the status of the crop diversity in understudied areas?

An additional challenge emerges in relation to the management of duplicated accessions in *ex situ* collections and its effect on collection data. Duplicated accessions have been removed from *ex situ* collections (a) to reduce overrepresentation of widespread landraces (i.e., those with a wide distribution range) and (b) to lower the cost of the management of collections. Such rationalization has led to landraces represented by a single representative accession with a single geographical coordinate. These factors, among others, may limit the value of gene bank collection data to elucidate spatial patterns beyond species and ploidy distributions (Spooner et al. 2010). Understanding the spatial distribution patterns of crop diversity at the intraspecific level (for cultivar groups and landraces) often requires a sufficient number of samples, together with prolonged and fine-grained mapping efforts. Several questions for future action can be posed:

- How can we create solid baselines of spatial distribution at the intraspecific level?
- How should we address this at multiple scales, given that varietal nomenclature changes rapidly, due to local and genetic marker systems?

It is important to note that spatial analysis can provide tools to identify environmentally homologous sites where landraces could be established, used, maintained, and therefore conserved. The intent of such an approach can have completely different outcomes: it can result in the conservation of landraces with farmers willing to manage and maintain landraces over time, but with farms located outside the centers of origin of the crop (Bazile et al. 2016; Ríos et al. 2007). It can also result in the diversification of agricultural production systems, which may lead to improved sustainability and ability of the systems to respond to extreme climatic and economic events (Kahiluoto et al. 2014).

An ultimate challenge is to expand spatial analyses so that patterns of crop diversity can be converted into models that elucidate how crop diversity is affected (positively or negatively) by different drivers and change scenarios. Such models would be useful to predict responses to human intervention and environmental events (e.g., floods, drought spells, variable rainfalls, land use change).

The use of spatial analysis in plant genetic resource collection and conservation has been broadly reviewed (Guarino 1995; Guarino et al. 2002; van Zonneveld et al. 2011). Over this twenty-year period, tremendous progress has been made in data availability (including location, environmental, and genetic data), analytical methods, and diversity of applications. Another clear change is that we have moved from paper-based computation (Guarino 1995), to specialized "GIS" software (Guarino et al. 2002), to the current situation where spatial data is no longer special and analysis is primarily done in a general data analysis environment such as "R."[1]

---

[1]   For information on R, a free software environment for statistical computing and graphic, see https://www.r-project.org/ (accessed Sept. 12, 2018).

We expect that spatial methods using genetic data ("landscape genetics") will become more common and provide new insights about the dynamics of crop diversity. Using genetic data together with archeobotanical data, environmental, and other data sources may provide additional insights into the study of crop diversity (see Kraft et al. 2014; van Etten and Hijmans 2010).

In terms of *in situ* monitoring, data capture can be improved by using remotely sensed data from satellites and "drones," crowdsourcing, and citizen science. Crowdsourcing techniques and citizen science (e.g., mobile technologies) offer the potential to involve local communities in the monitoring of crop diversity and the mapping of on-farm conserved crop genetic resources in real time. This could result in greater local community engagement and more frequent production of fine-scale georeferenced data.

Global Change and
Socioecological Interactions

6

Socioecological Interactions amid Global Change

Conny J. M. Almekinders, Glenn Davis Stone, Marci Baranski,
Judith A. Carney, Jan Hanspach, Vijesh V. Krishna,
Julian Ramirez-Villegas, Jacob van Etten,
and Karl S. Zimmerer

Abstract

How a group relates to agrobiodiversity differs greatly within and between user groups. This chapter explores the socioecological changes that are driven globally by migration and urbanization, agrarian change (de- and reagrarianization), market pressures, and climate. It introduces the concepts of *intentionality by default* and *conscious intentionality* to explore how two archetypical smallholder farmer groups, traditional/Indigenous and neoagrarian farmers, use agrobiodiversity. These groups represent the extremes of smallholder farmers for whom agrobiodiversity plays an important role in their lives. To increase understanding of how the use of agrobiodiversity can vary in response to the effects of global change, knowledge gaps and entry points are identified for different groups of actors (e.g., smallholder farmers, public breeders, private companies, NGOs, international organizations, and governments).

Current drivers of global change affect these groups on a local level in unique ways, and responding to them provides the potential for novel initiatives that can form the basis for a compelling overarching narrative to support the use of agrobiodiversity in multiple ways. Such a narrative would connect the wide diversity of agrobiodiversity users and provide a critical mass to reinforce ongoing efforts to find solutions to the challenges of global change. Important gaps in our knowledge remain to be considered by this new, integrative science, including the way in which participation and empowerment of vulnerable groups will be incorporated.

Group photos (top left to bottom right) Conny Almekinders, Glenn Davis Stone, Jan Hanspach, Vijesh Krishna, Marci Baranski, Karl Zimmerer, Julian Ramirez-Villegas, Conny Almekinders, Jacob van Etten, Judith Carney, Julian Ramirez-Villegas, Karl Zimmerer, Vijesh Krishna, Glenn Davis Stone, Marci Baranski, Judith Carney, Jan Hanspach, Conny Almekinders, Jacob van Etten, Glenn Davis Stone, Karl Zimmerer

Agrobiodiversity Functions, Values, and Interests

In whose interest is it to utilize or own agrobiodiversity? Tremendous profits have been made through industrialized agriculture, which tends to be low in agrobiodiversity and associated functions (e.g., resilience, food security, cultural identity). Policies associated with the industrialized agricultural sector tend to promote a regime of low functional agrobiodiversity, and expansion of this sector has pushed many farmers into production regimes with intended and unintended negative consequences for agrobiodiversity (IPES-Food 2017). A range of alternative production systems with different values and functions is currently emerging, also with intended and unintended consequences for agrobiodiversity. As a result, we see different groups and organizations with different value systems using agrobiodiversity in a variety of ways. While the relations of these actors to agrobiodiversity vary, they are all subject to global socioecological changes (Zimmerer 2010) that are driven by migration and urbanization, agrarian change (de- and reagrarianization), market pressures as well as climate change (see Chapters 7 and 8). The varied responses of diverse actors to these changes involve and affect their use and relation to agrobiodiversity, including different types and levels of agrobiodiversity. The ways by which agrobiodiversity functions change in the face of interacting global trends that need to be understood if we are to identify opportunities and threats to agrobiodiversity.

Agrobiodiversity has a wide variety of functions through which it contributes to human well-being (e.g., income provision, food security, resilience, absorption, and adaptation capacity), and the empirical relationship between agrobiodiversity and livelihoods is complex. Three types of values are associated with agrobiodiversity and can be distinguished on the basis of comparison with biodiversity: intrinsic, instrumental, and relational values (Chan et al. 2016; Ives and Kendal 2014). As in biodiversity more broadly, the intrinsic values of agrobiodiversity point to the value of it in its own right. Assigning intrinsic values to agrobiodiversity might seem less obvious, but the consideration is especially relevant for the conservation of associated biodiversity that supports agroecosystems. Instrumental values place emphasis on the utilitarian nature of agrobiodiversity and relate, for example, the contribution of agrobiodiversity to the production of food, fiber, fodder, and fuel (Almekinders et al. 1995). Relational values underpin the nuanced associations, interactions, and responsibilities that people can have with agrobiodiversity and describe the contribution of agrobiodiversity to personal and cultural identity.

In this chapter, we discuss how agrobiodiversity can be better used by diverse groups and organizations in newly emerging situations to contribute to improved human well-being in the context of different drivers and multiple pathways of change (Figure 6.1). We address sustainability and social justice issues across a wide range of the social, ecological, and agronomic sciences. In addition, we review institutional approaches, policy making, analysis, and

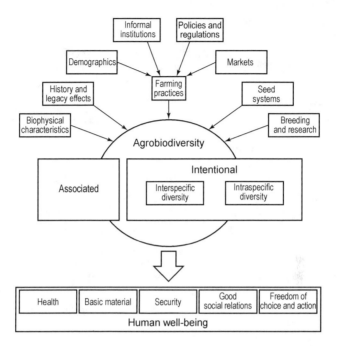

Figure 6.1   Conceptualization of agrobiodiversity showing how multiple pathways and drivers collectively influence agrobiodiversity, with farming practices assuming a crucial role. In turn, agrobiodiversity contributes directly and indirectly to different aspects of human well-being.

activism among the broadly defined public, which includes diverse civil society groups (Almekinders and de Boef 1999).

Archetypes of Agrobiodiversity Users

When identifying agrobiodiversity user groups (producers, breeders, consumers, and other stakeholders), one quickly discerns vast differences within and between groups (see also Chapter 8). Our focus here is primarily on farmers (i.e., smallholder producers, some extreme archetypes) as well as private and government organizations.

There is an almost endless variety of farmers, and how one defines the diverse types depends on the characteristics used to describe them. This variation is most evident among the world's estimated 2.2 billion smallholder farmers (Zimmerer et al. 2015). Defined as having an area of farmland that measures less than 2.0 hectares or that occupies the smallest quintile of farm size within a country or region (Graeub et al. 2016; Lowder et al. 2016), smallholder farmers do not only farm, many work the land part-time as they are involved in nonfarm or off-farm income-generating activities. The actual composition of smallholders is highly heterogeneous; socioeconomic, agricultural, environmental, and cultural characteristics vary widely within this group (Cohn et al. 2017; Zimmerer et al. 2018). While millions of smallholders reside in the Global North, including some who farm as a hobby, most are poor farmers located in the Global South who experience food insecurity and are vulnerable to trends and shocks in markets and their natural environment. Typically, they must combine agricultural strategies for market production with those for subsistence. Under diverse circumstances and rationales (explained below), smallholders maintain, produce, and consume the largest share of the world's *in situ* agrobiodiversity. In other words, even though they exist on the margins of the global economy, smallholder farmers disproportionately rely on agrobiodiversity. Likewise, relative to other farmers, agrobiodiversity tends to be disproportionately important to smallholders.

Two subsets of smallholders deserve being considered because of the agrobiodiversity they maintain and the relation they have with it, and the extremes they represent in the diverse gamut of smallholder farmers. The first group is comprised of *Indigenous* or *traditional farmers* (e.g., of the Americas, Africa, South Asia, Australia)—those who have managed to remain in their territories over centuries, if not millennia, despite sociopolitical change. This group is highly diverse and includes, for example, descendants of escaped slaves (e.g., maroons and *quilombolas* in South America) who practice an agrarian creolization, shaped by their African heritage and contact with New World native peoples (Carney and Rosomoff 2009). A distinguishing characteristic of this group as a whole is their cosmic worldview of nature, of which agrobiodiversity constitutes a specific part. The second group is made up of *neoagrarian farmers*—those who have established operations in recent decades in North America, Europe, and other regions. This group shares a philosophical rejection of industrial agriculture and a commitment to reforming it through humane, sustainable, and biodiverse production methods. It is estimated that farms matching this description in North America number in the tens of thousands and are in particular on the rise near large urban areas in Ecuador, Brazil, and Spain (where they are referred to as *neo-rurales*). Neoagrarian farmers aim to capture a premium by marketing farm products through short commodity chains. In this sense, they differ from most smallholder farmers in the Global South, who must frequently juggle subsistence with market production. Neoagrarian farmers stand in stark contrast to most of the world's 2.2 billion smallholders in terms of their (a) philosophical commitment to agricultural reform, balancing economic and ecological sustainability, and so forth; (b) dispersal across agricultural landscapes; and (c) organization in networks for exchange of information, seeds, and experiences (i.e., through seed saver groups).

Organizations also vary in their relation to agrobiodiversity. Interesting examples can be derived from plant breeding institutions. Although the public breeding sector tends to be lumped into one profile, Indian wheat breeders, for example, hold a plurality of views about breeding for broad versus local conditions (Baranski 2015a, b). However, the institutional structure of Indian wheat science (which is primarily public) allows for only a single view, focused on broad adaptation of wheat varieties. This view is mediated through a series of features, including both formal and informal incentives, policy (varietal testing and release system), and an institutional culture that values the widely adapted varieties of wheat. Thus, the wheat breeding system in India is implicitly biased toward less varietal diversity.

During the Green Revolution, tensions over the use of agrobiodiversity also existed among plant breeders in Mexico, in terms of the positioning and power of public sector versus private sector breeders (Bebbington and Carney 1990; Harwood 2009; Jennings 1988). Breeders who worked to strengthen the national plant breeding capacity in the Global South at the International Maize and Wheat Improvement Center (CIMMYT) offices were advocates of poor farmers, and they worked to develop breeding strategies relevant to them. They were not happy about policy changes induced by the U.S. government, which reduced public funding for research and weakened the capacity of national seed research. Yet there was little they could do to counter the concentration of seed development by corporations that relied on CIMMYT gene bank stock for developing their commercial hybrid varieties. Seed companies like Pioneer (subsequently purchased by DuPont) became increasingly in charge of setting research priorities that favored their own commercial goals. Later, many breeders also opposed CIMMYT's burgeoning repositioning of itself in terms of transgenic seed development.

Totally different breeding institutions are represented by farmers and their local breeding customs and networks. Take, for instance, smallholder rice growers in Sierra Leone who had selected and developed the hybrids of two rice species found in their fields, *Oryza glaberrima* and *Oryza sativa*, long before the Africa Rice Center formally developed and released Nerica rice in the region (Mouser et al. 2012). Examples of intermediate institutional forms of breeding include the participatory potato breeding programs with both conventional and organic farmers in the Netherlands working with conventional commercial companies and supported or mediated by semipublic research organizations (Almekinders et al. 2014).

It is important to keep the heterogeneity of views and values in breeding institutions in mind. There is increasing awareness among breeders as to the importance of public goods and the need to serve smallholder farmers. In many countries, public breeding efforts are progressing from a reduction of diversity to an increased use of diversity in response to smallholder demands, climate change, and production system rationales (Dawson et al. 2016a; Murphy et al. 2016). However, public breeding programs are on the decline. Currently, there are two prevailing paradigms: the productivist and alternative models. We anticipate that other breeding institutions and approaches will gain in importance, further diversifying this spectrum in the future. In this respect it is encouraging to note that a large number of public–private partnerships are emerging to develop and disseminate varietal technologies even in the major cash crops like maize.

Intentionality and the Use of Agrobiodiversity

Among smallholder farmers, intentionality around the use of agrobiodiversity varies greatly. It represents a spectrum that ranges between and combines what we call *intentionality by default* and *conscious intentionality*. Smallholders are highly aware and knowledgeable about agrobiodiversity. Each year they choose to work with certain varieties and species in crop fields and landscapes. In other words, significant skills and management inform their planting strategies. Still, their rationale for producing and consuming agrobiodiversity varies and is often multifaceted:

- Hardy crops may be needed in marginal, stress-prone growing environments to reduce the risk of crop loss (e.g., in response to extreme variation associated with climate change).
- Culturally familiar foodways may be the goal.
- Accessing seed at lower cost through seed saving and the informal seed sector may be a key motivation.

For these reasons, smallholders do not typically choose agrobiodiversity for its own sake but rather because it fits with underlying farming rationales or trait preferences (e.g., Almekinders et al. 1995). This type of agrobiodiversity management can combine one or more functions and is what we term "intentionality by default."

The functions of agrobiodiversity underlying "intentionality by default" are important when we consider the relationship with food security and income in smallholders, Indigenous farmers, and others whose land use and rights are based on collective or group identity and historical land ownership (e.g., the *ejidos* in Mexico). For instance, in the Amazon areas of Brazil, the production systems of many maroon (*quilombo*) communities are quite diverse, characterized by a mixture of seed, tuber, and tree crops. Because many men migrate to the gold fields of the Guianas, and children are in school, the agrobiodiversity-rich plots are often managed by the women. These plots provide maroon communities income benefits and enhance household food security. Cash earnings from the market can be reinvested into the farm to support continuous growth (in crops and agrobiodiversity) (Steward and Lima 2017). In other smallholder areas in the Global South, the poorest farmers may be forced to reduce agrobiodiversity by abandoning tastier and more nutritious crops in favor of higher-yielding varieties. A typical smallholder in Western Kenya may use more agrobiodiversity (number of crop/animal species) as she farms, but it may not bring tangible benefits, whereas an Indonesian small-holder may realize higher income benefits using less agrobiodiversity than the Kenya farmer (Sibhatu et al. 2015). Still, in parts of sub-Saharan Africa, the rural poor are deeply reliant on agrobiodiversity. In West Africa, for instance, farmers (often women) harvest semiwild yams and collect wild foods (notably vegetable greens and fruits) to supplement their diets. Such gathering strategies may involve active tending of wild resources and habitat manipulation to encourage wild plant progeny for future collection. It may also encourage ongoing selection, domestication, and crop evolution. In the Brazilian Amazon, peasant and Afro-descendant farmers recognize the fertility of dark earth (*terra preta*) soils created from cumulative swiddens by pre-Columbian Amerindians and they manage them differently to take advantage and manage the soil fertility of these plots. These plots serve as genetic reservoirs for specific cultivars and allow for agrobiodiversity maintenance (Glaser et al. 2003; Junqueira et al. 2016).

Many examples exist to illustrate the complex relationship between the use of agrobiodiversity and its possible benefits to the user: agrobiodiversity emerges out of processes associated with both (multi)functional and agrobiodiversity-conscious intentionality. These processes, however, may not always be present. For example, in most commercial agriculture, private seed companies produce an inadvertent increase in agrobiodiversity, based on number of varieties as well as a low functional agrobiodiversity due to the similar genetics of the varieties. Similarly, a farmer in India may plant different hybrid cotton varieties and brands to compare crop traits, but the effect on agrobiodiversity is tiny or even zero when the different brands are actually the same varieties (Stone 2016). The neoagrarian archetype occupies a highly distinct and limited portion of the smallholder spectrum (see Chapter 8): the neoagrarian tends to exercise more agrobiodiversity conscious intentionality. By making it an explicit part of their farming philosophy, neoagrarians are often focused on the marketing of agrobiodiversity while emphasizing the intrinsic value of it and prioritizing ecological sustainability over the profit motive. They are smallholders whose farm operations are based on a "moral economy" of capturing a market premium for farm methods that promote agrobiodiversity, whether marketed explicitly as such or as an implied "credence quality" to their clients (Darby and Karni 1973). For some of these producers, the use of heirloom varieties and the exchange of seeds is a core element of their philosophy. However, preliminary investigation of seed diversity employed by many of these small-scale "consciously intentional" growers in the United States indicates that seed sourcing is restricted to purchases from a limited number of companies. These sources are often presented as relatively small and local in their company histories; still, many sell nationally and receive a portion of their seed stock from large corporate conglomerates (e.g.,

Seminis). As a result, seeds used by neoagrarian farmers are not very different across the United States. Similar concerns pertain to the animals raised on these smallholder operations. For Joel Salatin, whose Polyface Farm is famous in the neoagrarian movement (although atypical in that Salatin actually grew up on the farm), the biggest moneymaker is called "Salad Bar Beef," in reference to Polyface's biodiverse pastures (Pollan 2006). While Salatin's pastures may be biodiverse, his farm animals are not: his chickens enjoy a diverse pasture diet but the flock is uniform Cornish Cross, the classic industrial bird. In other words, a local food movement is not necessarily a local seed movement or one that boosts agrobiodiversity. The relationship between the practices of these neoagrarian farmers and their belief system, and how they use agrobiodiversity, gives rise to interesting new research questions. Does the conscious intentionality of farmers, breeders, collectors, and seed exchangers, apart from providing good stories, really contribute to desirable agrobiodiversity outcomes?

The relationship between agrobiodiversity and intentionality is variable. It occurs in a socioecological context, mediated by farmers, organizations, values, and power relations. Intentionality is not a dichotomy; it operates along a continuum between and combines "intentionality by default" and "conscious intentionality." Within a particular smallholder system there are different degrees of intentionality that vary from place to place and are expressed in diverse ways along a farmscape. Among neoagrarian farming units in the southern United States, for instance, agrobiodiversity may not be so much evident at the scale unit of an individual field but rather across a farming landscape: smallholders may grow and raise different species but share a commitment to organic, sustainable, and diverse foodsheds. For this reason, the analysis of agrobiodiversity in temperate, smallholder zones demands a multiscalar approach—one that examines diversity at the field, farm, landscape, and regional levels. Neoagrarian farms may appear initially as a patchwork of alternative cultivation, existing in the interstices of a conventional industrial agricultural landscape. In tropical smallholder contexts, analysis of use of agrobiodiversity can require an entirely different approach since rural poverty is often associated with functional use of agrobiodiversity. For example, inorganic pesticides and herbicides being unaffordable, polycultural production with less insect predation and crop loss in year-round growing conditions continues to be more attractive. In such systems, in some cases, off-farm income opportunities may prove decisive for being able or choosing to maintain local knowledge of agrobiodiversity and shaping "intentionality by default." Certainly, this discussion of intentionality through two contrasting temperate and tropical zones underscores substantive differences in how distinct smallholder systems experience, and respond to, exogenous global agricultural trends. The institutions that mediate responses locally are thus situated.

The Intersection of Agrobiodiversity Use with Global Changes

Socioeconomic Trajectories and Scientific Models

Global trends such as climate change, migration, commodity market integration, population growth, and urbanization/deagrarianization exert highly varied impacts on the different archetypes of agrobiodiversity user groups, especially farmers (for further discussion, see Chapters 7 and 8; Zimmerer 2010). These differentiated impacts also modify the policy and action context of breeders, nongovernmental organizations (NGOs), and governments. In addition, health and nutrition needs (such as those related to noncommunicable diseases) provide important novel impacts. Many hold that climate change constitutes the most influential driver of global change. Climate change policy targets primarily food security through diversification, genetic adaptation of main crops, and different crop management practices (different planting dates, irrigation). Chapter 7 of the IPCC Fifth Assessment Report (Porter et al. 2014) comprehensively assesses the available literature on climate change impacts and adaptation for food production systems.

Although there is general consensus that climate change will reduce agricultural productivity globally, particularly across the Tropics (Challinor et al. 2014; Porter et al. 2014), biases are clear in the available literature. Specifically, the number of citations for livestock and fisheries is roughly one sixth of the number of citations for crops (Campbell et al. 2016). In addition, the types of adaptation strategies assessed in the climate change impacts and adaptation literature is biased toward agronomic management (e.g., irrigation and fertilizer optimization, varietal substitution, shifts in planting dates), whereas agrobiodiversity-related adaptation strategies are relatively poorly investigated (e.g., crop and varietal diversification, agroforestry, silvopastoral systems) (Porter et al. 2014).

It is not clear how farming systems will change and adapt to global changes. Farmers have many ways of dealing with climate change on their farm, and creativity is essential. Scientists, by contrast, tend to look for simple solutions that can be generalized across populations. For example, during the early Cold War era, U.S. national security interests framed population growth and communism as the problem, and promoted yield increases per acre through transferrable technologies as the solution; this led to the development of the global research institution, the Consultative Group for International Agricultural Research (CGIAR), and massive investments in the agricultural research that spawned the Green Revolution (Cullather 2010).

Currently, global socioeconomic scenarios use expected population growth and socioeconomic development to characterize future trajectories (Moss et al. 2010; van Vuuren et al. 2014). The results point to different challenges, in terms of adaptation and mitigation, both globally and regionally (O'Neill et al. 2014; van Vuuren et al. 2014). While mitigation challenges result from the need to keep greenhouse gas emissions at a specified level, adaptation challenges emerge from the warming that results from greenhouse gas emissions and concentrations. By definition, therefore, trade-offs and synergies exist between adaptation and mitigation (Lipper et al. 2014; Rogelj et al. 2016). Locally, however, adaptation and mitigation challenges are varied. For instance, projected reductions in the yields of maize, bean, and napier grass in the mixed crop–livestock systems in southern Kenya may imply modification of farm composition (e.g., number of livestock units) that are needed to cope with changes (Claessens et al. 2012). A different strategy would be needed for crop or livestock systems in other parts of the world (cf. Nendel et al. 2014).

Identification of Knowledge Gaps and Entry Points

Global changes create new situations that affect people's use of resources, their values and identities and, consequently, their production and consumption of food and use of agrobiodiversity. To be able to identify knowledge gaps and develop narratives that can help lead to better interventions, we assessed a series of agrobiodiversity-related characteristics for different user groups. We decided to focus on smallholder farmers, public breeders, private companies, national and international NGOs, and governments, which are discussed below. For smallholder farmers, we identified the archetypical neoagrarians and Indigenous farmers as being quite distinct from the large majority of smallholder farmers in this world. We based our exercise on the following questions:

- How does the group use agrobiodiversity (differently) in response to climate change and market integration?
- How can the use of agrobiodiversity by the group be characterized in terms of vision, values, and intentionality?
- What are the opportunities or entry points to support agrobiodiversity use for the group?

The results are summarized in Appendix 6.1 and discussed below.

Smallholder Farmers

Cuba provides a good example of how a political change created space for agroecological production and the potential for enhanced agrobiodiversity use. Creativity was needed to develop urban agriculture in Cuba (Chapter 8). In addition, global movements in gastronomy (taste and flavor) provide important incentives for farmers to produce agrobiodiversity, although this is only relevant to a relatively small number of smallholder farmers. Similarly, neoagrarian farmers and specific farmer groups are generating interesting new pathways. Still, such alternatives leave out the majority of the 2.2 billion smallholder farmers, who by themselves form a highly varied group in terms of vision, value, and intentionality. The future of this group, which includes many vulnerable populations, particularly in terms of how they will cope with global change, is unclear. Will smallholders need to leave the farm? If so, why? Will they be driven by other sources of income? What are the gender implications of rural-to-urban migration for maintaining agrobiodiversity? Will there be opportunities to use agrobiodiversity in novel ways, such as selling and shipping blue maize from Oaxaca to New York? What can the Fair Trade Movement offer? In addition, how can current examples of innovative agricultural production and marketing become more relevant for the large majority of smallholder farmers?

Indigenous and Neoagrarian Farmers

Although they are quite distinct, Indigenous Peoples and neoagrarian farmers both use agrobiodiversity more intentionally, compared to the rest of the aggregated but heterogeneous group of 2.2. billion smallholders. Nonetheless, how they use agrobiodiversity is also quite distinct. This diverse group of Indigenous Peoples holds the bulk of the globe's agrobiodiversity. For many Indigenous People, agrobiodiversity is an integral part of their being: it is embedded in their worldview and an element of what constitutes quality of life ("the living well movement," which has emerged in the Bolivian Andes). Although our knowledge is limited on how global drivers currently affect the livelihoods and agrobiodiversity use of Indigenous and traditional farmers, their capacity to sustain ongoing evolution has been well documented (e.g., Bonnave et al. 2015; Thomann et al. 2015; Vigouroux et al. 2011b): selection for new diversity has enabled adaption to new circumstances.

As commented above, the actual agrobiodiversity deployed by neoagrarian farmers may be limited and far less than that of Indigenous and traditional farmers. However, their practices and views represent interesting new opportunities. Many neoagrarian farmers acknowledge agrobiodiversity as a unique "credence quality" of their produce, which links to its intrinsic value and agrobiodiversity-conscious intentionality. More commonly, agrobiodiversity is an assumed (rather than explicitly advertised) credence quality, since pasturing livestock and agroecological practices generally require conditions for biodiversity integration. Two key issues emerge regarding the impact of neoagrarians on agrobiodiversity: (a) the type and quality of agrobiodiversity promoted on neoagrarian farms, and (b) the scale of the neoagrarian impact on agrobiodiversity. There are reasons to suspect that neoagrarian farms tend to be hotspots of biodiversity, but this is so far woefully understudied. Despite the explosive growth of farmers' markets and other channels, neoagrarian farms only account for a tiny proportion of U.S. and E.U. farmland. No solid measure of the acreage is available, but the U.S. Department of Agriculture provides a rough indication of the scale of neoagrarian farming: "144,530 farms sold USD 1.3 billion in fresh edible agricultural products directly to consumers in

2012" (USDA 2014:1). As this sector continues to grow, empirical research on its economic and ecological impacts will be necessary.

Certification needs to be mentioned as an opportunity to foster the value of agrobiodiversity produce. The process could be supported by an organization similar to the International Federation of Organic Agriculture Movements. On the other hand, people may be wary of yet another certification measure. Alternative forms of providing trust and legitimacy on short chains and direct consumer–producer relations should be explored.

Public Breeders

There is, when it comes to national agricultural programs and as discussed above, ample diversity in the vision, values, and intentionality among public breeders. Some public breeders have more freedom to engage explicitly with agrobiodiversity than others, but the goals of government funding may vary. We discern differences between modernized breeders and regenerational breeding: the former is more oriented to privatizing the benefits of the program and, consequently, in prioritizing the instrumental use value of agrobiodiversity. Public breeding programs as run by the CGIAR research centers have a logic of their own. With their dependency on fickle and unpredictive donor support, their goal is to fight hunger and poverty.

With the recent focus on breeding for climate change resilience, there are opportunities to reconceptualize the relationship between plant breeding and agrobiodiversity. For example, participatory and evolutionary plant breeding allows more user testing and engagement in the breeding process, which can lead to better adapted varieties (or mixtures) and quicker varietal turnover, both of which are crucial for climate adaptation. Creating a shift in the public plant breeding culture toward participatory, evolutionary, and location-specific breeding requires shifts in incentives and institutional values. Climate change, however, offers an opportunity space for organizations to reorganize their goals, incentives, and challenges.

Private Companies

Food manufacturers, insurance companies, and seed companies have obvious interests in agrobiodiversity as it creates new profit making and alternative opportunities. Above we discussed the potential for novel public–private sector partnerships in developing varietal technologies (see section on Archetypes of Agrobiodiversity Users). When there is a market, seed companies use agrobiodiversity to promote new crops (e.g., teff in California, quinoa in the Global North). Climate change can also open up alternative production systems for commodity production (e.g., the wine industry in new areas in Australia). Agrobiodiversity may become of interest to insurance companies concerned with production risks. Furthermore, many companies have established philanthropic organizations to finance research to support social responsibility and to create innovative opportunities (e.g., Monsanto and its carbon footprinting program). Since companies rely on data, and scientists may possess more comprehensive data, novel ways to support agrobiodiversity may be encouraged. Efforts to establish "sustainable sourcing" by multinationals may at times appear as "window dressing," but such initiatives also offer opportunities for collaboration.

NGOs and International Organizations

NGOs can employ agrobiodiversity for food security, which includes aspects of food quality and malnutrition, for adaptation to and mitigation of climate change as well as in relation to changing market opportunities. Many of the agendas that social and environmental justice NGOs promote are complementary to agrobiodiversity (see Appendix 6.1). Agrobiodiversity links into food and seed sovereignty, as shown in the work of the Action Group on Erosion, Technology, and Concentration (ETC Group) and La Via Campesina. They could also become practically involved in delivery of seeds.

There are many opportunities to tailor an agrobiodiversity message to and via NGOs, but there is also a knowledge gap in the link between agrobiodiversity and environmental and social outcomes promoted by NGOs. Agrobiodiversity is often complex and location specific, making it difficult to study and draw universal conclusions about the relationship between agrobiodiversity and environmental and socioeconomic outcomes. Targeted research on the links between agrobiodiversity and livelihood outcomes (see Figure 6.1) can help produce useful knowledge for NGOs and allow them to focus their efforts on critical agrobiodiversity-related interventions.

Governments

There are different ways in which governments can employ and support the use of agrobiodiversity in the context of global changes. Decentralization of land-use decisions has been adopted by developing country governments for the last two decades. In Indonesia, for example, the decision-making processes involved in land use of community landscapes has been turned over to the local governments. Prior to this, customary or common ("adat") lands were not recognized in Indonesia. On many occasions, this democratic decentralization has been found to have increased agrobiodiversity because local governments pay more attention to its constituents' needs. The governments of Brazil and India have instituted seed repatriation programs as a response to climate change. Peru and Bolivia have engaged in systematic monitoring of landrace diversity. In Europe, seed collections are directly available to farmers. Through information and communication technology, there are many new ways in which farmers can be given access to seeds and associated information. This does not, however, eliminate the need for farmers to experiment and adapt the provided seeds and information to their local conditions.

Global Drivers, Better Use of Agrobiodiversity, and Novel Institutional Arrangements

Different global drivers (e.g., climate change, population growth, urbanization and re- or deagrarianization and migration, market integration, and global market and food system transitions) generate new needs and create opportunities for better use of agrobiodiversity. Each driver requires a reconfiguration of the roles and linkages between groups of actors and organizations. The variation in values of breeders as well as among different groups of farmers and other agrobiodiversity users (e.g., governments, NGOs) raises the question of how these divergent archetypes of agrobiodiversity users, with their different values and uses, can be understood in relation to each other. In response, agrobiodiversity science can pursue three options:

1. Attempt to ignore this divergence by creating a "neutral" definition of agrobiodiversity.
2. Adhere to one particular set of beliefs about agrobiodiversity to the exclusion of others.
3. Try to explore how diverse definitions of agrobiodiversity systematically vary across different groups of people.

For us, the first option would be difficult or impossible to achieve and the second option fails to recognize the diversity in perspectives, even among scientists, and would thus be doomed to failure. The third option is attractive, as it provides an opening to reconcile differences that stand in the way of effective solutions.

A framework is required to analyze this variation in perceptions and their relations. One way to do this is by using ideas informed by grid-group cultural theory (for an introduction and review of this critique, see Tansey and O'Riordan 1999). This theory holds that people's value systems are embedded in social relationships in ways that are relatively consistent and predictable. For example, people who form highly bonded social groups will tend to uphold value systems that reflect this by emphasizing group solidarity and by measuring behavior against this standard. Beliefs are not only relevant to interactions with other people, but also with nature. A highly bonded social group will tend to emphasize external dangers, including environmental ones, as these dangers resonate most with the challenge of maintaining the social boundary between the inside and outside world of the group. Grid-group cultural theory originally proposes four archetypes of institutional culture: hierarchy (or top-down), individualism (egocentric), and isolation (isolates) are advanced as unique, consistent patterns of social relationships and value systems in addition to bonded groups (egalitarianism). In relation to agrobiodiversity, a gene bank, for example, needs an ordered structure with clear procedures; thus, a gene bank produces a hierarchical institutional culture based on an idea of agrobiodiversity as a classifiable set of entities. Entrepreneurs, on the other hand, tend to be embedded in a more individualistic institutional culture that is less concerned about producing shared and stable classifications, but views agrobiodiversity as a resource.

Different archetypes can be recognized in any social context, with varying degrees of divergence. What is of interest here is how the varied organizations that represent these different archetypes form stable institutional arrangements, which lead to more or less sustainable ways of using agrobiodiversity. Agrobiodiversity plays a role across these different organizational actors, who tend to have divergent views on what it exactly is and what role it plays, but need each other to sustain agrobiodiversity. To continue with the example, in order to obtain political support, gene banks need entrepreneurs to show the economic value of agrobiodiversity. In turn, entrepreneurs may rely on gene banks to introduce, for example, new variety or crop products. Although these actors' views on agrobiodiversity may conflict at times, shared understandings are needed if they are to collaborate.

Below, we explore the way each of these specific drivers of global change could lead to different opportunities for collaboration in the use of agrobiodiversity.

Climate Change

Climate change, as one of the principal drivers of global change, is responsible for negative impacts on the yields of agricultural systems on a global scale (Lobell et al. 2011; Tubiello et al. 2007). Initial projections a decade ago held that the worldwide yields of major crops would increase under conditions up to 2°C of global warming. More recent modeling estimates, however, now point to the probability of aggregate loss at this level of temperature increase unless significant adaptations are undertaken (Challinor et al. 2014). At the same time, biodiversity across a range of ecosystems is known to reduce the temporal variation of yield (Loreau et al. 2001). As a consequence, the biological diversity of agriculture and food systems is becoming a priority in the design of agroecological resilience to climate change (Altieri et al. 2015; Branca et al. 2013).

Specific insertions of agrobiodiversity into food production systems must be designed in response to the particular conditions of climate change. For example, the temporal diversification of medium-intensity production systems (e.g., crop species in rotational sequences) stabilizes yields under conditions of abnormal soil moisture (Gaudin et al. 2015). More generally, the modeling of crop yield under global climate change underscores the need for adaptations, including cultivar adjustments (i.e., new uses of existing varieties and the development of new varieties), that offer promising strategies for improved crop yield with climate change uncertainty (Challinor et al. 2014; van Etten 2011). These adaptations will depend on the strategic use and insertion of crop biodiversity.

Climate change effects and the capacity for adaptive responses are, however, most commonly experienced at the level of fields, farms, and communities in conjunction with the processes associated with other major drivers, such as global socioeconomic integration through markets and trade. In other words, climate change impacts and adaptive capacity are rarely felt as isolated impacts (O'Brien and Leichenko 2000). This insight must now be developed to encourage agrobiodiversity in diverse socioecological contexts. The idea of archetypes, as introduced and treated above, can be employed in this regard. Certain groups within the large and heterogeneous category of 2.2 billion smallholders, for example, are likely to use agrobiodiversity within the combined context of climate change and the transition to part-time farming associated with global trends of expanded labor migration, the growth of peri-urban areas, and deagrarianization (but not complete depopulation) in remote rural areas.

Climate change and other drivers of socioecological change present an opportunity space for new institutional arrangements and agrobiodiversity strategies. For example, public plant breeding in India is typically a top-down enterprise; farmers are not involved in adapting or finishing new varieties due to the agricultural policies and incentives structures of public plant breeding in the nation. Similarly, the National Bureau of Plant Genetic Resources (NBPGR) in India did not initially distribute germplasm directly to farmers for testing until Bioversity International worked with NBPGR to distribute 21 varieties of wheat in a pilot program in northeastern India for climate change adaptation (Mathur 2013). This and other scientific exchanges between the global CG-center, Bioversity, India's national NBPGR, and gene bank managers from other countries has led to a cultural shift in NBPGR that is more open toward farmers having access to germplasm. This multiscalar institutional network opens opportunities for collaboration with private organizations, such as the Swaminathan Foundation, that promote agrobiodiversity conservation, sharing, and use for climate change adaptation.

Population Growth

As a driver of global change, population growth has not yet stabilized in many low-income countries. Here, agrobiodiversity can play an important role in coping with challenges around food security and health, such as the availability of and access to nutritious and culturally appropriate food. Agrobiodiversity can contribute resilience through highly diverse and productive farming systems. For example, mixed cropping can result in higher and more stable yields in the face of environmental variation. To support the development of more agrobiodiverse farming systems, environmental schemes could be implemented in which a range of actors participate: farmers, NGOs, government agencies, and corporations could jointly be responsible for developing such schemes. By providing farmers better access to seeds and knowledge, joint initiatives between NGOs and public gene banks can use agrobiodiversity to improve the actual access to available food.

There is also evidence that agrobiodiversity can also be used to elevate the status of women in regions with high population growth and tenuous food security. It is generally accepted that raising the status of women and increasing their education levels leads to lower birth rates. To increase female status, one possible way is to train rural women to be local caretakers of agrobiodiversity and, at the same time, to increase their participation in the seed system in which they traditionally play an extremely important role in many societies. Side benefits of this approach could include better nutrition and increased economic security, and could bring the agendas of NGOs and governments together. More research is needed, however, on the connections between women, agrobiodiversity, and economic empowerment to determine where and how agrobiodiversity can contribute to the status of rural women. In addition, attention should be paid to the regionally specific nature and culture of women's agricultural work and social status (Carney 1993; Nuijten 2010; Zimmerer et al. 2015).

Urbanization, De- and Reagrarianization, and Migration

Urbanization and the intensification of adjacent peri-urban spaces for food production are currently a major global change driver shaping environment, land use, and socioeconomic changes across the world. The processes of urban and peri-urban growth are linked closely to processes of deagrarianization (rural out-migration), reagrarianization (e.g., the movement of exurbanites or "new farmers" into rural spaces), and migration across national borders and continents. Together, these changes are associated with influential yet complex impacts on agrobiodiversity as well as the actual and potential roles of institutional configurations. For example, urbanization has been strongly linked to the growth of new consumer cultures that favor agrobiodiversity through gastronomic cuisines among well-to-do consumers as well as through food movements among educated consumers who prefer organic, unprocessed food, and healthy eating. Rural in-migrant populations who retain preferences for inexpensive traditional foodstuffs and have cultural ties with specific food also induce demand for agrobiodiversity. These include not only rural peoples of the Global South who once farmed but also Global North retirees who have migrated to tropical countries. As international migration continues, so too does the demand for foods and condiments that are often unavailable in the host culture. Whether the demand for organic vegetables stems from retirees from wealthy countries relocating to the warmer climates in the Global South or from refugee populations resettling in the Global North, our contemporary world of people on the move offers multiple instances and possibilities to encourage and promote agrobiodiversity, with roles for formally and informally organized networks of actors.

Restaurants and food businesses are important actors in brokering the changes in urban consumer cultures. Urban and community gardens, as well as small farms, are a significant locus of this activity and can offer migrants support from governments (city, regional, national) and international agencies. The continued links of rural-to-urban migrants with their natal communities offer additional prospects for agrobiodiversity. Although these migrants often send remittances and other resources back to rural areas, which in many cases are deagrarianizing, such resource flows are often accompanied by a demand for seeds of traditional foods not easily sourced in urban areas, but which migrants can grow on city plots made available to them through community gardens and other governmental organizations. Such migrant resource flows can be used to support agrobiodiversity under the right combination of favorable circumstances (Zimmerer 2014).

Issues of gender relations, equity, and livelihood options are an integral component of smallholder deagrarianization. Migration patterns are often gendered, and the process frequently leaves women disproportionately behind as farmers in the sending areas. The "feminization of agriculture," reported in many smallholder farming areas of the Global South, is a globally significant feature of deagrarianization. While migration loosens household dependence on farming for family survival, it often results in pronounced gendering of land use and agrifood systems in out-migration areas (Radel et al. 2012). Female farmers and their household members must prioritize farming strategies and decide how to invest remittances from male migration. In Nepal, men's increased rate of out-migration has been reported as a positive influence on women's decision-making processes, including agrobiodiversity management (Bhattarai et al. 2015). Differential migration of household members carries profound implications for rural land use and agrobiodiversity. It can reduce the amount of land cultivated and diversity of crops planted, but it can also have the opposite effect: it may induce households to adopt new crops and practices into farming systems—a process that is evident today in many peri-urban areas of the Global South (Zimmerer et al. 2015). Proximity to emerging urban food networks and trends may encourage the adoption of new cultivars, including tree crops. In some rural locales with male out-migration, females create agrobiodiversity through informal seed networks that encourage risk-averse farming practices and flexibility in farm labor demands (Nuijten 2010). For example, women rice growers in The Gambia exchange seed varieties of differing maturities which are adapted to the diverse microenvironments that comprise a rice landscape. Seed diversity promotes subsistence security of a crucial dietary staple. Water retention in one rice microenvironment—the inland swamps—facilitates agrobiodiversity by permitting the planting of multiple crops and sequential cultivation. Women have established bananas on the bunds that enclose the plots and, after the rice harvest, use the plots' residual soil moisture to grow vegetables which are then subsequently marketed (Carney 1993).

Reagrarianization is another important process that encourages and promotes agrobiodiversity in rural areas. This refers to the patchwork of small farms that have appeared in the Global North (and in some parts of the South) in response to urban food movements. As in the "back-to-the-land" movement of the 1960s, we are again witnessing an interest in agriculture among young people with no previous farming experience. This is also encouraging those with farm experience to grow crops and animals responsive to new urban consumer demands for farm products that are produced and raised using ecologically sound practices. This convergence of trends has created the group of smallholder farmers we label neoagrarians. The reagrarianization of such landscapes is unfolding with an intentional use of agrobiodiversity, evident at several scales—within farm plots and across landscape levels—as well as through mixed cropping, agropastoral, and agroforestry methods. To what extent neoagrarian use of agrobiodiversity implies enrichment or conservation remains to be better understood. In any case, the repopulating of rural land with production techniques and crops in demand by urban "foodie" consumers is unfolding for the most part with weak institutional support. Other forms of organization are emerging as well, such as informal seed exchanges among this type of farmer to promote heirloom varieties or rescue ones that did not serve the expansion of the industrial food chain. Internet-based communication opens up an array of opportunities for disparate groups to connect.

There is a diversity of organizations involved in promoting agrobiodiversity in low-income communities of the Global North. Community-supported agriculture and farmers' markets are being increasingly encouraged by city governments. In Los Angeles public schools, school gardens are promoted to encourage better diets among low-income students, who live in urban food deserts. In areas affected by youth gangs, school gardens provide an opportunity to teach children at a young age the geographic origins of foods they like and to help gain an appreciation of the ethnic and cultural groups who developed the foods. These diverse food gardens have emerged as important sites of agrobiodiversity in urban areas, providing fresh vegetables and fruits where supermarkets are absent. Urban agriculture is also being increasingly supported by philanthropic foundations, which fund projects to encourage the availability of fresh produce for the needy, including the homeless, handicapped, and homebound senior citizens. Their efforts extend to involving at-risk youth and new immigrant groups in urban farming (Cockrall-King 2012).

The international migration of people from countries in the South to more prosperous countries in the North, for the purpose of finding better and safer futures, offers specific chances to link producers, entrepreneurs, and consumers, which may be supported or facilitated by government policies. Communities of migrants from the same country or region often find themselves in cities (e.g., in the United States, Canada, England, Sweden, and Germany). In this new environment and cultural context, the migrants have to construct new identities; in many cases, their original food culture plays an important role either as food for consumption or in generating income (e.g., restaurant, shop). The value of having access to produce from their homeland can open up new opportunities for agrobiodiversity grown by smallholders in these countries. Mexicans and other consumers in New York eating tortillas made from blue maize or Ethiopians preparing injera with teff from their home country are examples of migrant cultures that generate opportunities for new value chains of agrobiodiversity. These value chains tend to be thought of as dominated by small entrepreneurs (often from the migrant community) and as a direct linkage to producers in the homeland production areas. Whether such value chains contribute to the livelihood of farmers growing the agrobiodiversity or result in larger areas planted with valuable agrobiodiversity is not clear. A point of tension is the threat to such value chains when these agrobiodiversity crops or varieties are being picked up by producers in the country or region of these migrant cultures. Examples of this include identifying or developing teff varieties adapted to California, growing quinoa in Denmark and Holland, and popularizing the use of chili pepper diversity across the southwestern United States. Because this contributes to income opportunities for the producers, it potentially broadens the demand as well as the agrobiodiversity portfolio being used, leading to more types of maize and chili eventually being grown. When white-grain quinoa production expanded in the European Union and the United States, marketing from the Andean countries started to focus more on yellow, black, red, and pink varieties. "Protected designation of origin" labels governed by producer groups or NGOs could add value to the produce from the country of origin; advocacy by the migrants themselves could also assure demand for the produce of their home country. In addition, governments can facilitate trade through regulation or assume a protective role, as in Colombia, where the production of coffee other than Arabica is banned to protect the quality of the renowned Colombian shade-grown coffee, which also supports biodiversity.

Global Market Integration and Food System Transitions

Around the world, food supply chains and commodity markets are undergoing a drastic transformation directly linked with food system transitions (see Chapter 9). This change is typified, for example, by the increase in the number of supermarkets involved in food retailing in developing countries (Rao and Qaim 2011). The changes in the composition of food supply chains are expected to affect the scope of on-farm conservation of agrobiodiversity. The market institutions that structure the commodity value chains are important for two reasons: (a) they transfer consumer demand for agrobiodiversity to producers and farmers, and (b) they cater effectively to the consumer demand for diversity. These issues are particularly important for the conservation of perishable food products (e.g., vegetable, dairy, and livestock products), which are an integral part of agrobiodiversity in many farm portfolios. There are documented cases of traditional vegetable chains that are effective in conserving agrobiodiversity (Chweya and Almekinders 2000; Iskandar et al. 2018). Similarly, the demand for local products leads to the development of farmers' markets in developing and developed countries. Novel and emerging labeling and certification schemes (e.g., Happy Chickens, Roundtable on Sustainable Palm Oil) that embrace the consumer demand for nature-friendly production systems have a significant impact on on-farm conservation of agrobiodiversity. However, transmission of information on the agrobiodiversity conserved from farmers to consumers is often challenging in these commodity value chains.

Conclusion: Framing the Agrobiodiversity Narrative

Agrobiodiversity has a multitude of functions through which it contributes to the well-being of people in this world via different pathways. Agrobiodiversity and its functions in food systems transitions are subject to global change, driven by climate change, population growth, migration from rural to urban spaces, and market integration. The way in which agrobiodiversity can play a role in the world of tomorrow is likely to be different than it is at present. Historically, the dynamic nature of agrobiodiversity has allowed human–environmental relations to change adaptively. With the challenges of a warming world and agricultural uncertainty, it has the potential to continue adapting in contexts, however, where agriculture and food systems have been significantly transformed.

In this chapter we explored how current and future intentional agrobiodiversity use varies between and within different groups of users. Our examination demonstrates that current drivers of global change affect these groups locally in unique ways. Their response (in terms of value, intentionality, and use of agrobiodiversity) has created the potential for novel initiatives: user groups and organizations have been able to engage in new collaborations and diverse configurations at varying scales. Together, we consider that the experiences from these initiatives and analyses of opportunities offer the basis for a compelling overarching narrative: agrobiodiversity can be used in multiple ways by different groups of people, organizations, and networks to improve the well-being of people around the world. Such a narrative can connect the wide diversity of producers and consumers that are geographically dispersed, forming networks of personal and virtual interactions that provide a critical mass to reinforce ongoing efforts to find solutions to the challenges of global change. Farmers that seem now at the extreme poles of the gamut of diversity may connect and be of inspiration and support to each other, joining forces across geographic space amid myriad new forms of production and consumption.

The experiences and the narrative also point to important gaps in our knowledge which, when filled, should help to create space for alternative pathways of development and institutional strategies to enable adaptation and mitigation of the negative impacts of global change—perhaps by turning threats into opportunities. Different groups and organizations have different roles to play. Ensuring participation and empowerment of vulnerable groups will need to be incorporated into the responses to global change. New science integrating social, agronomic, and environmental knowledge is needed to support and analyze how initiatives of using and valuing agrobiodiversity fare, how these experiences can be connected or patterns discerned, and how experiences can be shared and linked to forge new institutions and organizational linkages.

Acknowledgments

The findings and conclusions in this publication have not been formally disseminated by the U.S. Department of Agriculture and should not be construed to represent any agency determination or policy.

Appendix 6.1

For seven different groups, we present an overview of how groups currently use agrobiodiversity and their associated vision, value, and intentionality. This information can provide entry points to strengthen future agrobiodiversity use.

1. For 2.2 Billion Smallholder Farmers

Group Use Related to Climate Change

- Forces people to shift to part-time farming or off-farm income
- Migration: seasonal versus permanent/forced versus voluntary
- Gender
- Varietal change
- Crop switching
- Mixed crop and livestock
- Crop diversification
- Intensification: skill and inputs
- Extensification: expansion (environmental cost)
- Social networks/information
- Not feeding food to animals
- Eating less or differently

- Crop insurance
- Agronomy

Group Use Related to Market Integration

- Create demand by linking with consumers to finance transition
- Increase cash crop production

Vision, Values, and Intentionality

- Highly varied
- Includes food as part of cultural identity
- Intentionality is often by default

Entry Points

- Link to agrobiodiversity institutions
- Embed activities into networks
- Policy to support small-scale initiatives based on creativity and alternative narratives

2. For Indigenous Farmers

Group Use

- Sparse knowledge on agrobiodiversity responses to global changes

Vision, Values, and Intentionality

- Intentionality by default
- Agrobiodiversity is an integrated part of livelihood sustenance, security, cultural identity, etc.

3. For Neoagrarian Farmers

Group Use

- Conscience intentionality: banking on "credence quality"

Vision, Values, and Intentionality

- Often a high level of intentionality toward agrobiodiversity considerations
- Short food chains with value for food and client relation

Entry Points

- Use knowledge as a product

- Support gastronomy that favors healthy tasting food
- Agrobiodiversity certification

4. For Public Breeders

Group Use Related to Climate Change

- Stress tolerance and biotic and abiotic
- Crop and variety substitution
- Conservation agriculture + resource use efficiency + agronomy
- Impact modeling $\rightarrow$ physiology

Group Use Related to Market Integration

- Food processing
- Entomophagy, shifting diets
- Productivity
- Commodification, intellectual property, seed market

Vision, Values, and Intentionality

- Modernized breeding
- Intellectual Property
- Profit oriented
- Instrumental agrobiodiversity use (e.g., drought or disease resistant), regenerational breeding
- Orphan crops
- Intrinsic (e.g., in centers of origin), relational (e.g., characteristics with cultural values), instrumental

Entry Points

- Increase public funding for location-specific breeding (more varieties)
- Increase public funding for orphan crops
- Change institutional incentives to increase the value of agrobiodiversity, away from wide adaptation
- Incentivize plant breeders to work with multiple stakeholders
- Open access to intellectual property
- Create alternative narratives

5. For Private Companies

Group Use Related to Climate Change

- Increasing yields, continuing Green Revolution (possible) trend for area-specific breeding if buyers concentrated; new crops (quinoa, teff)
- Production companies, land grabs, agrobiodiversity use in crops
- Food manufacturing, agrobiodiversity cell lines
- Insurance companies, big data approaches, maybe agrobiodiversity?
- Agrobiodiversity conditions insurance
- Food companies

Group Use Related to Market Integration

- Private companies mainly disuse agrobiodiversity for production processing, food system uniformity
- Novel products and markets, niche commodities

Vision, Values, and Intentionality

- Value: individualistic, for profit, create markets, "investment opportunities"
- Stakeholder (private wealth maximization)
- Intentionality: agrobiodiversity instrumentally for vision and values; agrobiodiversity for unique attributes of quality; patenting agrobiodiversity/intellectual property

Entry Points

- Niche markets, value chain
- Philanthropy
- Genetic resource use
- Corporate sustainability (e.g., Monsanto's carbon-zero strategy)
- Prosocial environmental responsibility (e.g., health and diets, Fair Trade)
- Voicing of societal concerns (related to market integration)

6. For NGOs, International Organizations

Group Use Related to Climate Change

- (Forest) Food security, food quality, food sovereignty
- Low-input agriculture (low fossil fuels)
- Diversification (livelihood diversity for adaptation)
- Advocacy
- Conservation (of intrinsic value)

Group Use Related to Market Integration

- Livelihood diversification/diet diversification
- Prolocal (against globalization/self-autonomy): food, seed systems, agricultural products

Vision, Values, and Intentionality

- Conservation (of intrinsic value), access to plant genetic resources
- Social justice, equality, collectivism (horizontal seed systems)/advocacy for rights, autonomy of Global South (e.g., property rights, intellectual property)/sustainability
- Agrobiodiversity as tool for advocacy of their values (related to market integration): self-autonomy and agroforestry/crop diversity for preservation, associated biodiversity

Entry Points

- In terms of market integration, agrobiodiversity can be a tool to advocate values
- Goal should be better linkages between NGO goals and agrobiodiversity and better tailoring of message, scientific output, and knowledge gaps
- Tailor the research agenda to knowledge needs
- Produce empirical evidence for agrobiodiversity use

7. For Governments

Group Use Related to Climate Change

- Payment schemes (agro-environment schemes), e.g., EU payments to farmers for traditional breeds
- Insurance schemes (e.g., for catastrophes):
  - ○ Productivity → public breeder, agricultural development
  - ○ Adaptation → breeding
  - ○ Mitigation → control of erosion, deforestation

Group Use Related to Market Integration

- Market price regulations
- Trade regulations
- Certification schemes

Vision, Values, and Intentionality

- Support for economic growth and stability
- Cultural heritage and identity, public health
- Diverging positions and goals related to market integration (i.e., profit and yield maximization vs. societal goals), food security, diversity, conservation, culture, etc.

Entry Points Related to Climate Change

- Repatriation

- Migration
- Development of value chains for new foods
- Rewards for agrobiodiversity stewardships
- Information/communication will raise awareness
- Biodiversity prospective
- Decentralization

Entry Points Related to Market Integration

- Removing perverse incentives and hidden subsidies to production and marketing practices that negatively affect agrobiodiversity
- Provide incentives to agrobiodiversity promoting production and marketing practices

7

How Do Climate and Agrobiodiversity Interact?

Jacob van Etten

Abstract

The interaction between climate and agrobiodiversity is framed in different ways by different scientific disciplines and researchers. These diverse frames inform climate action by defining the main questions that are being asked and the solutions that are attempted. This chapter explores these frames through select discussion of studies in archaeology, environmental, climate, agricultural, and social sciences. Archaeological and environmental studies frame the interaction between climate and agrobiodiversity as part of a historical coevolutionary process. Agricultural and climate sciences have focused away from systemic interactions between climate and agrobiodiversity, devoting limited attention to genotype–environment interactions and diversification. Another relevant frame is to see agrobiodiversity as an informational resource, which is undermined by climate change as local information about adaptation rapidly becomes obsolete. Knowledge generation then becomes the central engine of economic growth to counteract loss of information due to climate change. Climate action needs to confront climate change and agrobiodiversity management as "wicked problems"—problems that demand attention to the systemic nature of the problem, uncertainty, and the role of human values. Integrated scientific approaches are needed to design processes that explicitly address these aspects, contribute to climate action, and accommodate opposing values.

Introduction

Anthropogenic alterations of the Earth's atmosphere are changing the global climate so fast that this has created one of the most important challenges to humanity. To address climate change, wide-ranging responses are needed to reduce its negative impact, responses which are now often referred to as "climate action." In agriculture, climate action represents an important role for agrobiodiversity, and the intelligent use of biodiversity in agricultural land use is a crucial ingredient in addressing climate change. Mobilization of genetic resources is needed to address new stresses from heat, drought, and new pests and diseases induced by climate change. These genetic resources are needed to shift or create crop varieties and animal breeds that produce more, cope better with stresses, and contribute to reducing greenhouse gas emissions per unit of area or unit of product (Jackson et al. 2013). Climate change also requires the redesign of agricultural systems that can cope with higher temperatures or less available water supplies. The use of new combinations of biota that perform agricultural functions with biologically evolved effectiveness can increase resource use efficiency, sequester soil carbon, and decrease the use of fossil energy (Altieri et al. 2015; Branca et al. 2013). However, agrobiodiversity is not solely a tool for climate action: it is also a victim of climate change. Wild and cultivated species and intraspecies diversity may disappear as they lose their habitat to climate change (Bellon and van Etten 2014; Jarvis et al. 2008a).

To explore different views on the interactions of climate and agrobiodiversity, I bring together a range of different disciplinary and theoretical perspectives (see also Chapters 6 and 8): archaeological and environmental studies provide a long-term perspective on these interactions, whereas applied climate and agricultural sciences offer perspectives that focus on informing climate action. A recent generation of studies produces a better understanding of climate and biodiversity from a complex systems perspective. In addition, social science contributes to our understanding of how climate action is shaped by different social forces.

The scope of this essay is not that of a comprehensive review. I have selected studies that explain and illustrate concepts as well as others that provide potentially important insight into how the climate–agrobiodiversity interaction can be reframed. Frames, a term I use rather loosely, define the problem and eventually suggest the solution; hence, they inform climate action (Dorst 2015; Entman 1993). Frames have a strong normative aspect, so discussing research from this perspective involves necessarily a degree of subjectivity and eclecticism. More explicit frames will enable better choices to be made regarding climate action.

Long-Term Perspective

Agriculture emerged in different world regions over a prolonged period of time (Fuller et al. 2015). Whether or not agriculture emerged in response to climate change in different areas is still a matter of discussion. Even if climate change played an important role in specific areas, other aspects as well as demographic and sociocultural causes most likely played an equally important role. Archaeologists have learned that agriculture did not emerge as a sudden invention, but arose gradually from preagricultural land use as prehistoric people exercised selection pressure on plants that they collected and started to manage landscapes in ways that favored their own purposes (Fuller et al. 2015). Through conscious and unconscious selection of plants and animals, human populations started to alter a number of traits that favored their agricultural use, leading to a slow process of domestication. Prehistoric people expanded the range of domesticated species by taking plants and animals to new places and sharing them with neighboring human populations as well as through migration. Human populations proficient in agriculture started to grow demographically, expanding their demographic base and occupying more land. As a result, domesticated species radiated out of their areas of origin. The genetic geographies created by prehistoric crop dispersal are still recognizable in today's distribution of crop genetic diversity. As agriculture forced nomadic human populations to settle, they also became more reliant on highly localized resources and, in many cases, broadened the range of plant species they consumed (Brookfield 2001).

Agriculture expanded in Europe at an annual rate of 0.9–1.3 km (Fort 2012). This rate not only reflects the speed of demographic expansion of human populations and cultural transmission, but also the ability of crops to adapt to new climates. The dispersal of agriculture from Anatolia to northern Europe or the American tropics to North America is premised on a drastic climate adaptation process. Contemporary climate change can similarly be expressed in a speed rate, as climates shift away from the equator in response to heat accumulation. The prehistoric rate of expansion of agriculture exceeds the contemporary speed of climate change, estimated by Loarie et al. (2009) to be 0.48 km per year as a global average, although locally it could be much higher. In addition, for many systems the required speed of selection may be demanding without the aid of modern plant breeding.

Recent studies of the archaeobotanical record reveal some of the potential costs of climate adaptation. A fair number of species became domesticated and then fell into disuse. Allaby et al. (2015) argue that such false starts may be related to excessive selection pressure that prehistoric people placed on plant populations as they moved into new environments. Cultivated plant populations were often not prepared for human selection pressure as they had already gone through a genetic bottleneck compared to their wild progenitor populations. Prehistoric people removed much diversity as they only propagated small portions of wild populations; this led to genetic drift, the random loss of diversity. Also, they exercised seed selection for traits they found important, leading to so-called selective sweeps across the genome. Likewise, increased selection pressure on crop populations today may reduce their genetic base and the chances of finding traits that help crops adapt to new climates.

The prehistoric past sets the scene for agrobiodiversity in climate action in another way, too. Agriculture, one of the economic sectors most dependent on the natural environment, has typically been viewed as one of the main victims of climate change whereas another economic sector, industry, has been faulted with kicking off the massive fossil fuel use that has increased greenhouse gas concentrations in the atmosphere. It was presumed that agriculture contributes to accelerated climate change only to the extent that it is implicated in the Industrial Revolution. Consequently, a return to the "old" ways, with an increased role for agrobiodiversity, was seen as a way of bringing agriculture back on the right track: it would reduce fossil fuel dependency and even act as a redeeming sink, fixing greenhouse gases by converting them to soil carbon.

Recent research, however, challenges this view. Geologist Bill Ruddiman (2013) developed the hypothesis that agriculturally generated greenhouse gases before the Industrial Revolution were significant enough to stave off an Ice Age, which was bound to happen if normal climate cycles had followed their course. Prehistoric farmers cleared forests to plant crops, sending the carbon contained in the trees into the atmosphere. Also, they greatly increased the number of ruminants by domesticating them, keeping them in large herds and protecting them from predators. Given their specific digestion process, ruminants emit much methane, a potent greenhouse gas. Another important source of methane is irrigated rice cultivation. To grow rice, prehistoric people modified the landscape to increase the area that is periodically flooded, thereby creating the anaerobic conditions that favor methane emissions.

There is now a growing body of evidence to support Ruddiman's early anthropogenic hypothesis that agricultural emissions were indeed substantial enough to cause a notable warming of the atmosphere. Prehistoric emissions and concentration gradients predicted from archaeological data on the spread of agriculture match past atmospheric concentrations of greenhouse gases that can be inferred from air bubbles enclosed in Arctic and Antarctic ice cores (Mitchell et al. 2013). Obviously, the increase in greenhouse gas concentration due to preindustrial agriculture is dwarfed by the increase caused by industrial emissions, including those from modern agriculture, but it is not only the quantity of greenhouse gases emitted by preindustrial agriculture that matters. Its timing is also important. The climate system does not respond immediately to the upsurge in greenhouse gases in the atmosphere; it is highly inert and takes several decades to respond to change in greenhouse gas concentrations. The preindustrial buildup of greenhouse gases, although smaller, has had much more time than the industrial contribution to trap energy from the sun into the atmosphere. Simulations reported by Ruddiman (2013) suggest that preindustrial warming is somewhat greater than industrial warming.

To summarize, archaeological and environmental research over the last decade has markedly changed the scientific perspective on how climate and agriculture interact: agriculture is not only a "victim" in the story of climate change, it is also a "villain" contributing to emissions even before industry. Thus, agrobiodiversity is a hybrid product of human and natural forces, and is deeply implicated in anthropogenic environmental transformation over the last 10,000 years. Agrobiodiversity forms the tangible evidence of human environmental transformation over millennia. This transformation has brought us into the Anthropocene—the current geologic period defined by human impacts, such as elevated greenhouse gas levels as well as nitrogen and phosphorous levels, on soil and water systems, especially in estuarine environments (Ruddiman 2013).

These findings undermine the perspective of agrobiodiversity as a "natural" solution, because it predates the Industrial Revolution. Ruddiman's hypothesis and the concept of the Anthropocene suggest a change in our perception of our living environment, challenging the idea that we can go back to an equilibrium situation in which agriculture is "adapted" to its environment. Instead of a simple adaptive process in which agriculture slowly moves to an equilibrium, we are observing a nonlinear, coupled process which may not have a stable equilibrium or optimum (cf. Kauffman 1993). This "frame" suggests a far more active historic role for humans in environmental transformation and, at the same time, urges us to take responsibility for charting the course of the global environment in a forward-looking way. Human decisions about land use, energy, and food consumption will shape global climate. In all of this, agrobiodiversity is highly relevant and can potentially, but not inherently, be used as a positive tool for climate action.

Science and Climate Action

How do disciplines which directly set the climate action agenda in climate science and agricultural science frame the climate–agrobiodiversity interaction? The agriculture chapter of the fifth report of the Intergovernmental Panel on Climate Change (IPCC) summarizes the latest research and provides a sampling of the most influential research (Porter et al. 2014). What is evident, however, is the scant attention given to agrobiodiversity. Most of the agricultural research represented in the IPCC report is limited to the main staple crops, based on data from crop trials as well as crop model studies. Crop models are usually parameterized with data from on-station trials and then used to extrapolate the results to other environmental conditions. This approach is compatible with the model-oriented focus of climate science, but it hampers the understanding of wider interactions in which agrobiodiversity plays an important role (Thornton and Herrero 2015).

To explain the narrow focus of climate science, Demeritt (2001) makes the case that climate science is socially constructed. He does not imply that scientific findings are completely dependent on social relationships or that there is no reality outside of the interactions between scientists, but rather that microsocial relationships play an important role in making choices about what to include and what to leave out in the climate models, and what type of evidence is acceptable or not. For example, global circulation models are a very narrow abstraction of global physical reality and exclude a large number of features, such as cloud formation. This type of reductionism is beneficial because it makes reality analyzable and works toward a strategy that provides a clear focus on the importance of climate change as a global phenomenon. At the same time, social factors play a role in the decision-making processes involved in defining what allowable abstractions are in a given scientific domain, and these decisions may limit alternative epistemologies. Demeritt argues that social construction of climate science cannot be circumvented. Even though different mechanisms, such as peer review, can provide checks on the quality of science, trust in science cannot be solely generated with an appeal to correct procedures and abstract principles. In the long term, transparency about the social construction of scientific insights is necessary to generate trust in science, a clearer understanding of its limitations, and reflections on alternative scientific strategies.

Similar social analyses also apply to the agricultural sciences. In a recent study, Baranski (2015b) reexamined the role of environmental variation in the social construction of agricultural science behind the Green Revolution, which started in earnest with wheat breeding in India. She argued that the Green Revolution's focus on yields in high fertility and irrigated conditions was rooted in the institutional landscape of India, which had just centralized its agricultural research program. Of importance was also the belief that fertilizers would soon be widely available in developing countries such as India. Scientifically, the breeding approach was defended with the idea that "wide adaptation," or minimizing genotype–environment interactions, would also favor yield gains in more stressed environments. Baranski argues, however, that the photoperiod insensitivity and responsiveness of the new varieties to fertilizers did little to confer yield stability on the new varieties across a range of environments, including rainfed conditions. Using experimental data from the 1960s, Baranski demonstrated that under rainfed conditions, an important Indian tall variety was more stable and more productive in lower-yielding environments than the Mexican variety that was introduced. Despite this evidence and the more recent emphasis on breeding for drought and heat stress, the same views on wide adaptation still prevail in Indian science; these early choices and strategies are now codified into the wheat breeding program.

The problem with these views is that they obstruct the climate signal in plant breeding. Climate information is generally incorporated into plant breeding only indirectly, primarily through the definition of (widely defined) target production environments and specific insights coming from crop physiology about the relative importance of traits for climate adaptation. It could be argued that even though climate may not be a focus of explicit analysis, the climate signal will be picked up by plant breeding as it affects selection environments. Some breeders argue that a main adaptation response would be to keep a focus on broad adaptation, accelerate breeding, and shorten the time between breeding and farmers' use of the new seeds (Atlin et al. 2017). Ideas around wide adaptation, however, are increasingly being questioned. Desclaux et al. (2008) discuss how new demands placed on agriculture increasingly force plant breeders to rethink their approach to genotype–environment interactions. Some breeders have argued that breeding should be a more decentralized process, working with more diverse populations in "evolutionary breeding," which has been shown to work in favor of resilient crops in marginal conditions (Ceccarelli et al. 2010).

Desclaux et al. (2008) go a step further and expand the concept of genotype–environment interaction to involve other aspects of the environments in which crops grow. Climate change affects these environments not only by increasing plant stress but also by affecting other decisions by farmers on their cropping systems and farms. Breeders can no longer reduce a production environment to a few biophysical aspects. Crop management is not just a factor limiting the genetic potential of varieties but is important in itself and involves multiple trade-offs. In the future, crop physiology needs to play a more important role than just informing the overall setting of priorities. Currently, crop management and its interaction with diverse environments play an increasingly important role in breeding decisions, as environmental sustainability gains in importance, including reducing greenhouse gas emissions from fertilizer use. Desclaux et al. (2008) review several options to subject this expanded view of the environment to analysis and reflection. These authors find, for example, that a broader range of stakeholders needs to be involved in priority setting and that new demands placed on agricultural science can be addressed by increasing transparency and democracy. Demeritt makes a similar point for climate science (see above). New analytical techniques are being developed to tease out the influence of weather and other environmental factors in multiple environment trials. The aim of these analyses is to attempt to predict or address genotype–environment interactions rather than to remove them.

A stimulus for incorporating environmental data in breeding is the trend toward genomic "big data," which has not yet fully been paralleled with a similar effort for environmental big data. For example, Heslot et al. (2014) offer an approach to incorporate weather data into genomic prediction, a technique to predict phenotypic values from genomic information. They derived a range of environmental stress variables from weather data using simple crop growth models. Including these variables in the analysis improved the predictive power of models. This and similar approaches represent important progress, as they enable us to study the interaction between genotypic and environmental variation simultaneously, in a biologically meaningful way, to allow prediction across different environments. This opens up a range of new opportunities to generate physiological insights into plant environmental adaptation directly in breeding experiments.

The role of plant genetic diversity in breeding is likely to become even more important under climate change (see Chapter 5). Heisey and Day-Rubenstein (2015) predict that genetic resources will increase in value as climate change increases the demand for traits that confer stress tolerance. They cannot predict, however, which types of genetic resources will prove to be most useful in the future for breeding for climate adaptation: landraces, crop wild relatives, or nonplant genetic resources (such as useful soil organisms).

Nonetheless, gene bank collections have more value if information about their biology and environment of origin is available. Little hard data is available, however, about the economics of the use of plant genetic resources in breeding (for an overview, see Heisey and Day-Rubenstein 2015). Studies show that the costs of searching for accessions with biotic resistance traits are lower than the benefits these accessions give to breeding. In addition, they argue financial support for gene banks relative to the total revenue of the seed sector seems small. As a source of agrobiodiversity, the importance of gene banks will increase for climate adaptation; new breeding technologies will not undermine their function but rather expand it.

While climate is being factored explicitly into agrobiodiversity management for technology development and introduction, agrobiodiversity is also relevant for farm management strategies (see Chapters 6 and 8). Thornton and Herrero (2015) argue that interactions between components in farming systems can be managed to respond to climate change, with their study emphasizing interactions between livestock and crops. Branca et al. (2013) summarize some of the evidence for agrobiodiversity-based practices that are providing both adaptation and mitigation benefits, including agroforestry practices and nitrogen-fixing species. Still, little is known about the exact yield effects of many of these practices or how interactions within farming systems are affected by climate change. Consequently, farming system assessments have been largely absent from IPCC reports. Thornton and Herrero (2015) argue that to analyze the agricultural system, we need better models, scenarios, and indicators. Models are needed for a broader range of crops for which detailed crop models are currently unavailable. In addition, more data is needed to assess whole farm strategies. This requires consistent data across a range of metrics, including productivity, livelihood strategies, and nutrition. Thornton and Herrero argue that different data collection strategies are needed, using modern technologies and citizen data collection, to meet these needs. Hammond et al. (2016) present a relatively "light" electronic survey instrument that collects a large number of simple indicators. Agrobiodiversity is taken into account explicitly in this instrument (crop diversity, diet diversity) and can be analyzed in conjunction with indicators on greenhouse gas emissions, poverty, gender, and food security, among others, to assess the trade-offs and synergies in different agricultural systems and livelihood strategies.

Within agricultural systems, diversification is arguably an important risk management strategy and an important role for agrobiodiversity management. This is less controversial for mixed crop–livestock farming (Seo 2012) than it is for crop diversification (Lin 2011). For example, Barrett et al. (2001) have suggested that crop diversification is of limited importance as a strategy for risk management because yields of different crops are, in general, positively correlated. From plant modeling results, Gilbert and Holbrook (2011) suggest that diversification within grain crops does not contribute much to risk management, but that growing crops of different functional types or diverse physiologies (such as vegetables) does help to spread risk.

Empirical evidence on the role of crop diversification in risk management is rare because microstudies with detailed time series of crop yields require significant investments in repeated field data collection. One such long-term field study was reported by Matsuda (2013). He studied the upland farming systems of central Myanmar, which face very large interannual variation in rainfall. Myanmar farmers have a diverse agricultural system: pigeon peas, cotton, and sesame are the main crops. Each of these crops responds in different ways to climate variation, buffering the impact of drought. Yields between the main crops in central Myanmar farming systems showed a weak or even slightly negative correlation over the seven-year study period, which means that together they form a good portfolio for risk management.

Matsuda (2013) compared these findings with the farming systems in northeast Thailand, where interannual variation of rainfall is also high. In this area, rice is the predominant field crop and there are very few other crops, except in home gardens. Rice is grown under rainfed conditions and yields vary highly between years. In some years, the harvests are low or fail completely. Agrobiodiversity plays a very minor role in risk management in this area. Risks are mitigated between years through rice storage. Off-farm labor also provides an important buffer against income variation. This comparison demonstrates how risk management can take different shapes in different agricultural systems, depending on the actual configuration of farming systems, livelihoods, and institutions (see Chapter 8).

Padulosi et al. (2011b) argue that agrobiodiversity could play a larger role in climate adaptation through the expanded use of crop species that are currently underutilized. They hold that many underutilized species adapt to a wide range of environments and are tolerant to a range of stresses. Little information, however, is available on how these crops would function in new, adapted cropping and farming systems. From the perspective of climate action, prior evidence on the potential contribution of these species to production and livelihoods would be needed to justify targeted investments in research and development before their use is expanded. Systems analysis, however, has its own drawbacks: system modeling approaches often require a large amount of data, propagate uncertainties, and may provoke "analysis paralysis." Addressing this complexity is the main challenge.

Information, Networks, and Diversification

Scientific understanding of the interaction between climate and agrobiodiversity may have been limited by path-dependent historical choices in agricultural research, the need for closure in research strategies, and selective policy support. A number of alternative strategies have been suggested that are methodological in nature. What would be the corresponding theoretical perspective? To encourage a broader perspective, one approach would be to focus the discussion around concepts of *information*.

Utilizing the insight of Quiggin and Horowitz (2003), one of the main effects of climate change is to *destroy information*. For agriculture, this means that local knowledge about how to manage plant, animals, and environments becomes increasingly less valid, as conditions change. Under new conditions, well-known practices no longer work: species, varieties, and breeds are no longer adapted to local conditions; environmental observations no longer lead to reliable predictions; and well-honed skills become obsolete. Agrobiodiversity is arguably one of the most important forms of information in agriculture: it involves the genetic information contained in the biota that constitute agriculture as well as information (held by farmers, consumers, and institutions) about the functions of these biota that organize agrobiodiversity in time and space. Adapting to future climates and reducing emissions will involve creating new information at an accelerated pace. Even though the capacity of farmers to generate new information is important and needs to be strengthened, it cannot be assumed that the historical capacity of farmers to adapt to extreme climates or socioeconomic change will guarantee adaptation under highly accelerated, nonlinear, or disruptive change. While some aspects are visible to farmers (e.g., yield), others are highly invisible or not of immediate influence to their livelihoods (e.g., landscape degradation, emissions). Creating new information will be very challenging: it is often not codified but rather held as part of tacit knowledge gained through experiential learning; it is also embodied in artifacts or evolved populations of domesticated plants and animals. New information will need to be generated that is accessible at all levels.

In his book *Why Information Grows,* Hidalgo (2015) presents a new concept of information. Building on previous work by Kauffman (1993) on "complex systems," Hidalgo argues that biological evolution and economic growth processes are characterized by the growth of information. In this context, information is not the usual "entropy" of information theory (the Shannon index, which has no relation to meaning or function). Instead, information is defined as a measure of *order*, a concept in which function plays a role. This idea of information is actually much closer to the meaning of information as used in common language. Information growth in these studies is the product of functional diversification in both biological and economic systems.

An important implication of the theoretical work of Hidalgo, Kauffman, and others is that the growth of information or diversity depends on the size of interaction networks. In the industrial sector, for example, complex, less ubiquitous artifacts are created through the collaboration of a number of producers who specialize in certain components, which are then assembled. Economic growth is to a large extent the outcome of the ability to create and participate in such networks. In policy terms, this would mean that stimulating diversity, complementarity, and connectivity of economic activities are main ingredients of economic growth. Hidalgo (2015) shows the relation between economic diversification and economic development using data on international trade, but the theory would predict similar effects at other scales and for other sectors.

In parallel to this research, a small but growing number of studies explore how human networks underpin the ability to manage agrobiodiversity for climate adaptation. Bellon et al. (2011) show that for maize farming in a Mexican landscape with altitudinal gradients, farmers may have access to genetic materials adapted to future climates within the current geographic scope of their networks. Mwongera et al. (2014) did a detailed comparative study of two communities that occupy similar environments on Mt. Kenya. One community has been able to adapt to drought conditions much better than the other community. The community with better adaptive capacity has more intensive seed exchange with the drought-prone lowlands.

In a study on cotton seed acquisition behavior in Andhra Pradesh, India, Stone et al. (2013) show a case of maladaptation. They found that farmers' observations on the phenotypic performance of different cotton varieties hardly influenced their seed choices. Instead, farmers tend to imitate others and seek out new trends. As a result, there is a fast turnover of with cyclical fads sweeping through diffusion networks, but no detectable improvement of the varieties selected over time. Thus, varietal selection lacks environmental learning, which makes it very difficult for the seed system to pick up any climate signal. Stone et al. explain this as the result of a process of cotton farmers' gradual loss of agricultural skills of seed selection, which became superfluous when farmers started to depend on commercial seeds. The "wisdom of the crowd" principle points to the degree of independence between observations as a necessary property for networks to generate true information in response to external signals (Surowiecki 2004).

Complex agricultural systems can be regarded from the same perspective as complex industrial artifacts: both embody human interactions and knowledge. Agricultural diversification, including crop diversification, is important from a climate perspective (see above). Increased integration into financial and labor markets can, however, substitute the "natural insurance" function of agrobiodiversity, leading to the abandonment of certain species, varieties, and breeds (Baumgärtner and Quaas 2010). Partially replacing the natural insurance function of agrobiodiversity does not necessarily entail a net destruction of "information" from Hidalgo's (2015) perspective. The focus of climate information and financial service provision is on expanding the information network in which farmers are embedded. The effect of these interventions is to partially replace one type of information—agrobiodiversity in agricultural systems—with other types of information, embedded in new financial and information services. Brookfield (2001) introduced the concept of *agrodiversity*, which encompasses agrobiodiversity but also diversity of other elements that underpin agriculture institutions, management, and culture. It could be argued that within agrodiversity, different types of information or diversity may partially substitute each other in the "growth of information" during economic development. Likewise, different types of agrobiodiversity may replace each other. Duvick (1984) has described the development of a commercial seed sector as the replacement of diversity in space by diversity in time, from the diversity of landraces in different niches, to the quick turnover of crop varieties over larger areas. It is clear, however, that the substitutability between more and less "embodied" types of agrobiodiversity information is far from perfect. Material genetic resources (e.g., seeds in gene banks) and nonmaterial genetic information (e.g., gene sequences in digital format) are becoming more exchangeable as genomics, bioinformatics, gene editing, and developments toward synthetic life increasingly shape the way that breeding is being done. Even so, many aspects of agrobiodiversity as information remain embodied as they are associated with agricultural skills tied to specific places or knowledge about plant or animal performance and interact with managed environments that are difficult to transfer to new situations.

Even though different types of diversity can partially substitute each other, the theories discussed above suggest that a balanced climate action strategy should focus on diversification across various domains and scales, and that a narrow strategy of institutional and productive specialization should be avoided (see also Chapters 6 and 8). Highly developed agricultural systems are knowledge intensive and make smart use of agrobiodiversity, not only for its risk insurance function but also for many other purposes. For example, greenhouse horticulture in Europe, perhaps the most highly technological agricultural subsector, provides strong biological solutions to pest and disease problems, such that agrobiodiversity use makes chemical pest control no longer necessary (van Lenteren 2000). Agrobiodiversity creation should therefore keep pace with overall agricultural development and compensate the loss of information due to climate change. For example, modern crop variety development will only contribute to resilience if it has wide access to genetic resources, if it can periodically release genetically diverse, adapted varieties (cf. Duvick 1984), and if farmers have the capacity to purchase seeds and replace varieties over time based on reliable information (cf. Stone et al. 2013).

Current policies of developing countries and development donors, however, tend to stifle diversification. Pingali (2015) has documented the extent to which policies support staple grains to the exclusion of other crops. This reflects historical concerns about calorie supply; however, this policy focus fails to support diversification as a risk management strategy and is also incompatible with other policy goals, such as the fight against micronutrient malnutrition and child stunting as well as overweight and obesity. Pingali calls for a "crop neutral" policy instead, one that supports farmers' crop choices in response to market demand. Such policies would simultaneously allow farmers to respond to climate signals. Pingali argues that governments need to create a supportive environment for diversified agriculture through credit and infrastructure.

A specific stimulus for knowledge and information creation is needed, not only as an integrated part of the economic development process, but also to compensate for the loss of information due to climate change. There is now a growing body of scientific literature on knowledge creation, the role of networks, and diversification. As we look toward the future, we need to ask how these insights can be translated to climate action. In practice, diversification is often mentioned in climate action plans (Bedmar Villanueva et al. 2015), but scientific studies provide few actionable insights to inform such strategies. An emerging information "frame" could eventually provide an approach to evaluate diversification strategies.

Design

Recent scientific publications have characterized the challenges that agriculture faces as so-called "wicked problems." Most certainly, climate change and increasing pressure on natural resources cannot be addressed through agro-technological solutions alone. These problems are "wicked" because their solutions require changes in human behavior as well as the values held by agricultural producers, traders, consumers, and the institutions in which these actors are embedded. Scientists have responded to this "wickedness" by supporting a more systemic perspective and by stimulating co-learning processes for stakeholders to collaborate with scientists (Kristjanson et al. 2014; Struik et al. 2014). Both approaches have been suggested in relation to the interaction between climate and agrobiodiversity.

In the context of climate change adaptation and adjacent domains, the metaphor of "pathways" is often used to refer to decision-making processes that address the "wickedness" of problems. In their review of the emerging literature on pathways, Wise et al. (2014) critique the "predict-and-provide" and impact-analytical approaches that have been followed thus far by the IPCC, both of which "close down" the problem too early in the process. The pathways metaphor is deemed to be better at acknowledging (a) the connected nature of the climate adaptation challenge across space, scale, and organizational domains; (b) the inertia in many processes that result from path dependency; (c) the difficulty of monitoring adaptive responses; and (d) the way in which institutional culture enables or constrains social processes. Wise et al. (2014) suggest a number of deliberative and participatory methods that work well, such as scenario building, future visioning, and stakeholder forums.

The term "wicked problem," however, is rooted in an incisive critique of science itself. As one of the founders of the modern design movement (see Buchanan 1992), Horst Rittel coined the term to argue that there are problems that science is unable to address. Wicked problems are badly structured problems that are open to multiple interpretations, but with a single opportunity to find a solution. Scientists normally address relatively well-defined problems, find solutions that are either right or wrong, and can repeat their tests for as long as time and resources allow. Rittel suggested that design is not an applied science. Designers would never find creative solutions to problems if they simply applied scientific insights and did not produce novel knowledge themselves; design thinking uses problem-solving strategies that are different from those of science.

Climate change, agrobiodiversity, and their interactions have all the characteristics of wicked problems as put forth by Rittel. Archaeological and environmental studies characterize the climate–agrobiodiversity interaction as a historically contingent process. The climate change problem is not well defined; it challenges existing, narrowly defined models, especially when the role of agrobiodiversity comes into the picture. Climate action is an urgent, real-time challenge: only time will tell what is right or wrong. The climate change problem breaks down into many other complications.

Since the design community has engaged with these types of problems for a much longer time than other disciplines, it is important to examine the approaches developed by designers. The approaches discussed by Wise et al. (2014) share a large number of characteristics with elements emphasized in design literature, but there are also differences. In a recent book on design methodology, Dorst (2015) argues that a main distinguishing element of design thinking is that it places the creation of new frames at the center of the problem-solving process. Often, a design problem cannot be solved on its own terms; thus reframing the problem becomes essential to finding a solution. The connectedness of a problem should not be reduced by applying an obvious existing model to the situation just to make it manageable. Instead, designers treat the connections as a context that can lead to fresh solutions. They distil their findings as themes that are used to suggest different metaphors or patterns of relationships. These patterns are then tested to see if they provide new and inspiring suggestions toward a solution. One example from Dorst's book is a nightlife area in Sydney, where crime rates were invariably high, despite increased police presence. Designers suggested the frame of a "music festival." This frame proposed a large number of effective solutions that although commonplace in managing music festivals were never imagined when the problem was framed as one of policing and crime. Dorst describes "frame innovation" as a disciplined process of problem solving, outlines the principles that guide the process, and illustrates the process with successful examples (and failures) of redesign of public spaces, care provision systems, and retail experiences. Over time, designers accumulate experience and develop a particular set of skills and reasoning style that can be applied to a diverse set of problems.

Agricultural scientists have started to explore design approaches, especially in agroecology. Duru et al. (2015) suggest that to support the design and implementation of biodiversity-based agriculture, it is crucial for learning tools to be created. Knowledge bases are one type of support tool "that contain structured scientific facts and empirical information compiled from cumulative experiences and demonstrated skills and that enable biodiversity management to be inferred in specific situations" (Duru et al. 2015:1272). To improve adaptive management, they also emphasize the importance of easy-to-use monitoring tools to assess *in situ* the ecological dynamics of diversified farming systems and landscapes. Another approach is to use game-based modeling methods as learning tools to explore different dynamics. The new role for science in the perspective sketched by Duru et al. (2015) is to support design processes that generate locally specific, agrobiodiversity-based solutions to complex challenges.

Berthet et al. (2016) compare three different design methods used in agriculture: one is based on a modeling approach, another on a game, while the third is based on a collective design process after a methodology used in other economic sectors. The modeling and game approaches involve artifacts which focus the attention of participants on the design process and provide a useful learning experience in itself, but also constrain solutions in some way. The particular modeling approach chosen is especially appropriate to explore stakeholder conflicts. Of these three approaches, the collective design approach comes the closest to the frame innovation process described by Dorst (2015). Different experts were invited to talk about aspects of alfalfa meadows, the focus of the design process. The collective design approach, moving between existing knowledge and new concepts, was shown to be instrumental in drawing out the different potential functions of alfalfa meadows, and the process suggested new solutions that were subsequently implemented. Berthet et al. argue that the collective design approach could be especially useful in the first stages of an innovation initiative, while the game-based exercises may be more appropriate when solution directions are clearer.

Frame creation itself is not made explicit in the description of the collective design process (Berthet et al. 2016). Reference to a concrete, named frame may further help to bring about a common focus and motivate a collective vision from which more detailed solutions flow in a natural way to address the complexity of the problem at hand. Clearly, more design experience needs to be gained in the context of agricultural climate action and related to complex problems. It seems clear, however, that much can be learned from design experiences in other sectors.

Summary and Conclusion: Clumsy Solutions

From a historical perspective, climate–agrobiodiversity interactions have undergone a radical reframing. Crop diversity emerged out of landscape-use practices after foragers and hunters intensified their use of the land and became more sedentary and reliant on local resources as a result. Diversity was the "by-product" of processes that did not have diversity creation as their conscious purpose: prehistoric migration of agriculturalists and cultural diffusion.

Conscious selection contributed to diversity creation as well, but it was not a specialized process (see also Chapter 6). Climate adaptation was a result of trial and error while agriculture contributed to modifying the global climate. Modern breeding threatened crop diversity but at the same time needed plant genetic resources as a "standing reserve." This new framing of agrobiodiversity led to plant genetic resources conservation for breeding purposes, mainly in gene banks. Climate adaptation did not initially factor into this frame as breeding was premised on the radical transformation of growing environments through irrigation and fertilization that sought to diminish the influence of climate on production. Over the last two decades, framing has shifted. Agrobiodiversity is now viewed as a defining characteristic of ecologically and economically healthy networks of knowledge creation as well as both unintentionally and consciously designed agroecosystems or components. This corresponds to climate action as a broad response to a combined biophysical and social challenge.

This account is an oversimplification because these frames are not subsequent phases in the history of agriculture; they coexist with each other. Agriculture still holds much agrobiodiversity through *de facto* conservation, without an overall conscious strategy. Gene banks continue to play a clear role in conservation and breeding, but the ecosystems services value of *in situ* management has gained increased recognition. New attempts to create agrobiodiversity-based agriculture are still incipient and partial. Frames persist because social forces uphold them and a relatively coherent set of values underpin them. These frames will also be implied in the ongoing societal debate about climate action and agrobiodiversity. Both climate change and agrobiodiversity are highly contested areas with debates over multiple interlocking issues.

Verweij et al. (2006) note that opposing frames resurface continually in climate policy debates. Worldviews and social interactions tend to sustain each other, resulting in a limited number of stable institutional cultures. These cultures are organized around mutually exclusive principles but need each other, as neither will solve social problems alone. For example, markets typically need governments to function, yet problems surface when one of these cultures dominates. Verweij et al. argue that the Kyoto Protocol failed because it relied too much on a hierarchical culture that framed the prevention of climate change as a public good. The Protocol did not represent a balanced mix of institutional cultures. Specifically, "carbon trading" had little to do with real markets. Opponents of the Protocol emphasized values important in individualistic institutional cultures. Other opportunities exist for climate action and may more likely be able to garner support from different institutional cultures. Verweij et al. argue that renewable energy could be supported by public investment, at the same time providing opportunities for companies and constructive civic and local action. They call the arrangements that accommodate opposed institutional cultures "clumsy" solutions (Verweij et al. 2006:839):

Clumsy institutions are those institutional arrangements in which none of the voices—the hierarchical call for "wise guidance and careful stewardship," the individualistic emphasis on "entrepreneurship and technological progress," the egalitarian insistence that we need "a whole new relationship with nature," and the fatalists's asking "why bother?"—is excluded, and in which the contestation is harnessed to constructive, if noisy, argumentation. Clumsiness emerges as preferable to elegance optimizing around just one of the definitions of the problem and, in the process, silencing the other voices.

Climate action around agrobiodiversity would benefit from this type of "clumsiness." Through frame innovation, new frames could be identified that hold opposing institutional cultures in creative tension. As an example, a number of national gene banks, including those of India and Ethiopia, have started to transfer crop seed samples directly to local and farmer organizations as part of the "Seed for Needs" initiative to restore or introduce agrobiodiversity, often with a specific focus on climate-induced stresses. This reframes gene banks: they are no longer only a "standing reserve" of plant genetic resources to breeding. A new configuration emerges similar to how renewable energies are being managed. Renewable energy is difficult to store and requires careful, information-rich management of demand and supply in a so-called "smart grid." In the same way, gene banks cease to be the equivalent of long-term stocks of fossil resources and become the "batteries" or "supercapacitators" in a "smart grid" for agrobiodiversity that would also connect with community seed banks. In such a new configuration the management of flows of seeds and information is what would drive diversity, rather than the dispensation of resources from a central deposit.

Climate change poses one of humanity's main problems today. Diversity—agrobiodiversity as well as economic and institutional diversity—will play a key role in climate action. Ultimately, climate action will depend on our human capacity to innovate technologically, economically, and politically. An important challenge in all of this will be to channel human creativity to expand and support diversity.

8

How Do Global Demographic and Spatial Changes Interact with Agrobiodiversity?

Karl S. Zimmerer and Judith A. Carney

Abstract

This chapter seeks to identify the linkages between agrobiodiversity and global demographic and spatial changes. Using an interdisciplinary approach, it reviews research models and empirical studies that link demographic and spatial changes to socioecological interactions involving agrobiodiversity at different spatial and temporal scales. Concepts are employed from the frameworks of geographic synthesis, socioecological systems, global change science, coupled human–natural systems, agroenvironmental history, development studies, and political ecology. Seven globally predominant linkages are identified: demographic change and population effects; urbanization and peri-urban expansion; migration including refugee movements; agricultural trade, markets, and food systems; spatial and land-use planning, zoning, and territorialization; food security, food sovereignty, seed movements, sustainable intensification, ecological intensification, and agroecology; and ongoing historical, cultural, and social network influences. Conditions of the major drivers bear complex relations to agrobiodiversity that range from loss and genetic erosion to continuing utilization, the emergence of expanded or innovative new uses, and conservation contributing to the sustainability of food systems. Linking the causal drivers of change to the range of possible outcomes depends on assessing the context-dependent roles of intervening and intermediate-level factors rather than ironclad mechanisms. Several intermediate-level factors are evaluated for each topic and recommendations are offered for future policy-relevant scientific research.

Introduction: Defining Agrobiodiversity Interactions with Demographic Change and Spatial Integration

To understand current fluctuations in agrobiodiversity and equip us to meet future challenges, demographic and spatial changes require thorough examination. As primary drivers of agrobiodiversity change, both involve multiscale processes at various levels (from global to local). Thus, our overall framing must extend beyond the biological factors that have traditionally defined the core concept of agricultural biodiversity. The consideration of current population movements, both economic and refugee, and urbanization—whether rural to urban or displaced people increasingly concentrated in refugee camps—illuminates key geographic and socioecological factors that transcend classic disciplinary boundaries. By focusing on the interdisciplinary characteristics of demographic and spatial change, our goal is to examine the complex socioecological interactions involving agrobiodiversity at different spatial and temporal scales. We use the term "agrobiodiversity," which is already in common usage (see Chapter 1), to reflect this framing. This expanded conceptual framing is essential to delineating research directions in policy-relevant science and scholarship on agrobiodiversity amid dynamic changes (Bioversity Intl. 2017).

Our chapter extends well beyond conventional single-factor explanations by analyzing demographic and spatial changes in relation to population factors, political and economic forces, and production and consumption trends. The processes of demographic and spatial change include human population growth and decline, urbanization, migration (economic and political including populations in refugee or migrant camps), and the spatial integration of agricultural commodities and food systems across national boundaries and into international and global systems through institutions and policies as well as the everyday actions of migrants on the move. Much demographic and spatial change is closely related to global-scale drivers and the kinds of integration associated with globalization. Our referring to "global demographic and spatial change" also acknowledges that these processes often occur at the intermediate and local level.

We discuss the relationship between agrobiodiversity and sustainability, sustainable development, sustainable intensification, and ecological intensification in agriculture, food security, and food sovereignty as well as the regulation of food labeling. Each discussion includes the spatial and scale dimensions of these efforts; for a discussion of related agrobiodiversity interactions with plant breeding and industry, see Chapter 6. In addition, agrobiodiversity's interaction with demographic and spatial changes draws attention to the powerful influence still exerted by the past. Indeed, we highlight the relevance of historical precedents as well as specific past agrobiodiversity interactions, like the diverse food traditions enriched through pre- and early European trade across the Indian Ocean discussed below.

Agricultural biodiversity refers strictly to the biodiversity of genetic populations, cultivars and breeds, and species of both domesticated organisms and their wild relatives as well as agroecosystems (FAO 1999a). Various expanded versions of this definition incorporate knowledge systems, landscapes, and global environmental and socioeconomic changes (CBD 2000; Jackson et al.

2007; Zimmerer 2010). Recent research suggests bringing additional dynamic sociocultural and spatial processes into the broader conceptual framework of agrobiodiversity (Rangan et al. 2015; Zimmerer and de Haan 2017). In particular, we note that dynamic sociocultural and economic drivers, potentially resilient socioecological systems, and broad historical influences are also integral to agrobiodiversity. These additional dimensions have gained new importance in the context of current demographic change and spatial integration (Carney 1991, 2001; Pautasso et al. 2013; Zimmerer 2003a).

Our discussion focuses on the question: How do demographic and spatial changes interact with agrobiodiversity? We emphasize the predominant causality of past and current research, namely social and environmental impacts on agrobiodiversity. At the same time, we also recognize the potential importance of the converse: agrobiodiversity can and does affect demographic and spatial changes, although to date there remains little research on these less common impacts. Mindful of the coupling of socioecological interactions in complex environmental systems, we are keen to point out that agrobiodiversity itself can play an active and dynamic role in certain demographic and spatial changes, rather than the often assumed notion that it is a passive component in larger human systems and thus is strictly an outcome of socioenvironmental drivers. Agrobiodiversity's anticipated, significant influence on demographic and spatial changes, while nascent in research scholarship, leads us to highlight its importance to future research.

We discuss seven types of agrobiodiversity dynamics linked to demographic and spatial changes:

- Demographic change and population interactions
- Urbanization and peri-urban expansion
- Migration
- Integration of agricultural trade, seed and product markets, and food systems
- Spatial and land-use planning, zoning, territorialization, and social movements
- Food security, food sovereignty, sustainable intensification, and ecological intensification
- Historical, cultural, and social network considerations

For each dynamic, we survey current agrobiodiversity research and highlight potential directions for future research.

Demographic Change and Population Interactions

By demographic change we mean forces that range from global population growth rates to the expansion of conurbations into megacities and more recently, the accelerating growth of medium-size cities. This has also incurred population decline and transitions to part-time agricultural employment in rural areas of both the Global North and South, specifically in environments that are marginally productive. Simple causal relations are not evident between these demographic changes and environmental resource and management responses, of which agrobiodiversity is one factor. Even the agrobiodiversity associated with extensive forms of land use, such as swidden farming, has a complex relationship with population growth and decline (Blanco et al. 2013; Ironside 2013). Brookfield (2001) and others (Brush 2004; Zimmerer 1991a, 1997, 2010, 2014) have argued convincingly that these demographic processes are entwined with agrobiodiversity change in the fabric of large socioeconomic forces. Our perspective on this absence of simple demographic causation in global agrobiodiversity change also echoes the influence of prominent, wide-ranging treatises on environmental sustainability and degradation (Boyd and Slaymaker 2000; Fischer-Kowalski et al. 2014; Tiffen et al. 1994).

The majority of farmers cultivating and consuming agrobiodiversity belong to the group of 2.2 billion smallholders worldwide while other, distinct socio-economic groups—each with characteristic levels of agrobiodiversity—also play significant roles in the sustainability of this resource (Figure 8.1). The group comprised of smallholders consists of a highly heterogeneous group that encompasses both poorer segments, including much of the world's food-insecure population (HLPE 2017; IPES-Food 2017; Zimmerer et al. 2018), and slightly better-off populations as gauged by capital and resource levels (the two groups on the left side of the x-axis in Figure 8.1; see also Chapter 6). Our treatment of global populations important to agrobiodiversity also identifies two other broadly defined groups: medium-size farms and large farm enter-prises (the two groups on the right side of the x-axis in Figure 8.1). Each of the four groups is treated as a general category or archetype per methodological advances in the socioecological sciences (Bennett et al. 2005; Janssen et al. 2006; Oberlack et al. 2016; Sietz et al. 2017).

As shown, the food growers corresponding to the slightly better-off small-holder population—the sociocultural and demographic groups comprised of somewhat more well-to-do peasants, Indigenous or traditional peoples as well as neoagrarians and "back-to-the-landers"—generally produce and con-sume the highest levels of agrobiodiversity of any of the four main groups. Figure 8.1 illustrates that the poorer smallholder group, which includes many resource-scarce and food-insecure farmers, is by far the largest.

Demographic change is widespread among both the largest group of small-holders and the slightly better-off group. Large-scale socioeconomic processes, such as the increased shift of rural livelihoods in the Global South to part-time farming and land use, are significant drivers of that change. This shift is typically incompatible with agrobiodiversity because the agricultural empha-sis becomes centered exclusively on cash-cropping monocultures. Similarly, this shift is frequently counter to the input needs of agrobiodiversity, such as the critical labor–time shortages that frequently undermine agrobiodiversity

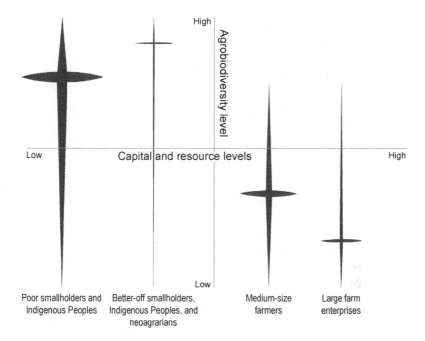

Figure 8.1 Global demographic groups based on resource-level archetypes (thickness represents relative demographic importance) and corresponding ranges (y-axis) and relative extents (filled area within each shape) of agrobiodiversity. Figure drafted by Sophie Najjar and used with permission.

production and consumption systems (Zimmerer 1991a, 2010; Zimmerer and Vanek 2016) as well as seed systems (Bellon 2004).

This shift of livelihoods to part-time farming can, however, potentially coincide with continued agrobiodiversity use when other conditions are present, like polycultures, mixed farming, adequate resource access, and institutional support (Zimmerer et al. 2018). Specific examples of this compatibility, and the conditions required, are addressed below. Still, even scenarios that are favorable for agrobiodiversity conservation as part of persistent land use, livelihood, and consumption strategies may prove inadequate amid global changes broader in scope, thus requiring additional external organizational and institutional support (Reenberg et al. 2013).

Demographic change often intersects with sociocultural and economic differences in the processes affecting agrobiodiversity. One of the most common examples involves the demographic dimensions of increasingly common migration and part-time farming, examined in detail below. In these circumstances, farm labor and knowledge is often gendered; women are likely to be most immediately involved with agrobiodiversity growing and food preparation (Chambers and Momsen 2007). Other social differences, such as race, ethnic group, and age, may also become actively incorporated into demographic shifts such as migration (Zimmerer 2014).

Urbanization and Peri-Urban Expansion

Recent, rapid urbanization throughout the world, especially in tropical regions, is the result of powerful global drivers and their effect on the environment. This is causally linked to changes in food systems and agriculture that impact the growers of agrobiodiversity and consumers' access to it via food markets (Bioversity Intl. 2017; HLPE 2017; IPES-Food 2017; Zimmerer 2010). Partial persistence of smallholder food and farming systems involving agrobiodiversity is possible in the expansive peri-urban contexts of the Global South (Ávila et al. 2017; Eakin et al. 2014; Lerner and Appendini 2011; Poot-Pool et al. 2015; Prain et al. 2010; Zimmerer 2013). Gardens, small fields, and informal spaces, such as river margins and rented or vacant land within urban areas, provide diverse opportunities for food production and continued use of agrobiodiversity. At the same time, many peri-urban agrobiodiversity growers are engaged in highly diversified livelihood strategies that depend principally on off-farm employment.

Ethnicity and socioeconomic interactions also influence the spatial dynamics of food production in urban contexts that incorporate agrobiodiversity. The encouragement of dedicating space to growing food varies widely from city to city, and support is typically partial at the most, making community, neighborhood, producer, and consumer groups vital. Approaches include urban and peri-urban farming in backyards, abandoned lots, community gardens, under power transmission lines, in traffic medians, and along sidewalk curbs as well as new designations like peri-urban agrarian parks (Pinna 2017; Pirro and Anguelovski 2017; Serra et al. 2017). Immigrant gardens are important sites of agrobiodiversity-growing in both private lots and in public spaces that are either informal or, in other cases, supported by city and municipal governments (Baker 2004; Imbruce 2007; Taylor and Lovell 2014).

Widespread local organic growing and direct producer-consumer arrangements, such as community-supported agriculture (also known as community-shared agriculture), as well as more restricted innovations like "seedy Saturdays," which may or may not support high agrobiodiversity, also have widespread potential. Such initiatives feature food that can be grown and consumed locally; it can be either marketed to consumers with high purchasing power or, in some cases, may be priced affordably.

Urbanization also reveals the potential of abbreviated food supply chains, innovative food marketing, and cultures that value local and artisanal food traditions. A good example of innovative marketing is Bioversity International's projects that promote diverse African leafy vegetables in Kenya and other high agrobiodiversity foodstuffs for their benefits to human nutrition and food security (Bioversity Intl. 2017; HLPE 2017; IPES-Food 2017). Research analysis of such initiatives needs to evaluate not only the shortness of the supply chain but also the impact of demand, and which markets are linked to agrobiodiversity.

This emphasis and the role of urban-based trends can also promote high agrobiodiversity foods and the consumption of diverse wild, gathered, and caught kinds of food (Aubry and Kebir 2013; Reyes-García et al. 2015). Research on urban food cultures needs also to consider the global marketing strategies of the industrial food sector and its massive domination of food supply and consumption trends to date (Cockrall-King 2012). These trends can seem to embrace agrobiodiversity, at least at first glance, while actually undermining it, as discussed below (see section on Agricultural Trade, Markets, and Food System Integration, which reviews findings on the global quinoa boom).

Even the current success of fresh food, farm-to-table or soil-to-plate retailing, the business and marketing boom for "healthy" products, and the innovations of celebrity chefs, restaurateurs, and television shows promoting experimentation with diverse foodways may unfold in unanticipated ways. On the one hand, the urban-based food movements involving consumers and chefs who embrace high agrobiodiversity foods, such as has occurred in Lima and in other urban areas of Peru, can lead to demand for more agrobiodiversity as part of the valorization of new eating experiences. The Slow Food movement, which is active across many countries, vividly illustrates these trends and how they suggest resistance against the homogeneity of food system integration.

On the other hand, the current interest in novel foods can reinforce existing categories of social difference, disadvantaging Indigenous Peoples' diets when some of their traditional foods (e.g., quinoa, açaí, amaranth, chia, maca, rooibos, goji berries) become a fad in the Global North because of presumed nutritional benefits. This can move crucial subsistence staples to urban areas or from one region of the world to another, negatively affecting the farming practices and diets of specific groups based on ethnicity, race, gender, or age (Skarbø 2015; Winkel et al. 2016).

To determine the impact of peri-urban and urban farming on agrobiodiversity requires also an assessment of whether the shortened food chains concentrate on producing a few specialized crops or incentivize diverse foods that contain high levels of nutritional diversity, agrobiodiversity, and agroecological functioning. Future research is needed that will measure the impact of new labor and environmental linkages on farms, in communities, and across the geographical networks that link producers to consumers (McMichael 2013; Shaver et al. 2015).

An example is the impact of urban "gastronomic booms" that encourage the demand for agrobiodiversity and that can potentially offer benefits to poorer growers (see Chapter 15). There is also evidence, however, that price premiums, to the extent they exist, are often captured by more well-to-do agriculturalists. Problems like malnutrition, massive dietary shifts among young people, and the entrenchment of socioeconomic drivers that remain disadvantageous to poorer growers persist (García 2013; Tobin et al. 2018). The demand derived from booms is generally short lived, triggering unsustainable value chain interactions and land use change.

Research on demographic changes and spatial integration associated with urbanization, and their significance to the balance of human populations and food systems is also needed. In particular, the worldwide growth of urbanization and the concomitant, increasing influence of urban on rural spaces (the so-called "urban in the rural") as well as the expansion of global, industrial food systems must be seen in relation to the role of smallholders. Within the heterogeneous group of 2.2 billion smallholders, more exist in the peri-urban and urban-in-the-rural contexts than those whose livelihoods can be characterized as remote rural (Zimmerer et al. 2015).

Migration and Movements of Refugees and Displaced Persons

The mobility of people on our planet has reached unprecedented levels. The United Nations reported that in 2015 the number of international migrants—persons living in a country other than where they were born—reached 244 million. This represented a 41% increase for the world as a whole compared to 2000 (UN 2016). Included in this estimate of mostly economic migration are more than 20 million refugees who were forced to relocate. Long-distance trade and economic migration has long been important, even prior to the period of European colonization (see below), but its scale has vastly increased in this century. Tens of millions of people now migrate across borders each year. The estimate of migration within countries is even larger. Furthermore, a new generation of temporal migrants, frequently part-time farmers moving back and forth between the countryside and cities, has emerged.

Significantly, many migrants from each of these groups are smallholder farmers who have traditionally (and disproportionately) relied on agrobiodiversity in their farm plots and gardens. Garden areas are especially important to migrants since they involve lesser amounts of land and field labor while containing moderately high levels of agrobiodiversity (Galluzzi et al. 2010). The capacity for innovation is great since even "traditional" home gardens have been proven to be impressively versatile when adjusting to changing circumstances (Trinh et al. 2003).

Migrants often maintain connections to seed networks through family members and extended households that remain in their "source areas" (Alexiades 2012), but sometimes migrant households lessen their levels of agrobiodiversity. Individual migrants can be viewed as diversifying the livelihood strategies of their source households by extending such linkages into the areas to which they move. This vast expansion of international migrants combines with local population movements that affect agrobiodiversity, like the common practice of a relocated spouse bringing seed from their former home to their new location in exogamous marriage. In special circumstances, the level of knowledge of migrants can even be higher than those from the "source areas" (Vanderbroek and Balick 2012). Migration also intersects deeply gendered processes, often contributing to the significant "feminization" of on-farm work activities that can include agrobiodiversity (Carney 1993; UN 2016; Zimmerer 2014).

While such strategies for livelihood diversity can negatively impact agro-biodiversity, under certain conditions, the levels of agrobiodiversity are maintained and agriculture may even be intensified (McCord et al. 2015). As illustrated in Figure 8.2, a household's access to land, water, and seed resources as well as specific demographic, socioeconomic, and political–economic factors are often the key conditions influencing agrobiodiversity outcomes in livelihood diversification contexts (Zimmerer and Vanek 2016). This illustration illuminates the many human–social and environmental dimensions that resource access encompasses (Ribot and Peluso 2003) along with multifactorial influences on agrobiodiversity outcomes. Determining the conditions of specific trajectories through empirical and predictive models is a crucial next step for research and policies guiding environmental sustainability (Gray 2009; Gray and Bilsborrow 2014).

Refugees' flight is forced, involuntary, and currently extensive in scope, and an important contrast and counterpart to the economic migration described above. Both forms of mobility are especially important to agrobiodiversity in current global contexts. The contemporary turmoil in Syria and Iraq has instigated an influx of refugees into Europe of epic proportions; over a million refugees have arrived in recent years. In sub-Saharan Africa, refugee camps are disproportionately made up of women and children who have fled conflicts

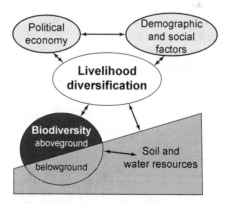

Figure 8.2 Conceptual diagram of the linkages between livelihood diversification and human–social factors as well as agrobiodiversity and related environmental factors. The lower arrow represents the pathway of impacts occurring via the linkages of below-ground agrobiodiversity and natural resources, principally soil and water resources. Adapted from Zimmerer and Vanek (2016), an open access article distributed under the terms and conditions of the Creative Commons Attribution (CC-BY) license.

and sought safety elsewhere in their country or in neighboring countries. The capability of regional food systems and food aid to feed these impoverished and uprooted people in concentrated settings is already stretched and worsening as years go by and families are still unable to return to their homeland (e.g., Somalis living in Kenyan camps, Rohinga refugees in Bangladesh and Malaysia, Myanmar's Karen people in Thailand). The necessity of feeding people through means other than food aid will grow as intractable conflicts turn such camps into quasi-permanent settlements.

The land surrounding refugee camps, many of which are located in rural areas of Africa, Southeast Asia, and the Middle East, could be providing food, given the number of camp inhabitants with little to do. In addition, such sites allow for the reassertion of cultural traditions that inform specific foodways. Growing the condiments and foods specific to their culture has always been a way for migrants to reestablish meaning and identity in exile (Carney and Rosomoff 2009). Among current refugees, the combinations of these cultural concerns and foodway practices together with the need for food security and health—and nutritional security in particular—has spurred a number of institutional initiatives worldwide that involve the diverse sourcing of seed for food plants and gardens among refugee populations. The 15th Garden, an initiative supporting seed availability and gardens in refugee camps in Syria, is an example that receives support from displaced Syrians and Europeans (Zimmerer 2017a).

The likelihood that refugees will remain in these areas for many years, indefinitely over generations, or will be resettled in cities, needs to be foremost in how we think about self-provisioning, agrobiodiversity, and resilience in the coming decades. Experts on projects in this domain underscore the complexity of agrobiodiversity interactions in refugee contexts. Paul Richards and colleagues point out the promise and applicability of agrobiodiversity to enhance food sovereignty among refugees and displaced persons worldwide and especially in several countries in Africa (Richards and Ruivenkamp 1997). Their work also cautions how such prospects can be severely compromised in unequal, agrarian social orders that exert persistent influence on resource access and knowledge systems in general and agrobiodiversity use in particular.

The works of Louise Sperling, Shawn McGuire, and others highlight the potential of seed aid as a potent force for impacting agrobiodiversity initiatives among refugees and displaced persons (McGuire 2008; McGuire and Sperling 2008; Sperling and McGuire 2010). McGuire recently noted that "in practice much seed aid comes as direct (free) distribution of seed, usually of a handful of formal sector varieties....[S]uch forms of seed aid commonly target large-scale displacement and resettlement situations (e.g., northeastern Nigeria at the moment)" (McGuire, pers. comm.).

Human migration related to agrobiodiversity also incorporates the biotic movements of plants, animals, and agroecosystems, a general perspective that has been widely applied to environmental history (Carney and Rosomoff 2009;

Crosby 1986; Rangan et al. 2015). Populations of these organisms typically undergo diversification through migration and genetic-level integration as illustrated by the peninsular oases of California (De Grenade and Nabhan 2013). Other results of migration, such as genetic bottlenecks, can severely restrict the biodiversity of plant and animal populations as well as limit the capacity for subsequent evolution. More generally, it is important to understand the interactions between agrobiodiversity and human migration as well as other forms of demographic change and spatial integration by employing the perspective of crop metapopulations rather than the narrowly defined notions of varieties and species (Alvarez et al. 2005; Orozco-Ramírez et al. 2016; van Heerwaarden et al. 2010; Zimmerer 1998).

Agricultural Trade, Markets, and Food System Integration

Agricultural trade growth and spatially integrated food and seed systems in the context of world population growth and globalization have fueled the connectivity of world markets. The resulting flows of standardized seed and immense quantities of food into global agrobiodiversity centers of production and consumption have reduced the genetic diversity of crops and agrobiodiversity in staple food agriculture and weakened associated smallholder farming systems (Lockie and Carpenter 2010). Food imports include frequently subsidized processed foods and food staples representing a narrow range of biodiversity that are inexpensively produced in the Global North.

The impact of this trade-dumping is mostly negative on the agrobiodiversity of staple foods, although these linkages need to be systematically investigated. Similarly, the market imports of seeds, such as vegetable seeds, seem for the most part to represent the ascendance of a narrow suite of crop species and varietal types at the expense of a broader, existing agrobiodiversity. Increased global integration of the trade of agricultural seeds and its basis in exclusive patenting and intellectual property rights has been highlighted in recent work as offering some of the most potent threats to food security and agrobiodiversity in Africa and other world regions (AFSA 2017). Here, too, there is the need for careful, well-designed studies, including ones based on historical research frameworks (discussed below).

Seen across geographic scales, the variation of agrobiodiversity outcomes—ranging from the global trend of a sharp reduction in staple food diversity to a continuing emphasis on diversity in certain local markets—is notable with regard to the impact of the food trade. The impacts of agricultural trade combining low-agrobiodiversity imports with more complex local food systems does enable agrobiodiversity to prevail in some markets (Bellon and Hellin 2011). It is not the dominant pattern, but tends to result in the availability of agrobiodiversity of moderate levels in rural-migrant neighborhoods in the Global South (e.g., Van Andel and Fundiko 2017). One of the key features of these markets—many of which are informal markets—is that they tend to entail well-established social relations and complex offerings of at least moderate levels of varietal diversity within staple foods as well as less common species. This feature is important because the continued marketing of intraspecific varieties (also known as local cultivars or farmer varieties) enables the integration of consumer demand and production that is vitally important to both the agroecological and sociocultural functions of sustainability involving agrobiodiversity (Kawa et al. 2013; Temudo 2011; for a contrasting interpretation, see Van Dusen and Taylor 2005). Moreover, this feature is a contrast to many global food trends that can appear to stimulate agrobiodiversity and potential sustainability benefits, but that in reality rely on a narrow range of varieties within a few novel food species.

The global "Quinoa Boom" is a good example of this phenomenon of the spatial integration of markets involving agrobiodiversity albeit with complex and often unanticipated consequences (Oberlack et al. 2016; Walsh-Dilley 2013). It has resulted from the rising demand of consumers—principally in the United States and Europe—together with responsive and highly integrated supply chains involving producers, distributors, and investors. These conditions have led to the rapid and widespread expansion of quinoa production concentrated in the southern Altiplano of Bolivia and extending through the Andes of Bolivia and Peru and, to a lesser degree, of Ecuador, Colombia, and northern parts of Argentina and Chile. Much of the global market involves a single variety known as Quinoa Real, and production often entails simplified agroecosystems relying on monoculture and continuous cropping. In addition, it has led to the severalfold increase of market prices: 600% between 2000 and 2008 (Ofstehage 2012). As various researchers have discovered, however, the home consumption of quinoa remains vibrant amid many Indigenous peasant farmers who are among the main producers (Kerssen 2015), even in the core producing areas of the Quinoa Boom (Oberlack et al. 2016; Walsh-Dilley 2013). This fact, along with conflicting interpretations of environmental data, has stimulated scientific debate and still unresolved conclusions regarding the environmental impacts remain, including agrobiodiversity loss as well as the social consequences (Jacobsen 2011; Winkel et al. 2012).

The spatial integration globally of industrial food systems has recently begun to be measured, and estimates show pronounced impacts on agrobiodiversity (Khoury et al. 2014). This new research seeks to estimate agrobiodiversity levels at multiple scales (including countries) and, in some cases, to determine the relationship between food system biodiversity and the dietary diversity of different socioeconomic groups within a country. It promises empirically derived policy information and insights that can be used to clarify and address the challenges of food and nutritional security in general and dietary quality specifically that are faced in the ongoing changes of agrobiodiversity in global food systems as well as those at smaller geographical scales (AFSA 2017; Bioversity Intl. 2017; HLPE 2017; IPES-Food 2017).

Even so, the expanding spatial integration of agricultural trade worldwide may be more complex than imagined. Recent U.S. agriculture missions to Cuba include a congressperson who advocates continuation of the country's policy support for its organic farming system rather than a return to conventional large-scale farming. This shift to organic food production began in the early 1990s following the breakup of the Soviet Union and the loss of the Soviet oil subsidy, which had encouraged the export of traditional monocrops. As Cuba's oil-dependent, mechanized, large-holding sector collapsed, the country reorganized agriculture, turning away from monocultures (especially sugarcane and citrus) grown for export to organic farming of basic foodstuffs on smaller farm units to feed its population. The export monocrop agriculture model has long limited Cuba's political options. Prior to the Cold War, large-scale farms produced tropical export crops for the U.S. market.

Following the Cuban Revolution and the ensuing nationalization of large landholdings into state farms, the Cold War political realignment oriented the traditional export sector to the Soviet Union. This continued the country's dependency on imported food while doing little to improve regional food availability. The dissolution of the Soviet Union brought the historic pattern to an end, forcing Cuba to prioritize feeding itself. This was achieved by turning the large and centralized state farms into small and decentralized cooperatives with an emphasis on food sovereignty. With more than twenty years' experience growing food organically, Cuba is now at a critical, historical crossroads. The recent détente with the United States holds considerable promise for a new export model based on organic food. While organics represent a small proportion of total food purchases in the United States (5% on average), it is the fastest growing food sector. Cuba's export of organic fruits, vegetables, coffee, sugar, and citrus could potentially increase the trade of intermediate levels of agrobiodiversity while earning valuable foreign exchange (Severson 2016).

Spatial and Land-Use Planning, Zoning, Territorialization, and Social Movements

Spatial and land-use planning, zoning, and territorialization in towns, cities, individual countries, and regional political and economic unions often bring about spatial changes. These activities commonly comprise the spatial integration associated with globalization, whether economic, political, or environmental in nature. Often, they also concurrently create or exacerbate spatial differences. Territory-based and territory-focused social movements play an important role in drawing attention to these differences. Such movements include many peasant and Indigenous groups who perceive the powerful entwining of territorial rights and claims to agrobiodiversity production and consumption. Their combined claims to territorial rights and efforts to protect and control resources, including agrobiodiversity and corresponding seed systems, constitute the powerful linkages created by local groups and global organizations (de Wit 2016; Zimmerer et al. 2017).

Modern land-use planning, zoning, and territory-based resource management have tended to place negative pressures on agrobiodiversity through state-led paradigms of improvement and governability. The spatial impact of the current leading global initiatives for food security and agricultural modernization is explicit and implicit (e.g., the "New Green Revolution," "Next Green Revolution," and "Agriculture for Africa"). Various other sustainability and agricultural modernization campaigns, especially but not only in Africa, have also left powerful spatial imprints by combining direct and indirect designs (discussed below). The European Union's reformed Common Agricultural Policy is another important example that expects benefits to agrobiodiversity whilst relying on land-use planning (Overmars et al. 2014).

Land-use planning at the national level is presumed to exemplify this negative impact on agrobiodiversity but merits additional in-depth study (Dawson et al. 2016b). More generally, the growth of sustainability goals, participatory approaches, and enhanced territory and property definitions during the past decades are mentioned more often in new government-based planning projects (though far from a priority in the latter) as well as nongovernmental initiatives.

The impacts on agrobiodiversity of these initiatives are still uncertain, since even village-level land-use planning tends to exert little influence on agricultural decision making (Lestrelin et al. 2011). Nonetheless, these new projects and policies have the potential to create a positive influence on local access to resources, especially land and water. In turn, such enhanced resource availability impacts agrobiodiversity. Such projects and policies vary widely and important critiques have been applied from the still-expanding *ordenamiento territorial* in Latin America (Zimmerer 2017b) to the *gestion des terroir* approach in West Africa (Bassett and Gautier 2014). The combination of social participation and territorial approaches can potentially become beneficial for local resource access and agrobiodiversity outcomes.

A specific focus on agrobiodiversity has also been incorporated into zoning and hotspot mapping. These forms of spatial integration are being used for planning the zonation of land use and the prioritization of efforts aimed at the continued use and conservation of agrobiodiversity. One national-level example is the initiative spearheaded by Peru's *ministerio del ambiente* (MINAM). MINAM is currently involved in building a program for the application of zonation and other planning models to enable the sustainability of agrobiodiversity. Among the goals of MINAM is to create zonation models that will enable the continued use and conservation of agrobiodiversity while protecting such areas from potential gene flow associated with the production of genetically modified types of crops.

At the same time, these planning and zoning decisions alone need to be seen as a necessary but not sufficient means of contributing to local governance and sustainability initiatives that favor agrobiodiversity. In other words, specific factors occurring in conjunction with planning and zoning (such as access to land and water, especially among smallholder and Indigenous or traditional populations) are particularly important to agrobiodiversity outcomes. In sum, the evolution and adoption of planning, zoning, and territory-based resource management have expanded considerably in the past couple of decades. These approaches are likely to continue to grow, increasing their power to influence ongoing spatial integration. The propelling forces include sustainability goals concerned with the world "running out of farmland" and the associated increase in land investments and "land grabs."

Food Security, Food Sovereignty, Sustainable Intensification, and Ecological Intensification

Spatial integration also occurs via the policies and knowledge systems—existing and planned—that combine sustainability with global food security and sovereignty. Research on ways to promote food security through ecological (Tscharntke et al. 2012) and sustainable intensification (Garnett et al. 2013) has identified the central role that biodiversity, including agrobiodiversity, plays in providing agroecological functions and resilience that can close the yield gap without incurring major environmental impacts. In particular, ecological and sustainable intensification recognize that agroecosystems' biodiversity furnishes much-needed agroecological services and agroecosystem regulatory functions. According to Vanek (Chapter 4), landscapes with greater levels and scope of biodiversity and agroecological functions are generally more productive, self-regulating, and resilient than those with lesser levels (see also Delaquis et al. 2018; Jackson et al. 2007; Jarvis et al. 2007).

Global-level ideas and potential policies for food security and sovereignty both extend their emphasis on agroecological functions and traits to encompass human diets, nutrition, and health. Food security policies interlink the goals of human nutrition and health with income generation and access to high-quality foods. Food sovereignty ideas and movements, which have been growing in global significance, meld the aforementioned functions with prioritizing the need to forge the independence of agrifood systems by controlling access to their agrobiodiversity, its use and benefits (see also Chapter 14). The broad spectrum of goals within food security and food sovereignty suggest ample scientific and policy opportunities to examine and encourage agrobiodiversity.

The sociocultural and ecological processes inherent in global change drivers are integral to research on food security and food sovereignty's prospects and pathways. Creating and sustaining viable, high-quality local seed systems, for example, must respond to the need for valued foodstuffs and foodways while replenishing productive capability and providing adaptive capacity in the face of climate change and other drivers (see also Chapter 7).

These areas of research highlight the contextual and spatial variation in ecological and sustainable intensification that include high-yield and medium-intensity agriculture. This creates new research opportunities for intensity-related incorporation of agrobiodiversity and agroecological approaches (Jordan and Davis 2015). For example, the functions of agrobiodiversity under the socioeconomic and environmental conditions of medium-intensity cropping can involve seeding with larger numbers of varieties, thus reducing risk and also producing higher yield than less diverse cropping (Di Falco et al. 2010). Other functions include provisioning specialized food and market products.

In addition, cultivation for enhanced agrobiodiversity can be compatible with crop commodity specialization in medium-intensity production (Flachs 2015; Turner and Davidson-Hunt 2016; Zimmerer 2013). Analysis of this compatibility amid intensification will require new research and the application of concepts and models of agricultural and economic development that address farm-level rationales that combine risk management and poverty alleviation via income generation (Barrett et al. 2010).

Intensification approaches (both ecological and sustainable) incorporating agrobiodiversity in medium-intensity agriculture may lead to increased insight into ecology, cultural practices, local knowledge systems, and continued co-evolution. This evolutionary ecological research is using molecular-level tools and techniques to vastly expand understanding of the deeply entwined cultural and ecological complexity of agrobiodiversity (Bradbury and Emshwiller 2011; McKey et al. 2010a; Orozco-Ramírez et al. 2016; Vigouroux et al. 2011a). With a new era of discovery around plant and animal microbiomes just having started, the potential to generate new insights with practical implications for ecological and sustainable intensification is substantial (Köberl et al. 2015; Sessitsch and Mitter 2015).

Finally, agrobiodiversity's socioecological processes in ecological and sustainable intensification suggest functions that are both similar to and yet distinct from those belonging to the general field of agroecology. Further research is needed to clarify and strengthen their complementary roles, as well as our vision of the possibilities for agrobiodiversity and agroecology.

Historical, Cultural, and Social Network Considerations

Historical and cultural perspectives provide additional insight into the interactions between agrobiodiversity, demographic change, and spatial integration. In the past, many combinations of trade, migration, and landscape change involving food and food traditions significantly enriched agrobiodiversity and supported cultural practices and religious belief systems. One specific example is the agrobiodiversity that emerged in African and South Asian agrifood systems connecting pre-European (and pre-Islamic) trade networks around the Indian Ocean, roughly between 800 and 1500 C.E. (Rangan et al. 2015). Counterexamples are also abundant, as in the case of European livestock and pasture flora that were spread across many of the world's temperate zones during colonization in the nineteenth and twentieth centuries, devastating native species (Crosby 1986). Within Britain, agrobiodiversity has been affected by a range of forces, including but not limited to seed commodification in Europe, allotment growing, seed saving, and de-skilling of the contemporary workforce. Such factors have resulted in agrobiodiversity loss or, at the least, shown complex and potentially unfavorable interactions with farm management and food production (Aistara 2011; Gilbert 2013).

These considerations can be extended to more extreme examples and also to present-day demographic change and spatial integration. For example, enslaved African rice growers in the United States and other locations were able to use marginal environments of little value to slaveholders to grow this crop for subsistence (Carney 2001; Carney and Rosomoff 2009). This example highlights the importance of cultural knowledge systems, social interactions, and power relations within and among groups (slaves, slave communities, plantation managers and owners) as well as new forms of intensification that can result from heterogeneity among smallholder farmers.

In our own times, a confluence of enabling factors can lead to favorable levels of agrobiodiversity in landscapes of migrant or refugee agrobiodiversity growers and consumers (Zimmerer 2014). More generally, the dynamic contexts of demographic and technological change have created circumstances where some households and communities have been able to continue the use of agrobiodiversity while they intensify production. In these cases, the practice of sustainable and ecological intensification has demonstrated emergent compatibility with the continued utilization of agrobiodiversity (Flachs 2015; Turner and Davidson-Hunt 2016; Zimmerer 2013).

Whether individual producers and consumers and their communities continue to conserve agrobiodiversity in contexts of migration depends on such factors as their capacity to participate in social networks, the amenability of the host government to help them do so, and different forms of seed exchange and acquisition. Where agrobiodiversity transfer is facilitated and plant health is protected, the social linkages in migration can facilitate further coevolution of agrobiodiversity through connecting the networks of otherwise separate seed systems. While past plant exchanges were the result of extraordinary journeys (Roullier et al. 2013a; Van Andel et al. 2016), globalization and increased connectedness has accelerated germplasm flows. Social networks and the other aforementioned factors are similarly important in the transitions to part-time and peri-urban farming that have created increasingly new and different conditions for agrobiodiversity during the past couple decades (Ellen and Platten 2011).

Conclusion: Directions of Future Research on the Interactions of Agrobiodiversity with Demographic Changes and Spatial Integration

Drivers of demographic change and spatial integration are vitally important to understanding the challenges facing agrobiodiversity. Current agrobiodiversity outcomes, ranging from genetic erosion and the creation of new diversity to continued use and conservation, are highly dependent on the interactions with these drivers. Specific, major drivers include human population changes, urbanization, globalization (e.g., migration, agricultural trade, and food system integration and transitions), spatial planning, potentially widespread approaches to food security and sovereignty, and a broad swath of historical, cultural, and social network considerations. Several of these drivers have powerful historical antecedents that also have importance to agrobiodiversity, both in understanding the past and processes that continue to exert influential legacy effects.

A number of the current drivers of global change tend to bear indeterminate relations to agrobiodiversity that depend on intervening and intermediate-level factors rather than enacting ironclad outcomes. Drivers such as population growth, migration with continued part-time farming, urbanization, and narrow plant selection responses to climate change can and often do weaken and undermine agrobiodiversity. Conversely, certain supporting conditions are comprised of factors potentially occurring together, such as resource access, innovative knowledge systems, and continued culinary value and use at the household and individual levels. These combined conditions can function as intermediate-level modifiers of global change drivers that support the continued demands for agrobiodiversity and the potential future expansion of its utilization.

Engaging policy and practitioner communities is a high priority (e.g., AFSA 2017; Bioversity Intl. 2017; HLPE 2017; IPES-Food 2017). Our chapter is keen to expand interactions with several of these communities involved in the Second Global Plan of Action for Plant Genetic Resources, the Convention on Biological Diversity (CBD), and the Intergovernmental Science-Policy Platform on Biodiversity and Ecosystem Services (IPBES) as well as other robust policy networks. Researchers need to be partnered with organizations to ensure that much-needed scientific analyses and scholarly understandings of global drivers and their impacts on and interactions with agrobiodiversity across multiple scales are able to be channeled into concrete policy insights, creating potentially new opportunities for more sustainable food futures worldwide.

We must recognize, moreover, that expanding the intermediate-level understandings of the varied processes inherent in agrobiodiversity change will increase the complexity of research. Therefore, the development and application of change and complexity-sensitive concepts is urgently needed in agrobiodiversity research. Conceptual frameworks such as geographic synthesis, socioecological systems, global change analysis, complex adaptive systems, coupled human–natural systems, and historical agroenvironmental studies have the potential to be particularly well-suited to understanding this complexity. In addition, future studies are needed to ascertain whether, or to what degree, the complexity of agrobiodiversity systems can be addressed through promising action research and design perspectives (see Chapter 7).

One crucial area of policy-relevant research is historical analysis that leads to a better understanding of the specific conditions that support agrobiodiversity amid dynamic demographic changes and spatial integration. Future agroenvironmental historical research can help address the aforementioned challenges of complexity in agrobiodiversity change. It is needed, for instance, to gain insight into the processes and patterns of recent refugee movements that have incurred both agrobiodiversity use and disuse (e.g., Vietnamese and Laotian refugees in the southern United States). These broader historical research themes will need to be complemented by specialized studies. For example, historical information needs to be uncovered and clarified that can serve as baseline data for in-depth temporal comparisons which in turn provide the foundation for potential monitoring systems (Pereira and Cooper 2006; see also Chapter 5).

Another area of crucial future research is to take the multiple dimensions of demographic change and spatial integration surveyed here and apply them to the actual dynamics of agrobiodiversity. Having recognized potential compatibility, future studies must reconsider agrobiodiversity research topics that previously focused primarily on so-called traditional agriculture in seemingly remote cultural contexts. Take, for example, the topic of seed networks. New research is urgently needed on the seed exchange networks of large groups of agrobiodiversity growers who are living in urban and expanding peri-urban areas amid migration processes, including refugees.

At the same time, novel research is needed to examine the characteristics of agrobiodiversity in these new, expanding contexts. It should consider agrobiodiversity-focused factors, such as whether and how varietal and species-level diversity and agroecosystem functions are being altered together with demographic change and spatial interactions. More generally, innovative agrobiodiversity research that implements the insights identified above should be able to make comparisons across space and time between agrobiodiversity systems. Such comparative research will certainly need to be mindful of variations, both conspicuous and subtle, that are contained in diverse manifestations of agrobiodiversity knowledge among countries, cultures, and time periods.

Acknowledgments

The authors gratefully acknowledge the insights and helpfulness of the extensive review of Shawn McGuire and additional anonymous comments. In addition specific gratitude is owed for the insightful comments of Stephen Brush and Jacob van Etten and the editorial guidance and inputs of Julia Lupp and Aimée Ducey-Gessner. The paper was drafted and circulated prior to the Forum. Our revisions, particularly the clarification of demographic groups illustrated in Figure 8.1, benefitted from discussions within our working group at the Forum (Socioecological Interactions Amid Global Change) and its co-members: Conny Almekinders, Glenn Davis Stone, Jacob van Etten, Marci Baranski, Jan Hanspach, Vijesh V. Krishna, and Julian Ramirez-Villegas. Figures were drafted by Stephen Zimmerer, Sophie Najjar, and Steve Vanek.

From Food and Human Diets to Nutrition, Health, and Disease

9

Agrobiodiversity and Feeding the World

More of the Same Will Result in More of the Same

Anna Herforth, Timothy Johns, Hilary M. Creed-Kanashiro,
Andrew D. Jones, Colin K. Khoury, Timothy Lang,
Patrick Maundu, Bronwen Powell, and Victoria Reyes-García

Abstract

Food systems large and small around this planet are changing more quickly and more profoundly than ever before in human history. If the same processes and priorities continue, we can expect more of the same results: the last fifty years of a productionist paradigm have resulted in increased production of a small set of calorie-dense crops, increased calorie availability, and increased global homogeneity of diets, while environmental sustainability, human health, and equity issues remain unresolved. Food system sustainability is threatened by soil erosion, fertilizer pollution, water overuse, tropical forest degradation, climate change, and genetic uniformity in agricultural production. Meanwhile, access by all to healthy, diverse, and safe food choices is far from realized, and food-related noncommunicable diseases such as type 2 diabetes, obesity, and heart disease are now epidemics as the world increasingly partakes in a diet high in sugar, fat, and salt. There is reason for hope, as eaters on every continent are demanding healthier, more diverse, safer food. This chapter argues that agrobiodiversity will help to improve sustainability, equity, and nutrition outcomes in food systems. We briefly review the current evidence on the linkages between agrobiodiversity and sustainability, equity, and human health and nutrition, differentiating between linkages at different

Group photos (top left to bottom right) Anna Herforth, Timothy Johns, Hilary Creed-Kanashiro, Colin Khoury, Patrick Maundu, Bronwen Powell, Andrew Jones, Victoria Reyes-García, Timothy Lang, Hilary Creed-Kanashiro, Bronwen Powell, Anna Herforth, Timothy Johns, Patrick Maundu, Timothy Lang, Andrew Jones, Victoria Reyes-García, Timothy Johns, Anna Herforth, Patrick Maundu, Victoria Reyes-García

geographical and temporal levels. We next identify research gaps in understanding the impact of agrobiodiversity on health. Because of the urgent need for action to create more sustainable, just, and nutritious food systems, we further propose tasks for the public sector as well as strategic alliances that support agrobiodiversity's contributions to sustainability, equity, and human nutrition.

Feeding the World: Battling Narratives

The productionist paradigm, which dominated mid- to late twentieth-century agricultural and food policy, has been successful in creating higher outputs of a few key crops and feeding more people. The model wobbled in the 1960s and early 1970s (during the first energy crisis) but was revamped and modernized by the continued Green Revolution with great effect (Chapter 6). Still, it was shown to be in a fragile state during the banking and oil crisis of 2007–2008 and was, yet again, polished and promoted by calls for more environmentally benign technical changes to tackle what analysts said would be the ultimate challenge: accommodating even more population growth and increasing dietary expectations as well, only this time in an era of climate change and the challenges of agriculture-related pollution and biodiversity loss. Consistent throughout the revamping of this model has been the perspective that problems can be resolved by producing more food (primarily staple grains, oils, sugar, and animal products) through ever more refined and sophisticated methods (Foley et al. 2011).

What can be predicted from proceeding down this familiar path?

First, more food. To clarify, more food of a certain type: more cereals, starchy root crops, meat and dairy, oilseeds, and sugar. Often emphasized in agricultural investments, these are the only foods tracked in the Food and Agriculture Organization of the United Nations' (FAO) food price index (FAO 2013). From a nutritional perspective, however, the emphasis on securing future consumption of these particular foods is increasingly puzzling. A diet comprised only of these foods would increase risks for negative long-term health outcomes. Sugars and fats, including saturated fats found largely in animal-source foods, are cited in a majority of dietary guidelines as components to limit because of their harmful relationship with health if eaten in excess (Herforth 2016). They are also the foods for which production and consumption have increased dramatically over the last fifty years (Khoury et al. 2014). Their increased consumption is a key driver of the nutrition transition and the global obesity epidemic (IFPRI 2016; Popkin 2004). Indeed, the projection of future food consumption is based to a large extent on trends in past food consumption—the same trends which brought with them social, health, and environmental resource impacts.

Second, in addition to more of the same kind of food, the dominant food narrative will bring more of the same kinds of malnutrition: obesity and diet-related noncommunicable diseases (e.g., cardiovascular disease, diabetes, cancer) will continue to increase, alongside persistent undernutrition and micronutrient deficiencies. This is a triple burden found in the same countries, communities, households, and even individuals within a life course (IFPRI 2016; Popkin 2004). Diet-related diseases have become a top risk factor in the global burden of disease (GBD Risk Factor Collaborators 2017). These dietary risks are due to the low consumption of fruits, vegetables, whole grain fiber, nuts, and seeds as well as high intake of sodium, processed meat, red meat, and sugars, including sugar-sweetened beverages. Diabetes, overweight, and obesity have risen in all regions and are projected to rise the fastest in Africa (e.g., IFPRI 2016).

Third, the current path supported through the dominant food narrative will produce more carbon emissions (Macdiarmid 2013) from the very production systems that are supposed to be designed as "climate smart" (Chapter 7). Myers (1997) and Hedenus et al. (2014) conclude that limiting global warming to a 2°C increase cannot be achieved without reductions in meat consumption, while others assert that we need to increase meat production to meet future food needs, particularly in low- and middle-income countries (McLeod 2011). A continued drive to produce more will also continue to exhaust natural resources (e.g., water, phosphorus), erode arable lands, and be a leading cause of the global decline in biodiversity, including pollinators, soil microorganisms, traditional farmer varieties and crop wild relatives, and other organisms that support the human agricultural endeavor (Bodirsky et al. 2014; Castañeda-Álvarez et al. 2016; Cordell et al. 2009; Foley et al. 2005, 2011; Matson and Vitousek 2006; Phalan et al. 2011; Rockström et al. 2009).

Fourth, the dominant food narrative's current production focus will continue to exacerbate wealth inequality, social and environmental injustices, and the power disparity between urban and rural areas as well as to devalue farmers and rural labor in general. Continuing the current production system will worsen the inequalities and injustices between urban and rural settings, with urban populations obtaining most of their food and energy from rural areas and returning their waste (Gracey and King 2009). The pressure to control the effects of industrial agricultural systems is lessening because the urban majority does not experience the environmental degradation and social injustices that affect rural populations (Coimbra et al. 2013). Moreover, the current agricultural production model places farmers and the farming profession at the bottom of the social ladder (Avila-Garcia 2016). This is intensified by the aging populations of farmers in high-income countries and increasing migration to cities in low- and middle-income ones (e.g., Martinez-Alier et al. 2016; Toledo et al. 2015).

These trajectories demonstrate that the status quo is unfit to improve the global nutrition situation substantially or to meet the sustainable development goals (SDGs) (UN 2015:15ff). Issues related to sustainability, nutrition, and health challenge current food systems and policy in most countries and institutions. The status quo is neither environmentally nor socially sustainable, rendering it economically unsustainable as well.

We envision that the following actions are necessary to support an alternative food narrative promoting sustainable, just, and nutritious food systems:

- Meet human nutritional needs and help to protect against noncommunicable diseases.
- Provide stable access to adequate food everywhere.
- Be resilient, that is, remain productive under changing and increasingly challenging environmental conditions.
- Conserve soil, water, and other natural resources; protect (agro)biodiversity; provide ecosystem services; and mitigate climate change.
- Minimize health risks and hazards, such as exposure to toxic chemicals and infectious diseases.
- Support social well-being and mental health.
- Provide culturally appropriate taste and variation and thus increase quality of life and demand for diverse species, varieties and breeds.
- Engender dignity, autonomy, and respect for all people.

Agrobiodiversity is one factor central to all of these desired goals. For the purpose of this chapter we refer to agrobiodiversity as "the variety and variability of animals, plants, and other organisms that are used directly or indirectly for food and agriculture, including crops, livestock, forestry and fisheries. It comprises the diversity of genetic resources (varieties, breeds) and species used for food, fodder, fiber, fuel, and pharmaceuticals. It also includes the diversity of nonharvested species that support production (soil microorganisms, predators, pollinators), and those in the wider environment that support agroecosystems (agricultural, pastoral, forest, and aquatic) as well as the diversity of the agroecosystems" (FAO 1999a). In a broad sense, agrobiodiversity is clearly essential to food and nutritional security and to human health (Friel et al. 2013; Frison et al. 2006; Graham et al. 2007; Johns and Sthapit 2004; Jones 2017; Negin et al. 2009; Powell et al. 2015). Yet this invaluable resource is threatened, including by the very agricultural systems that depend on it (Khoury et al. 2014). Alternative scenarios exist, however, where agrobiodiversity is able to flourish for the benefit of public health and the environment (Chapter 11), and such scenarios provide models for ways to transition toward more sustainable, just, and nutritious food systems. Adapting the words of President Bill Clinton (First Inaugural Address, January 20, 1993): "There is nothing wrong in [food systems] that cannot be fixed with what is right in [food systems]."

The "food shortage" paradigm (contained within the dominant food narrative) that arose in the twentieth century provided moral backing for productivity increases of a few major crops. Today, however, it has been asserted that the overarching food system problem is one of a nutritious food shortage (HLPE 2017; World Bank Group 2014). If food were equally distributed on the planet (which of course it is not), everyone would be able to satisfy or exceed their calorie needs, but it would still be impossible to fulfill their recommended dietary and nutritional needs (Herforth 2015). Specifically, not enough fruits, vegetables, and legumes are produced on the planet to meet the nutritional demand for those foods, while animal-based foods often are available only to the wealthy and not the undernourished who would most benefit from access to them. This nutritious food shortage stems from a focus on a narrow set of crops and livestock, which has effectively (socially, politically, and biologically) outcompeted a wider range of foods that provide diverse nutritional attributes.

Offering a true alternative food narrative requires understanding agrobiodiversity as one of the keys to the world's current food system problem—the systemic mismatch of humans, biosphere, and food supply that has narrowed the diversity base of agriculture and produced the modern scourges of obesity and diabetes while failing to resolve hunger, food security, and micronutrient deficiency. Table 9.1 compares the alternative food narrative described above, in which agrobiodiversity plays a central role, to the current dominant food system paradigm.

Table 9.1 Moving toward a new agrobiodiversity-based paradigm for the food system.

	Current Food System Paradigm	**New Food System Paradigm**
Environmental considerations	Climate change	Agrobiodiversity conservation and many other environmental considerations in addition to climate change
Social considerations	Undernutrition and famine; poverty	Women's livelihoods and empowerment, smallholder farmers' livelihoods, just and decent livelihoods for workers along the food chain, nourishment for all
Economic considerations	Maximize profit from economies of scale, product homogenization for global trade	Potential for higher earnings from greater utilization of diverse, high-value crops
Primary solution	Produce more of the same	Increase diversity of production systems, food, and diets
Results	Continuation of current dietary trends Widening gap in access to staples relative to other diverse plant foods Increased or insufficiently decreased carbon emissions, accelerating climate change Continued increases in diet-related noncommunicable diseases	Increased access to a diet that matches dietary recommendations, healthier diets Improved food security Increased equity and justice in the food system Reduced carbon emissions Increased adaptive capacity to climate change Decreased incidence of diet-related noncommunicable diseases

We argue that agrobiodiversity is an essential concept and actuality for the transition to a new food system paradigm. The ideas presented in this chapter are the outcome of one of the four groups in the week-long Ernst Strüngmann Forum, which discussed and evaluated the linkages between agrobiodiversity and sustainability, cultural equity, dignity, and viable livelihoods, and human health and nutrition for the purpose of identifying future research needs. We present here what is currently known about these linkages, and what needs to be better understood in the future. Because of the urgent need for action to create more sustainable, just, and nutritious food systems, we also propose tasks for the public sector and strategic alliances that support agrobiodiversity and its consequences (environmental sustainability, social equity and justice, and human health) as well as a research agenda.

What Do We Know? Current Evidence about the Impact of Agrobiodiversity on Diets and Health

Growing attention is being directed to the importance of agrobiodiversity for human health and nutrition at the global, landscape, community, farm, and household level. We have sought to summarize and evaluate current scientific knowledge applied in policy making and management as well as the perceptions, open questions, and controversies at each of these different levels. We highlight how the relationships between agrobiodiversity and human health and nutrition are shaped by geography, culture, policy, and power differentials.

Global Level

Over the past 50 years, evidence from national food supply data suggests that human diets across the world have become more diverse in terms of lower dependence on starchy staples and more food groups consumed; at the same time, they are more homogeneous in a "global standard" diet dominated by a relatively small number of major commodity crops (Khoury et al. 2014, 2016). These crops have substantially increased their share of the total food energy (calories), protein, fat, and food weight provided to the world's human population. The most prominent plant foods include wheat, rice, sugar, maize, soybean, and palm oil. Such globalization of food supplies is associated with mixed effects on food and nutrition security, including reduced undernutrition in some regions alongside diet-related diseases caused by overconsumption of macronutrients (Khoury et al. 2014). Although reasonably successful in providing sufficient calories to all but 700 million of the world's population, global food production does not supply everyone with the foods and nutrients aligned with dietary guidelines (Herforth 2015). The environmental "foodprint" of the simplified production model is also unsustainable (Tilman 1999; see also

Chapter 11). Inadequate focus on production, distribution, and availability of affordable fruit, vegetables, nuts, and seeds clearly contributes to these gaps.

The increased global homogeneity of diets is accompanied by a corollary decline in the importance of local or regionally important crops. Concern with regard to the global decline in agrobiodiversity and in particular the loss of crops and livestock and their many traditional varieties and breeds has been raised for over a hundred years, at least since botanist N. I. Vavilov traveled the world in search of plants useful for cultivation in his Russian homeland (Vavilov 1926b). He noticed that diversity was disappearing in the cradles of agriculture—places where crops and livestock had been cultivated continuously for thousands of years (Khoury et al. 2016). The alarm was sounded even louder by agricultural scientists fifty years ago, during the Green Revolution, when farmers in some of the most diverse regions of the world partially displaced their many locally adapted wheat, rice, and other grain varieties with fewer, more uniform, higher-yielding, professionally bred varieties (van de Wouw et al. 2009, 2010). At the same time, smallholder farmers in many parts of the world, particularly in rainfed, traditional, and marginal settings, have continued to maintain diversity as an integral part of food systems (e.g., Brush 2004; Perales and Golicher 2014).

While there is a global, scientific consensus of concern regarding the rate of loss of agrobiodiversity over the past century, this has been difficult to quantify (cf. Mekbib 2008; Shewayrga et al. 2008). Most of the statistics on agrobiodiversity loss can be traced back to a handful of reports (FAO 2010b) and books (Fowler and Mooney 1990) that reference a few studies which have been challenged (Heald and Chapman 2009). Many estimates likely represent inflated, generalized statements about the global state of crop diversity loss, the most common being some variation of "75% of diversity in crops has been lost" (FAO 1999b). Quantification of erosion is additionally complicated by the lack of data and inconsistent methodology used historically to measure diversity. The best evidence assessed changes in genetic diversity within cereal crop varieties associated with the Green Revolution and showed a decrease in diversity when farmers first replaced traditional varieties with modern types, but a more complicated relationship subsequently (van de Wouw et al. 2009, 2010). Further studies here are certainly needed, including qualitative studies (Chapters 2 and 3). It is important to emphasize that although the scientific community may lack the tools or the drive to show definitive changes in agrobiodiversity, in many locales its loss is abundantly clear to farmers and scientists working with farmers. Qualitative work can show that many varieties have been lost within farmers' lifetimes, or are simply not accessible to the farmer any more when one or two preferred high-yielding varieties are promoted, wiping out dozens of varieties within a span of a few planting seasons or years.

At the global scale, agrobiodiversity is clearly necessary for basic food provisioning and for avoiding catastrophic crop failure. Reliable access to food staples is important to nutrition and societal stability. For these aims, genetic diversity is needed to underpin the production of major staple crops. Gene bank (*ex situ*) systems are essential to the conservation of this genetic diversity and to ensure its availability to plant breeders via formal seed systems, but they do not sufficiently conserve all agrobiodiversity, neither do they provide access to all producers in food systems nor permit the ongoing evolution of agro-biodiversity via interaction with pests, diseases, and climatic change. Other strategies, in particular *in situ* conservation, are complementary to large gene banks, providing ongoing evolution and access to diversity at the community level (Chapter 2).

Landscape and Community Level

The idea that agrobiodiversity at the landscape scale (Amend et al. 2008; van Oudenhoven et al. 2012) might impact diet quality aligns well with the concept of the "food environment" (Herforth and Ahmed 2015). If better access to nutritionally important foods within the food environment shapes dietary choice in urban food deserts (Ramirez et al. 2017), it should also shape dietary choice in rural areas (Powell et al. 2013). However, little is currently understood about the ways food environments impact dietary choice in low- and middle-income countries, especially in rural agricultural landscapes (Herforth and Ahmed 2015; Powell et al. 2013). Cultural mindsets influence which type of market, farm, or wild foods people consume (De Schutter 2011; Vadi 2011), but dietary choices seem to be very easily influenced by structural elements, such as social status and gender, urbanization and food industry marketing, and trade policies (Nelson and Chomitz 2011; Toledo et al. 2015). This has largely been the case in many low- and middle-income countries where changes in food cultures, and the profound effect that structural factors have exerted on dietary choices, have led to the nutrition transition associated with rising rates of obesity and diet-related chronic diseases (Popkin 2004).

Presumably diversity within the landscape and foodshed (Horst and Gaolach 2015) supports access to a diversity of affordable foods in markets. The role of agrobiodiversity in a community setting was well-illustrated by Stanner (1969) in his research on Kitui Kamba markets, *ndunyu*. Many districts are notoriously poor in crops which grow well in a neighboring district. Stanner notes that maize grows well in *Migwani* (a village) while *Mutitu* (a neighboring village) which is dryer produces good millet, but is too dry for maize. He notes that "the wider distribution of these [market] commodities to rectify local deficiencies is undoubtedly a primary function of the *ndunyu*." The *ndunyu* is therefore of the greatest use in equalizing such local productive variations and arbitrary climatic strokes.

Recent research has shown that diversity at the landscape scale can support better dietary quality through access to wild and agroforest foods, and reductions in seasonality of food availability (Powell et al. 2013). The importance of wild or forest foods from diverse agricultural landscapes seems to be highly site specific. A review of primary research papers that assessed the dietary contribution of wild foods showed high variation in the importance of wild foods for diets and nutrition among studies (Powell et al. 2015). In a number of sites, wild foods made up a significant portion of nutritionally important food groups, including vegetables (between 83% and 43% of vegetables in diets were from wild sources in studies from Tanzania and Vietnam, respectively) and meat and fish (between 88% in the Brazilian Amazon and 0% in a pastoral community in Kenya were from wild sources). The contribution that wild foods made to total energy intake was low in most studies: despite this, wild foods accounted for a large portion of micronutrients consumed at a number of sites (Powell et al. 2015). Similarly, a recent study (Rowland et al. 2016) that examined the dietary contributions of wild forest foods relative to agricultural foods to various food groups across 37 forest adjacent sites in 24 tropical countries found high variation across sites in the proportion of fruit, vegetables, and animal foods from the wild. Forests contributed an average of 14% of the total supply of fruits and vegetables (sites ranged between zero and 96%) and meat and fish (between zero and 92%). The reasons for variation among sites are currently poorly understood, but are likely linked to landscape diversity as well as agricultural practices, including agrobiodiversity management or candidate foods available in a complex, distinct ecosystem. Studies have shown that a shift from less intensive subsistence (swidden) agriculture to more intensive (sedentary) agriculture at both the household and the landscape scale is associated with less wild food use (Broegaard et al. 2017; Schlegel and Guthrie 1973).

One example of landscape-level diversity is the presence of forests or agroforests within the agricultural landscape. In the last five years there have been a number of papers showing a relationship between tree or forest cover at the landscape level and various indicators of diet quality (Ickowitz et al. 2014, 2016; Johnson et al. 2013; Powell 2012). For example, Ickowitz et al. (2014) used Demographic and Health Survey (DHS) data to show a positive relationship between tree cover and children's dietary diversity in 21 African countries. They also found that consumption of fruits and vegetables increased with tree cover up to a peak of 45% tree cover and then declined.

While these studies suggest that landscape-level diversity in agricultural systems may be associated with improved diet, the pathways between tree cover and diet remain a "black box" (Powell et al. 2015). Income is unlikely to explain the relationship: although forests and forest products may contribute to income that can support food security (Angelsen et al. 2014; Pimentel et al. 1997), communities that live close to forests are often poorer than those who live further away (Angelsen and Wunder 2003; Sunderlin et al. 2008). Other pathways are plausible:

- Tree cover impacts diets through the mechanism of farming systems that include more trees (and are more agrobiodiverse), providing a greater diversity of cultivated and wild foods as well as foods that contribute to healthy diets, such as foods from swidden agriculture and agroforestry (Ickowitz et al. 2016).
- Trees (agroforests) produce food groups important for healthy diets, such as fruits and vegetables (Powell et al. 2013).
- Agricultural systems with more forests and biodiversity are better able to provide the ecosystem services needed for the production of nutritionally important foods including pollination (Eilers et al. 2011; Garibaldi et al. 2011) and microclimate variation.

Temporal Level

The fact that different varieties mature at different times can confer improved food security and resilience. Agrobiodiversity, like other types of diversity, is insurance for more vulnerable communities. This is especially true among communities that depend on small land holdings, rainfed agriculture, and small-scale irrigation. In dry environments, rainfall can be erratic. Similarly, water resources are limited in many small-scale irrigation systems. Within a species, some varieties or types are better at withstanding adverse weather such as drought than others. Therefore, a farmer growing several varieties of one crop stands a better chance that at least one or more varieties will be favored by the prevailing weather.

In the Andes, where there is extensive agrobiodiversity of native potatoes and other species, such as maize and beans as well as other tuber crops (oca, mashua, olluco), the different varieties and species stretch production across different times of the year. This gives continuity of supply and pushes back the vulnerabilities of seasonality, enhancing food security (Graham et al. 2007). Some potato and tuber crop varieties, for example, come to maturity within three months while others take longer—up to eight months (Moscoe et al. 2017; Rodríguez et al. 2016). Even without staggered planting, the supply of this staple is extended by diversity in maturation time. The varieties also have different storage and processing characteristics, with some fresh tubers storable up to three months while others can be stored for up to six months. Freeze-dried potatoes remain edible for several years (de Haan et al. 2009, 2012a). Potatoes' availability can thus extend year round. Indigenous Peoples in Peru recognize and know this function of agrobiodiversity. The naming of some varieties clearly suggests cultural recognition of earliness. The availability of Andean maize varieties ranges similarly from three to eight months, providing much needed resilience in response to rainfall and irrigation uncertainty among smallholder farmers that include Quechua Indigenous People (Zimmerer 2014).

Furthermore, agrobiodiversity can prolong the period of nutrient availability. In Kitui County of Kenya, for example, five varieties of mangoes may be found on one farm. The variety called Kakeke ripens from October to December; Kikamba (the traditional variety) from January to February; Dodo from February to March; and Boribo and Ngowe from March onward. Mangoes represent an important source of vitamins A and C, and in this case the traditional variety (Kikamba) would have provided nutrients for only two months, but the additional varieties extend the period to at least six months. Protracted availability is also seen in pigeon peas and cowpeas, two important legumes in Kitui. A diversity of wild fruits, each with its ripening period, can be viewed similarly (e.g., Kehlenbeck et al. 2013).

Farm and Household Level

Recent reviews indicate that household-scale agricultural biodiversity (i.e., crop species richness) is consistently associated with higher dietary diversity among farming households (Jones 2017; Powell et al. 2015). However, this association is small in most cases (i.e., four to ten additional crop species are needed to increase household dietary food group diversity by one food group). Furthermore, the association is not linear but rather an "inverse U" relationship such that the association between crop species richness and dietary diversity is higher among households with low crop species richness and lower among households with high crop species richness. It is unclear if differential access to markets influences these dynamics. In most contexts, the association between crop species richness and dietary diversity does not vary across farms of differing market orientation. While diversification of production may limit opportunities for specialization that could increase access to niche markets, greater diversification may also increase opportunities for farmers to spread risk and to introduce emerging cash crops into their production systems. Regardless of the market orientation of farms, most farming families still rely on markets for a large percentage of their food purchases. Therefore, maintenance of agrobiodiversity is important in supplying markets with diverse foods, and the diversity of food available in these markets is important for shaping the quality of farming households' diets. Yet, maintenance of on-farm crop species richness for subsistence consumption remains an important strategy for maintaining household diet diversity even among more market-oriented farming households (Chapter 10). Horticultural crops cultivated in homestead gardens may be especially important for preserving this "safety net" of diversity for home consumption.

Market Interventions

Although marketing channels have been implicated for the reduction of agrobiodiversity when they demand conformity, homogeneity, and specific shipping-friendly properties, markets can also provide a strong incentive for conservation of agrobiodiversity when underutilized crops are cultivated and sold. Several studies demonstrate efforts to support agrobiodiversity through markets (Chapter 15). The following case studies show how markets have been harnessed to support agrobiodiversity.

Andean Grains, Bolivia

Commercialization, value chain, and demand limitations very often stem from the stigma of *food-of-the-poor* that accompany traditional crops, including Andean grains. Consistent efforts by a project supported by the International Fund for Agricultural Development (IFAD) was undertaken to popularize the consumption of Andean grains in ways that would create a positive image of the Andean grain (Giuliani et al. 2012). The most successful intervention of this nature was the strategic partnership developed with the Bolivian private coffee shop chain "Alexander Coffee." This joint venture resulted from a collaboration between the PROINPA foundation, the Bolivian NGO "La Paz on foot," the Italian NGO UCODEP (today Oxfam-Italy), and Bioversity International. This alliance launched promotional campaigns for underutilized species. Customers of Alexander Coffee shops across Bolivia were exposed to the nutritional benefits of Andean grains through attractive leaflets, table tents, posters, and tasting of attractive and novel Andean grain-based modern food recipes. The snacks, biscuits, and other food items developed with the support of Alexander Coffee's chefs were a great success and are now very popular items in the network of this catering chain with spillover effects in other shops. At the same time, this initiative promoted the establishment of direct linkages between Alexander Coffee and poor farming communities from Lake Titicaca for the supply of grains (for a discussion on economic value chain approaches to agrobiodiversity use and conservation, see Chapter 15).

From Neglected to High-Value Vegetables: The Promotion of Traditional Vegetables in Kenya

In Kenya, vegetables are an important accompaniment for the main staple dish called *ugali* or *sima* (*nsima* in Zambia and Malawi). Green leafy vegetables are cheap and thus readily affordable to many people in rural, peri-urban, and urban areas (Chweya and Eyzaguirre 1999). Being accessible to low-income communities, they play a crucial role in food security and in improving the nutritional status of poor families. Despite these beneficial attributes, leafy vegetables have generally been neglected by both researchers and consumers as resources for consumption and as a source of income. Vegetable diversity especially for urban consumers had narrowed significantly since colonialism. Local vegetables were stigmatized as associated with poverty and the past. Cabbage, kale, and occasionally Swiss chard (locally known as spinach) were the modern vegetables of choice with the diversity of African leafy vegetables being notably threatened in the 1980s.

From 1996 onward, a consortium of institutions led by the former International Plant Genetic Resources Institute (now Bioversity International) researched and promoted traditional vegetables in Kenya. The Traditional Food Plants database of the Kenya Resource Centre for Indigenous Knowledge (KENRIK) at the National Museums of Kenya registers 210 species of local traditional vegetables consumed by more than 55 ethnic groups in the country. Ninety percent of these varieties grow in the wild or appear spontaneously in cultivated lands where they are managed. With the help of farmers, scientists prioritized 24 species. The following decade saw about a dozen promoted as high-value traditional vegetables, changing the vegetable landscape in both formal and informal markets in Nairobi. Production, consumption, marketing, and market demand for African leafy vegetables increased over the ten years of the program. This transformation was due to concerted efforts, including selection of seeds with preferred characteristics, determination of nutritional and agronomic qualities, capacity building of community groups, development of local seed systems and market linkages, food fairs, and cooking demonstrations. Efforts were supported by media campaigns and even street demonstrations.

By 2006, the consumption of traditional vegetables in Kenya no longer carried the stigma it once had. The choice of vegetables to grow or purchase became much wider with increased opportunity to sell any local vegetable (Gotor and Irungu 2010; Moore and Raymond 2006). Farmers who grew African leafy vegetables and their children tended to eat a greater diversity of vegetables, with positive impacts on diet quality and nutrition (Herforth 2010).

Ecuador: The Importance of Flavor and Taste

Flavor and taste encompass the physical, chemical, and neurophysiological aspects of food, while taste can be further understood as a multimodal and multifaceted social concept which may include how people come to perceive, value, and identify with gastronomy and other sensorial encounters. Social movements that are centered on gastronomy and flavors have shifted the appeal for a transition to regenerative food production from agriculture to *food*. They have thereby created space for "those who eat" and opened up the doorway for a broader public to join the traditionally rural peoples' movements of agroecology and food sovereignty. In 2015, the *Colectivo Agroecológico* in Ecuador initiated a provocative campaign to recruit 250,000 families (5% of the population) for "responsible consumption" (Sherwood et al. 2017). This campaign aimed to capture about USD 650 million per year of the present-day financial resources invested in food in the country and use it for alternative purposes. Through a collapse of dichotomies between rural–urban, producer–consumer and poor–rich, the campaign has grown in both size and intensity. The experience generated through the campaign reveals how the sensations, associations, and entanglements of food and its taste can have social outcomes that range from new forms of producing, distributing, and preparing foods, to involvement in transnational initiatives such as civil society-based efforts to overcome violence in the Amazon and to address pandemic overweight and obesity. The campaign has united practitioners from different and sometimes competing ethnic, cultural, and social traditions around a common purpose and cause: food enjoyment. In the same vein, other research has shown that consumer taste preferences are strongly linked to sustainable cultivation practices that are based on agrobiodiversity (Ahmed et al. 2010), and that taste preferences seem to be the most relevant motivation for those who continue to consume wild food plants in rural areas of high-income countries (Serrasolses et al. 2016).

Power and Culture

Landscape- and species-level diversity have had much to do with cuisines and identity among individuals, communities, and cultures (Chapters 11–13). Three examples of the connection between food cultivation, culture, and power are presented here. The first is a story of how food norms and culture can shift based on the introduction of varieties that were originally introduced from a different place and culture. The second proposes the concept of "*dietary keystone species*" for species that are central to a particular food culture. The third story is about power and how cultivating one's own food and sovereignty over that process can be empowering, regardless of any other more normative or biological outcomes. We direct the reader to more comprehensive compilations of case studies on this theme for other examples of the intersection between food, culture, and power (e.g., Posey 1999a).

Sunflowers, Russia, and the Americas

Sunflower is one of the few native North American crops. It reached Europe early in the Columbian Exchange and was found in the 1500s in European botanical gardens (as an ornamental) (Putt 1997). Peter the Great may have had an influence in spreading its importance in Russia, as he was an advocate of botanical gardens. In the Orthodox Church, diets were quite constricted during Lent, forbidding the more "developed" life-forms and foods (animals, butter, fat, and even plant oils like olive). For these reasons, and because it was not on the prohibited list, sunflower became the preferred oil in the region, eventually becoming the dominant oil year-round. The oil crop expanded across Eastern Europe, where it is still very important today, as well as in the Mediterranean, where it is second in importance only to olive oil. In the 1930s, Eastern European Mennonites who resettled in the New World (Canada, United States, Argentina) brought sunflower varieties with them, instigating major industries present in these countries today. Argentina remains a major producer with the crop providing the most important oil in the diet.

Dietary Keystone Species

The concept of a "keystone species" in ecology refers to species that have a disproportionately large effect on their environment relative to their abundance (Paine 1995). In the ethnobotany literature, "cultural keystone species" are those central to the traditional livelihoods of a given cultural group (Garibaldi and Turner 2004) and are often foods. "Dietary keystone species" would perhaps comprise foods central or critical to a nourishing traditional diet. A keystone species, if it were to disappear, would cause loss of a whole repertoire of dishes containing an important suite of food taxa and a profound dietary shift. For example, central Mexico is the birthplace and center of the agrobiodiversity of maize. Mexican cuisines are built around maize, including countless recipes that call for particular racial complexes and varieties. For example, a kind of fish stew requires a specific variety of large-kernel maize; if that variety were unavailable, it is questionable whether the soup could still be prepared. If the maize variety were to disappear, the entire recipe, including the flavor and matrix of nutrients available in it, could be lost. Likewise, among the Mijikenda of coastal Kenya, women mix up to seven species of vegetables: there is typically a main one, while the others are called *kitsanganyo* meaning "for mixing" with the main one. The purpose of *kitsanganyo* species is to moderate the texture, taste, flavor, and even appearance of the main vegetable. Each of the *kitsanganyo* species plays a specific role (Maundu et al. 2011).

Urban Agriculture, Lima, Peru

A study on the outskirts of Lima exploring the role of urban agriculture on nutrition and food security showed no change in diet, but indicated the social and psychological benefits of producing one's own food, in addition to a material contribution (Prain and Dubbeling 2011). Families mentioned that having home production saved them money that they could spend on other foods or other necessities as well as serving as a safety net by "having food on hand." Women producers stated that while it was very important to buy food, it was equally important for them to have a space to plant their own vegetables. By sowing their own crops or rearing animals, the women felt they were caring for the environment as well as for their own families' health by eating fresh and uncontaminated foods. They also consider the physical activity in itself healthy and contributing to their sense of well-being and relaxation. In this sense they felt that urban agriculture enhanced their quality of life (Chapter 8).

What Do We Need to Know? Research Gaps in Understanding the Impact of Agrobiodiversity on Health

Clear changes in crop-species level diversity, both in agricultural fields and in per capita national food supplies, have been documented over the past fifty years, but the impact of these on human health and nutrition remains to be fully explored (see also Chapter 11). Accordingly, variation in food diversity availability, access, and utilization must be examined throughout food environments. Factors for analysis include changes in agricultural research policies and innovations; international trade regulations; international food aid policy; multinational, national, and local food companies' product penetration; markets, particularly supermarkets; demographics, particularly urbanization; and economic development, with particular emphasis on increased consumer purchasing power in many regions globally. Key elements of the analysis are cultural norms and changing dietary expectations, including increased demand for animal products, high fat, sugar and salt foods, and other "Western" foods; impacting these trends are celebrity chefs, organic agriculture products, health food industry products, and renewed emphasis on locally produced and traditional foods. Particular attention should be paid to the impact of agrobiodiversity interventions with the potential to resolve undernourishment and improve diet quality.

While strongly advocating agrobiodiversity's importance to nutrition and health as the central impetus for this paper, we consider here the state of the evidence and key gaps organized around the functions food systems should provide. These functions of agrobiodiversity include potential for improving nutrition and diets; for food security and resilience; for protection of health against risks and hazards; and for protection of dignity, autonomy, and quality of life. Inconsistencies and deficiencies in methodology or the ancillary nature of relevant research to date calls for a more systematic approach to refining and validating methods for directly assessing agrobiodiversity as it informs nutrition and health questions. Such approaches will necessarily merge quantitative and qualitative information, and can cut across both the natural and social sciences.

Potential for Improving Nutrition and Diets

While the link between excessive consumption of carbohydrates, protein, and fat in diets and noncommunicable diseases is well established, how changes in agrobiodiversity at the plant and animal species level have contributed to increased consumption of macronutrients is inadequately understood. The often high micronutrient content of underutilized species is also well known, but benefits of biodiversity with regard to major micronutrient deficiencies (vitamin A, iron, iodine, zinc) are not well elucidated. Furthermore, almost nothing is known about the impact of differences at the level of crop variety and animal breed diversity on human health at the global scale, and even less about the impact of simplification on both intestinal and oral microorganism diversity associated with dietary change (Obregon-Tito et al. 2015; Sonnenburg and Backhed 2016). Research oriented toward the identification of dietary keystone species for fulfilling micronutrient requirements, preventing noncommunicable diseases, and maintaining healthy microflora populations within the human body, while also supporting the social and cultural importance of diet, would be novel and welcome.

Data from various, mostly circumstantial sources support contributions of agrobiodiversity to positive nutrition outcomes and demonstrate how the protection of wild and cultivated diversity prevents undernutrition or noncommunicable diseases on a case-by-case basis. Nonetheless, more systematic evidence is needed to fully define these relationships or to predict how and when agrobiodiversity can be utilized for better health outcomes. Investigations from the several perspectives already discussed in this paper can continue to strengthen understanding of the relationships between agrobiodiversity and nutrition outcomes within different contexts ranging from populations adhering to more traditional patterns of resource use and exchange to those fully integrated into modern production and market economies. More research is needed to better delineate the contextual relationships and pathways responsible for the importance of wild foods for human diets and nutrition (Powell et al. 2015; Rowland et al. 2016).

As food systems are increasingly global and market oriented for a majority of the population, traditional foods will be increasingly accessed through markets. Some underutilized crops have become market commodities of regional or global distribution (e.g., quinoa, açaí, finger millet), sometimes marketed as "superfoods." From a health perspective, agrobiodiversity's contribution in such systems needs to be examined in relation to noncommunicable diseases. Do production and consumption of more diverse plants reduce noncommunicable diseases? Although evidence exists that more diverse plant foods aid risk prevention, research that establishes thresholds of minimum diversity, or proposes optimal diversity levels to reduce risk, will be useful.

Community-level foodshed diversity is an important scale yet to be more adequately understood as socioeconomic realities shift (Horst and Gaolach 2015). As an ever-greater proportion of farming families interact with urban environments, purchase some food in markets, and earn at least part of their household income off farm, research into the availability of and access to agrobiodiversity by communities through different mechanisms is warranted. While urban food deserts (Ramirez et al. 2017) are characterized by issues of availability and access, questions arise as to whether improving proximity to fruits and vegetables necessarily improves diets and nutrition. Equally, rural food deserts characterized by monotonous diets on farms where foods are grown predominantly for often distant markets deserve attention.

Research that is defined by contemporary case studies and observational data has distinct limitations. Importantly, the few studies that have examined the association between household-scale crop species richness and dietary diversity among farming households have used many different approaches and indicators to assess the relationship, and all of the studies have relied on cross-sectional analyses (Chapter 10). Such analyses do not capture any potential longer-term nutritional benefits of agricultural biodiversity and also preclude the drawing of causal inferences to understand the relationship. Furthermore, changes in agricultural biodiversity may have impacts on dietary diversity and quality over longer time periods, or on populations outside of the producing families themselves. Therefore, methodological approaches applied to date to examine these dynamics may misalign with those needed to properly assess the potential impact of agricultural biodiversity on diet outcomes. Impact on diet and nutrition can be observable, albeit often small, but the more critical question could be related to loss of cultivated and wild species and varietal diversity from a whole food system. What effect would this have at the community level and beyond? Relevant prospective research requires continued compilation of empirical evidence, but also extension into modeling of system transformations.

Which policies and interventions support access to diverse food resources and the ability to maintain agrobiodiversity within production systems? Undoubtedly, different messages and policy applications are needed for different regions and social groups. What is appropriate and what works are themselves issues which urgently need to be resolved.

While in mainstream North America food culture is being reinvented, in many other regions such as Southeast Asia, East Africa, or the Mediterranean, food culture has never been lost (Johns and Sthapit 2004). Maintenance of biodiverse traditional food cultures provides a powerful tool for moving forward. Under what conditions do farmers stop or continue to eat their local and traditional foods when they enter markets? What are their economic, nutritional, and social vulnerabilities of engaging in specialized production systems (Johns et al. 2013)? Intermediation in markets is an essential aspect of policy-guided research focused on producers, consumers, and supply chain.

Potential for Food Security and Resilience

Population growth, coupled with the virtuous objective of raising living and health standards of the billions living in poverty, challenges sustainable food security in unprecedented ways, and agrobiodiversity's role in this context requires examination.

The specialization of farmers linked to markets may come at the expense of the resilience that agrobiodiversity offers, which can be examined as a determinant of a long-term livelihood strategy. How and where agrobiodiversity is important for specific groups, including women and children, needs better understanding. In places where people rely on agrobiodiversity for their subsistence and well-being, microstudies focused on mechanisms, whereby agrobiodiversity brings benefit, are important because that is often where interventions are targeted. Research focused on issues such as the seasonal importance of biodiversity to health and nutrition or understanding of how dietary keystone species, landscape, ecology, and sociocultural factors affect the relationship will continue to draw on comparative case studies from which generalizable patterns and insight emerge. Undernutrition remains a primary focus for such research.

Food security in the twenty-first century assumes a requisite response to climate change. Agrobiodiversity, as it comprises variation in crop microadaptation to variable temporal and spatial conditions in temperature, moisture, and other characteristics that have been exploited for millennia by farmers, offers a key resource. The resource capacity of agrobiodiversity in response to climate change (Chapter 7) is also linked to local and global mediation of future change (Brondizio and Moran 2008) that has potential social and gender dimensions (Bhattarai et al. 2015).

The role of formal and informal seed systems in mediating the connection between food security and agrobiodiversity needs better documentation (Mucioki et al. 2016). Similarly worthy of fuller examination is in-field diversity: Why is it important and to whom? Modern plant breeders might assert that spatial diversity of traditional systems has been effectively replaced by temporal diversity of modern systems (i.e., varieties partially replaced or renewed every few years). Indeed, there have rarely been widespread full crop failures based on this model over fifty years. However, power relationships come into play in the dissemination of improved varieties and their adoption, with diminishment of the role of agrobiodiversity in smallholder communities for market opportunities, seasonal home consumption, and social function (Chapters 8 and 13).

Potential for Protecting Health and Minimizing Health Risks and Hazards

Beyond direct, consumption-related impacts on diet and nutrition or on food production, potential pathways that link agrobiodiversity and health include reduction in the use of external inputs with detrimental effects on the quality of air, water, or soil (Chapter 11). Landscape diversity, structure, and management is related to vector- and foodborne disease transmission as well as use (and misuse) of agricultural chemicals (Bianchi et al. 2013; WHO/CBD 2015). Agrobiodiversity's role in minimizing exposure to pesticides and other agrochemicals can be further supported. Likewise, documentation of the ecology and environmental determinants of human disease in relationship to agricultural systems as evidenced by the examples cited above extends understanding of the role of agrobiodiversity.

Although the World Health Organization (WHO) includes mental health as an essential aspect in the definition of health, related research exploring links with the natural environment is lacking. Some potential pathways between agrobiodiversity and mental health could relate to access (or lack of access) to both sufficient food and food that is considered culturally adequate, or to pertinence to social networks which provide access to resources (e.g., seeds, information), social influence (e.g., spread of nutrition related behaviors), social engagement, the provision of social support (both perceived and actual), and the enjoyment of life through reduction of monotony (further discussed in Chapter 11).

Potential for the Protection of Dignity, Autonomy, and Quality of Life

While culture's potency as a mediator of human behavior has been recognized above in relation to the market link of agrobiodiversity and nutrition, food culture can be examined directly as a desirable outcome. What is the effect of agrobiodiversity loss on food culture and dietary keystone species? Conversely, it is important to understand food culture as a driver of conservation. Agrobiodiversity can be distinguished in relation to sensory qualities (organoleptic); anthropological research might further examine aspects of food enjoyment with both intrinsic and economic value (Ahmed et al. 2010).

Dignity, autonomy, and respect for all people define a principle rather than strictly a research topic, but this can be examined in relation to policy and decision making and self-reflectively as an influence on research agendas. The need for studying relationships and outcomes between agrobiodiversity and human health beyond nutritional indicators (e.g., food taste, cultural foods, cuisine, preparations, gender) calls for a more integral methodological approach. Such an approach will be developed not by going to the field with refined research tools to collect data on additional variables, but by building community-inclusive research agendas and developing trustful relations with communities. In this sense, communities should be addressed as full participants in the identification of local foods, varieties, and preparations of traditional, historical, and current relevance. Communities need to participate in finding workable solutions to problems emerging from the relationship between agrobiodiversity, food, and health. An example of a guide to research protocols is "Documenting traditional food systems of Indigenous Peoples: international case studies guidelines for procedures" by the Centre for Indigenous Peoples Nutrition and Environment (CINE) (Kuhnlein et al. 2006). In research involving Indigenous traditional knowledge and practices associated with the collective use of plants, animals, and insects, the participatory process should be recognized and attributed (cf. WHO 2003).

Advocacy Gaps and Opportunities

Perhaps larger than research gaps are discrepancies in awareness and advocacy about the problem of declining agrobiodiversity and its potential for positive impacts on nutrition. At this Ernst Strüngmann Forum, we discussed and debated why the international community is not more engaged in agrobiodiversity as a central issue of our time. We believe that gaps in scientific evidence do not fully explain this lack of engagement. Rather, the lack of a coherent and compelling story, and the failure of scientists to communicate it, is a larger gap. There is a need to make the multiple stories clearer to a wide audience. What are the stories of agrobiodiversity in a world of over-, mal-, and under-consumption? The narrative of continued use by producers and consumers and the embeddedness of agrobiodiversity in society is less visible compared to the "doomsday" and "gene banks" story line.

The current international policy agenda on food and environment is dominated by climate change. It has taken 30–40 years of hard work and consistent, coherent evidence to achieve top-level policy engagement on critical global issues, including obesity and climate change. The international commitment to those topics is now manifest in the 2014 United Nations (UN) statement on noncommunicable diseases and the 2015 Paris Climate Change Accord (now ratified by over 55% of governments).

The lesson is that one needs a combination of good evidence, clear simple messages, and good organization to engage the policy agenda. The role of the Intergovernmental Panel on Climate Change (IPCC) is exemplary, but it has been enormously helped by the huge sociopolitical infrastructure of the NGO community, which has been active, noisy, and persistent on climate change. This mix of "inner circle" of respectable science and "outer circle" of nimble, noisy, active civil society is essential. One targets decision makers and formal institutions; the other does the same but by harnessing public attention. On obesity within health and food agendas, input comes from more "voices" and organizations—a good mix of inner and outer circles—and the message has been consistent, despite there being no one IPCC equivalent.

Environmental science offers many more policy challenges than just those posed by CO_2. Biodiversity and agrobiodiversity compete for policy attention with soil, water, land use, air, and the general concern about food supply in a time of climate change (see, however, African and other regional assessments of the Intergovernmental Science-Policy Platform on Biodiversity and Ecosystem Services; IPBES 2018). The food system as a whole is widely agreed to be in stress (Gladek et al. 2016; GLOPAN 2016). However, concerns over biodiversity and agrobiodiversity do not rise as high as climate change, notwithstanding the reality that a major consequence of global warming for humanity is ultimately the loss of biodiversity (Rockström et al. 2009).

We recommend a proper review and advancement of the most effective strategy for enhancing the profile and importance of agrobiodiversity for nutrition and health. This should consider different options, different scales (global to local), and the role of different policy actors, including governments, companies, consumers, scientists, and civil society.

Meanwhile, one option is to attach agrobiodiversity more clearly to specific immediate threats. What does it have to offer to climate change adaptation or noncommunicable diseases? Are there any links? It always helps to get an issue onto the policy agenda if protecting and enhancing it can help resolve other problems for society and the planet. In short, agrobiodiversity needs to build the right mix of problem, solution, organization, and message to ensure greater attention at the policy level.

Accountability to Existing Commitments

There is a need to enforce and act on existing, signed international policies (Chapter 14). The following policies and initiatives are the most significant with regard to agrobiodiversity conservation, sustainable use, and human health:

- The Convention on Biological Diversity (CBD) requires parties to develop National Biodiversity Strategies and Action Plans (UN 1993). Agrobiodiversity is an integral part of both the Aichi Biodiversity Targets (Target 13) and the Global Strategy for Plant Conservation (Target 9). The Global Environmental Facility has supported various related initiatives, for example, the UNEP/FAO implemented the multicountry Bioversity for Food and Nutrition project.
- The International Treaty on Plant Genetic Resources for Food and Agriculture, in alignment with the CBD, outlines specific responsibilities with regard to plant agrobiodiversity (FAO 2009).
- Initiatives such as the CBD's Cross-Cutting Initiative on Biodiversity for Food and Nutrition, which was adopted at the 8th Conference of the Parties (COP8) under the Millennium Development Goals. The updated SDGs strongly consider agrobiodiversity conservation (Goal 2.5), environmental sustainability (Goals 7, 11, 12, 13, 14 and 15), and the need for improved human health (Goals 2 and 3).
- The World Health Organization, through initiatives such as the recent joint report "Connecting Global Priorities: Biodiversity and Human Health: A State of Knowledge Review" (WHO/CBD 2015). Among the thematic areas featured in the report are (a) agricultural biodiversity, food security, and human health and (b) biodiversity and nutrition.

Linking Agrobiodiversity to Related Agendas

While agrobiodiversity features prominently in some international agreements such as the CBD or the International Treaty on Plant Genetic Resources for Food and Agriculture, this is much less the case in relation to health policy decisions (see COP 12 Decision XII/21: Biodiversity and human health). The following examples show linkages between agrobiodiversity and identifiable global agendas of relevance to food systems and nutrition.

Food Security Agenda

Simply put, food security entails consistent access to diverse food to meet nutritional needs, as has been agreed to by all signatory nations for over twenty years (FAO 1996). Greater agrobiodiversity is undeniably necessary at the species level to make food security a possibility for all. At the varietal level, evidence needs to be astutely portrayed so that it becomes clear in what situations, and for whom, varietal diversity contributes to year-round food security through reduced seasonality and enhanced resilience.

Climate Change Agenda

Agrobiodiversity offers adaptation potential. For example, different native potatoes can be grown at different altitudes in the high Andes according to the climate, thus handily providing climate change adaptation. Also, complex knowledge systems are associated with the adaptive capacity of agrobiodiversity and are at risk of loss (Chapter 7).

Nutrition-Sensitive Agriculture

Agrobiodiverity is at the intersection between nutrition, agriculture, and environment because it offers a means toward nutrient adequacy, reduced seasonality, gender equity, and resilience. To some extent it is already embedded in the nutrition-sensitive agriculture conversation. FAO (2015:5) states that "diversified production systems are important to vulnerable producers to enable resilience to climate and price shocks, more diverse food consumption, reduction of seasonal food and income fluctuations, and greater and more gender-equitable income generation." FAO recommends interventions and policies that facilitate diversification and increase incentives for availability, access, and consumption through environmentally sustainable production as well as the trade and distribution of nutrient dense and safe crops and animal-source foods (e.g., horticulture crops, legumes, nuts, seeds, small-scale livestock, and fish—foods that are relatively unavailable and expensive, but nutrient-rich and vastly underutilized as sources of both food and income). Thus nutrition-sensitive agriculture assumes a coherent nutrition and ecosystem focus. The Scaling Up Nutrition (SUN) Movement Strategy and Roadmap (2016–2020) identifies agriculture and food systems as essential in making diverse, nutritious food more accessible to everyone, and supporting small farms as a source of income for women and families.

Health and Nutrition Agenda

The UN Decade of Action on Nutrition (2016–2025), led by the FAO and WHO, embraces the current policy consensus that eating a variety of foods, including fruits and vegetables, is important for health. Abundant diversity of plant foods is one of the common characteristics of international and national dietary guidelines and epidemiologic research that protect against noncommunicable diseases (Herforth 2016). Researchers have yet to illustrate clear linkages between agrobiodiversity and people's access to a diversity of plant foods to protect health.

Sustainable Development Goals

While SDGs as a broad agenda indicate better food systems for sustainable diets, progress toward improved diets demands specific strategies and greater accountability. Despite the importance of health in 70 of the SDG targets related to food, none are tied to indicators that measure dietary intake. FAO and Bioversity International have articulated a concept of sustainable diets that directly links agrobiodiversity and nutrition (Burlingame and Dernini 2010). Although the discourse on sustainable diets and associated research has accelerated the emphasis on diversity for food and nutrient adequacy and resilience (Fischer and Garnett 2016; Gustafson et al. 2016; Jones et al. 2016), it has yet to realize the impact needed. Agrobiodiversity needs to be more effectively understood as integral to SDG 2, 3, and 5 at least. Goal 2 calls for food security, improved nutrition and sustainable agriculture. Goal 3 (UN 2015:18) calls for healthy lives and the promotion of well-being for all people at all ages. Goal 5, which strives to achieve gender equality and empower all women and girls (UN 2015:18), converges with nutrition and health priorities as underlined in the SUN Movement Strategy and Roadmap. Policies that are gender sensitive are more likely to value underutilized species and other components of agrobiodiversity that are typically grown and harvested by women (cf. Bhattarai et al. 2015; Johns et al. 2013).

Biodiversity Conservation Movements

The closest ally for agrobiodiversity may naturally be the conservation community. The World Wildlife Fund's "Metabolic Report" says food systems are essential for conservation (Gladek et al. 2016). But this rationale is often used as an argument to simplify and intensify systems ("Growing more intensively to spare"). How can it be reshaped to protect and support agrobiodiversity? A major shift in philosophy over twenty years has led to the understanding that conservation in diverse regions cannot succeed without the approval, buy-in, and participation of diverse local cultures. Perhaps it is not so different in the case of agrobiodiversity: conservation of agrobiodiversity needs both people and places to flourish. It cannot only exist in seed banks.

Where Next? (Conclusions)

Agrobiodiversity is an essential part of both the storyline and the mechanism for solving the world food problem. It underlies a shift in paradigm from merely a Green Revolution-era "food shortage" to an updated realization of a "nutritious food shortage"—which is a reflection of production systems that are lacking in diversity, sustainability, and equity with visible consequences on nutrition and health. The recent regain in food system thinking and practice reflects the more holistic take now being adopted by donors and policy makers when balancing nutrition security and the socioeconomic and environmental imprint of agriculture on people and planet. There are roles for the support and revalorization of agrobiodiversity at every level.

At the global level, the UN approved the SDGs in 2015 and the Paris Climate Change Accord, as well as updated biodiversity targets via the CBD, and more member countries have signed onto the International Treaty on Plant Genetic Resources for Food and Agriculture. In 2016, Habitat III addressed issues of the urbanized world (UN 2017a). These intergovernmental commitments and aspirations need to be translated into concrete, precise, and coherent actions at the local, regional, and national levels. Although nutrition is implicit in many of the SDGs and targets, precise actions are often lacking. Agrobiodiversity has much to offer. We want to see all member states create new SDG-informed commitments at their national and local levels. New public engagement is essential. Eating differently to protect and enhance biodiversity requires consumers and the food industry to change. More use of existing diversity, breeding, production, and trade, particularly in fruits and vegetables, is needed for building diversity in the field, not just in parks or forest edges.

Global trade policies and agreements have profound diet-related consequences (Friel et al. 2013). Since they impact access to healthy and unhealthy foods through global and local supply chains, they need to be more responsive to ensuring the benefits of agrobiodiversity for desirable health outcomes.

Civil society organizations worldwide are aware of the importance of biodiversity and some see agrobiodiversity as worthy of support. We urge them to give higher priority to the following:

- Support farm and food systems which do not "mine" the earth.
- Restrict herbicide and agrichemical use.
- Reinvigorate skills sharing and training to farm well by building on local knowledge.
- Link agrobiodiversity to youth engagement, education, and revalued local identity.

- Conduct annual "state of nature" farm and food reviews to hold governments to account.

Supporters of "nutrition-sensitive" agriculture should consider how to be both nutrition and ecosystem sensitive. A new generation of extension and advisory services needs to halt the degradation of ecosystems and enhance their protection. Gene banks have their place, but agrobiodiversity *in situ* is a useful means for retaining a pool of genetic diversity in the field, spread across regions and growers. Hundreds of millions of people are already caretaking biodiversity through their livelihoods, even if they do not conceive of themselves in that way.

The most sensitive issue of all concerns the role of the public. Consumers eat the environment. The global rise in meat and dairy consumption is widely agreed to be a major driver of ecosystem threats, notably of climate change. Different messages are needed for different regions and social groups. Little advantage rests in asking consumers in low-income societies to eat less when they desperately need to have access to more and better diets. Yet, there is a value to promulgating the message to eat less but more sustainably to high-income society consumers. What is appropriate, and what works where, are themselves issues which urgently need to be resolved.

Our final appeal is to fellow scientists. First, agrobiodiversity must be more consistently and openly supported. The complex relationship of biodiversity, food, and nutritional health requires us to speak out coherently and with united and clear voices. Scientists and researchers often relish the incompleteness of their tasks: there are always new questions to ask, new avenues and connections to explore, new data and insights to absorb, new pathways to map. The connections between biodiversity and health are no exception to this, nor is the role of agriculture and food lacking in fascinating, complex issues to explore. The debate, however, about what agrobiodiversity has to offer for improving public health and nutrition in an era where diet is the factor with greatest impact on noncommunicable diseases—the most significant but not the only source of twenty-first century ill-health—is of importance not just to scientific journals but to everyone. The real world of farms, fields, and food systems has entered a new era, the Anthropocene, in which human activity is both driving and being shaped by the consequences of our collective actions. These are well-known and documented by science—climate change, ecosystems stress, demographic change, the nutrition transition and more. Scientists in all our organizations must come together to give clear, coherent, evidence-based advice and advocacy. We cannot expect public opinion and behavior to adapt to the Anthropocene if we add to the policy and cultural cacophony, or worse, keep silent waiting for yet more research to answer our questions. Despite the need for more and better knowledge, we already know enough to speak up and out for the value of agrobiodiversity for human health, joy, and indeed, survival.

Acknowledgments

We would like to acknowledge Gabriel Nemogá, Matthias Jäger, and Steve Sherwood for contributing to the writing of this manuscript. We would also like to thank all participants of this Ernst Strüngmann Forum on Agrobiodiversity who contributed to a rich discussion.

10

Agricultural Biodiversity and Diets

Evidence, Indicators, and Next Steps

Andrew D. Jones, Gina Kennedy, Jessica E. Raneri,
Teresa Borelli, Danny Hunter, and Hilary M. Creed-Kanashiro

Abstract

This chapter synthesizes key findings on how agricultural biodiversity influences diets, and, based on this evidence, provides both policy recommendations as well as priorities for a future research agenda that can help to inform the promotion of diverse food systems for healthy diets. Empirical evidence is reviewed of the linkages between terrestrial agricultural biodiversity, both cultivated and wild harvested, and the diversity and quality of human diets. Further, the principal pathways through which agricultural biodiversity may influence diets are identified. An assessment is provided of the research challenges inherent in linking agricultural biodiversity and nutrition. Diet diversity and quality indicators are reviewed and analyzed relevant to understanding the relationships between agricultural biodiversity and diets. The chapter concludes with a set of policy recommendations for driving change at global and country levels to inform policy aimed at producing more diverse foods and improving diet quality through the mainstreaming of biodiversity into overall development objectives.

Introduction

Global cereal production and inventories are expected to reach record high levels in 2017 (FAO 2017). Long-term trends in global cereal production, which for decades have demonstrated growth since the Green Revolution, have rendered such record-setting achievements almost common place. This unprecedented abundance, however, may obscure the less conspicuous trend that food supplies worldwide are becoming increasingly homogenous (Khoury et al. 2014). Indeed, of the tens of thousands of edible plant species on the planet, only three crops provide the majority of calories in human diets: rice, maize, and wheat (FAO 2010b). The declining diversity of agricultural production systems worldwide is worrisome for multiple reasons. Agricultural productivity is fundamentally dependent upon the supporting and regulating ecosystem services that are provided by species diversity (Hooper et al. 2005). Biodiversity within agricultural systems may also provide resilience to climate-related shocks, as well as help to preserve cultural identity (Johns et al. 2013; Mijatović et al. 2013).

Agricultural biodiversity may also play an important role in contributing to diverse, healthy diets (Lachat et al. 2018). Poor-quality diets are the largest risk factor in the global burden of disease (GLOPAN 2016). This burden is composed of both undernutrition (i.e., it is largely driven by micronutrient deficiencies associated with low-quality, monotonous diets) as well as obesity and diet-related chronic disease (e.g., type II diabetes, hypertension, cardiovascular disease) (see also Chapter 9). To be certain, malnutrition is a complex challenge and has numerous underlying causes, not all of which are related to diets (Scrimshaw and San Giovanni 1997). The promotion of healthy diets through diverse food systems, however, may uniquely contribute to addressing the multiple burdens of malnutrition currently faced by many low- and middle-income countries.

In this chapter we synthesize key findings on how agricultural biodiversity influences diets, and, based on this evidence, provide both policy recommendations as well as priorities for a future research agenda that can help to inform the promotion of diverse food systems for healthy diets that also consider consumer demand and the role of the private sector.

Agricultural Biodiversity and the Diversity and Quality of Diets

Empirical Evidence of the Linkages

The most comprehensive review to date found that in 19 out of the 21 studies reviewed, there was a small, positive association between agricultural biodiversity and household-level diet diversity (Jones 2017). If interpreted as causal, the magnitudes of these associations indicate that four to ten additional crop species would need to be added to household agricultural production systems to increase the food group diversity of household diets by one group.

Though this assessment included only studies that adjusted for factors which also contribute to differences in household-level diets (e.g., wealth and education), the cross-sectional nature of nearly all of the reviewed studies makes it difficult to assess whether such a large number of new crop species would actually need to be introduced to affect diets. Policies or programs, for example, that strategically incentivize or introduce new crops into farming systems may require fewer such crops to influence the diversity of diets, particularly if incorporating behavior change efforts into the program (Berti et al. 2004). Furthermore, the association between agricultural biodiversity and diet diversity was not always monotonically increasing, such that marginal increases among low-biodiverse farms were associated with greater increases in diet diversity than increases among intermediate or high-biodiverse farms. This relationship makes intuitive sense, particularly when considering food group diversity of household diets as the outcome of interest. If the crops produced on high-biodiverse farms contribute directly to household consumption, these farms are most likely providing a large number of different food groups as compared to low-biodiverse farms, which may produce only a small number of staple crops that are likely to be considered part of the same food group (e.g., grains, roots, tubers). If one considers intraspecies variation, it is possible that there would be a distinct association. However, almost no studies have examined the association of varietal diversity with diet diversity. One study that did examine this association found no difference in the association of crop species and varietal diversity with diet diversity (Jones 2016). Thus it is likely that species diversity is a more efficient strategy for improving diets (Berti and Jones 2013). Yet, where intake of specific staple crops are high (e.g., in centers of crop origin such as Southeast Asia for rice and the Andean region of South America for potato), increasing intake of micronutrient-rich varieties of key staples may be an effective complementary approach.

Because nearly all of the studies in the above-mentioned review assessed only food group diversity of diets and not diet quality (e.g., micronutrient density), it is not clear how changes in crop species richness, at any level of existing agricultural biodiversity, might influence the availability of limiting nutrients in diets. As discussed below, glaring gaps in our understanding of the composition of underutilized species, or of diverse varieties of common species, is another challenge to examining the contribution of species richness to diet quality. One study that assessed daily intakes per adult equivalent of several macro- and micronutrients within households found that crop species richness was positively associated with energy, protein, iron, zinc, and vitamin A intake (Jones 2016). Nonetheless, limited data are available to make generalizations about the relationship between agricultural biodiversity and diets. It seems likely, though, that not all forms of agricultural diversification will have equivalent impacts on diets. Diversification efforts that increase access to nutrients deficient in the diets of vulnerable groups, such as women and children, especially iron, vitamin A, zinc, and folate, may have a more pronounced public health impact than other approaches (Jones 2017). In addition to the type of species, the dietary impact of agricultural diversification will ultimately depend on the use of the crop by the household (i.e., as food or cash crop).

Pathways

Raneri and Kennedy (2017) identify two principal pathways through which agricultural biodiversity may influence diets:

1. Direct consumption by households of self-produced plants and animals (or wild harvested/caught species).
2. The sale of agricultural crops for income which may then contribute to diets indirectly through the purchase of foods.

The nutritional importance of the second pathway may be even more important than the subsistence pathway. Households with more market-oriented farms can have more diverse diets than those with less market-oriented farms (Jones 2017). In addition to providing income that can be used to purchase diverse foods in markets, food crops destined for market may also be kept in part for own consumption, thus contributing directly to diet diversity. However, as Jones (2017:8) indicates:

> Despite the importance of market-oriented production for diet diversity, the relationship between agricultural biodiversity and diet diversity appears to be consistent across farms with varying degrees of market orientation. This observation is consistent with evidence that suggests that greater diversification, especially in highly subsistence settings, may reflect greater, and not foregone opportunities for market engagement by smallholder farmers who maintain a foundation of subsistence staple crop production, but have also diversified into one or more cash crops.

Therefore, any assessment of the potential for agricultural diversification to influence diets must examine the potential for

*   new and different crops to contribute to diets via these distinct pathways;
*   gains or losses to income either through specialization or diversification;
*   synergies between increased diversity for own consumption and for market; and
*   the influence of a host of other factors, including market access, gendered control of decision making, land size, labor availability, food price-to-wage ratios, and consumer preferences.

Furthermore, changes in landscape-level biodiversity (i.e., aggregation of changes across multiple households or large-scale changes in a region) may facilitate additional pathways for dietary change beyond the scale of individual households. Landscape-level agricultural diversification, for example, may influence the diversity of foods available at regional markets, thus increasing the likelihood that income generation can lead to more diverse diets. The ecosystem service functions provided by enhanced biodiversity at the landscape scale may also initiate positive feedbacks on total farm productivity and contribute to household resilience (Raneri and Kennedy 2017). The dietary contribution of sustainable forests and wild-harvested foods, at both landscape and household scales, is another important pathway to consider between agricultural biodiversity and diets.

Wild Foods

An insightful and comprehensive review by Powell et al. (2015) indicates that while the contribution of wild foods to energy intake among most populations is low, such foods may constitute 20–50% of dietary intake of essential vitamins and minerals that are commonly lacking in the diets of low-income households (e.g., vitamin A, iron, calcium, riboflavin) (Powell et al. 2015). In many contexts, wild vegetables and fruits—commonly harvested in and around agricultural lands near to homes—make up more than half of the total vegetable and fruit intake of a population (Powell et al. 2015). Powell and colleagues caution, however, that wild food availability and use are not analogous, and that such foods are lacking entirely from diets in many regions, including in parts of the Amazon and East and West Africa (Powell et al. 2015). Perhaps more importantly, studies of the nutritional contributions of wild foods, as well as the extent to and mechanisms by which they may influence diets (including their possible role in mitigating seasonal food shortages), are few and sporadic such that there are numerous knowledge gaps remaining in the research literature. For example, recent work indicating positive and negative associations of tree cover and deforestation, respectively, with consumption of animal-source foods, fruits, and vegetables among children are intriguing, yet do not provide mechanistic insights into how tree cover may influence diets (Ickowitz et al. 2014; Johnson et al. 2013; Jones et al. 2017). Forest cover may be a proxy for more diverse agricultural production systems associated with proximity to forests, or forests may provide food directly or provide ecosystem services important for enhancing production of fruits and vegetables. Well-designed research studies that are focused explicitly on elucidating these and other potential mechanisms are required to better understand these dynamics.

Challenges and Priorities for Research and Policy

Research Challenges of Linking Agricultural Biodiversity and Nutrition

Several aspects make it difficult to derive lessons on the impact of agrobiodiversity and nutrition. Both agrobiodiversity, as an input, and nutrition, as the outcome, are the most commonly hypothesized pathways tested. Both are very broad and complex concepts. For example, nutritional status is influenced not only by food intake, but also by health and care practices, which in turn are influenced broadly through sociocultural norms and political and economic conditions (UNICEF 2015). Similarly, the use of agrobiodiversity for nutrition is mediated through a range of factors including affordability, availability (which can be either season, landscape, or market based (e.g., Cruz-Garcia and Struik 2015), and acceptability which is influenced by culture, tradition, nutrition knowledge, and food preferences (see Chapter 12). There has been a lack of research that tests the reverse relationship, how consumer demand can drive changes in agricultural production or diversification. The way in which these multiple agroecological, market, social, and cultural contexts shape the relationships between agricultural biodiversity, diets, and nutrition is important yet difficult to measure through commonly used research methods. Masset et al. (2012) have highlighted many of the challenges in assessing the impact of agriculture interventions on nutrition. Many of these factors are applicable here: study design, measurement of participant exposure to the intervention, and attention to metrics. Additionally, there are methodological gaps in how we measure dietary intake, in order to understand the role of agrobiodiversity in diets, as well as scientific gaps in documenting the actual nutrient composition of biodiverse foods (Kennedy et al. 2017b).

Typical impact studies in the field of nutrition focus on one specific intervention and one outcome (e.g., supplementation with vitamin A capsules to reduce vitamin A deficiency). In real life, it is both difficult and costly to test more complex relationships between agrobiodiversity and nutrition using a classical clinical-style randomized controlled trial model. Thus, cross-sectional studies are often used to try to understand some of the relationships, particularly at the level of production or market diversity and diversity of diets. Currently, scientists lack consensus on the appropriate metrics for measuring on-farm diversity as well as diet diversity. The biological significance of findings relating these concepts is also not clear.

One important critique is the mismatch between the agricultural biodiversity measurement and the nutritional meaning of the outcome measure chosen. Many disparate "count" metrics are used to measure agricultural biodiversity on farm or within landscapes, with no standardization to date. Agricultural biodiversity measures include counts of the number of plant species and varieties produced on farm, counts of plant and animal species, and, in other cases, the number of wild foods gathered. Agroecological concepts of species evenness and richness, such as the Simpson, Shannon, and Margalef indices, have also been applied. On the nutritional side, simple aggregate measures of dietary diversity are the most common outcome chosen; these metrics, however, do not usually reflect dietary biodiversity. Despite this, studies vary between measuring this at household, woman, or child level. A strong critique of the available evidence base is that dietary diversity, when measured at the household level, is not a nutrition outcome (Verger et al. 2016). Often, there can be a mismatch between the count used to measure agricultural biodiversity and the count used to measure dietary diversity. Even when individual level scores are used to test the relationship, there can be inconsistency in the number of food groups and the reference period (i.e., 24 hours or 7 days) used to derive the dietary diversity score (Jones 2017).

Another challenge for understanding the impact of the full range of agricultural biodiversity is that efforts to map inverse relationships between production diversity and diet diversity often neglect to include the contribution of wild and semi-wild foods and markets. Both wild plant and animal foods contribute to diet quality and may have important socioeconomic and cultural values to communities (Powell et al. 2015). These foods, even if not cultivated, are often collected from or around farms, and opportunities to leverage domestication possibilities can be missed when interventions are considered to diversify production systems. With increasing feminization of agriculture and increasing migration from rural areas, essentially resulting in part-time farming, the complementary role of markets to household agrobiodiversity production is often undocumented. Research to better understand how markets in local food systems can be utilized to leverage the agrobiodiversity available at a landscape level may offer more insight into the production, market, and diet nexus of biodiversity in local food systems.

Finally, analysis of the nutrient adequacy of diets strongly depends on the public availability of food composition data. National food composition tables often include data on the most commonly consumed foods by the majority of the population. This means that underutilized and biodiverse foods, which are often integral parts of traditional food systems, can be missing. Many studies document significant differences in nutrient content of varieties within species (Burlingame et al. 2009). Average values of the nutrient content of a species, when not considering differences among varieties, could incorrectly estimate the roles of nutritious biodiverse foods in the diet and further obscure important connections between agricultural biodiversity and nutrition (Lachat et al. 2018).

Importance of Diet Diversity and Quality Indicators

Despite the difficulties of evaluating the role of agricultural biodiversity in diets, the body of evidence is beginning to grow, and further detailed guidance is available on how to consider agricultural biodiversity in study design for dietary intake assessment (Kennedy et al. 2017b). It is important that we bear in mind the limitations of the various indicators most commonly used and how they may highlight or blur biodiversity–nutrition relationships. In addition, recent calls have been made for more attention to studies of diet quality that tackle issues beyond dietary diversity and that include considerations of both balance and moderation (Herforth 2016).

Individual diet diversity indicators have been validated for micronutrient adequacy in the diets of infants and women of reproductive age (FANTA 2006; Martin-Prevel et al. 2015). Household measures of diet diversity—such as the Household Diet Diversity Score (HDDS) (Swindale and Bilinsky 2006) and the Food Consumption Score (FCS) (World Food Programme 2008)—have been validated for energy intake and are intended to be used as food security indicators and not as nutrition indicators (Hoddinott and Yohannes 2002; Wiesmann et al. 2009). Recent validations of the HDDS and FCS concluded that despite its common application, these household-level indicators may not be viable indicators of household food security, and improvements seen cannot be used to assume improvements in household diets (Lovon and Mathiassen 2014; Vellema et al. 2016).

When applying household measures of diet diversity as nutrition outcome indicators, a main concern is that they do not account for intrahousehold dynamics, which often leave women and young children vulnerable to malnutrition, even when households have access to nutritious foods (Dang and Meenakshi 2017). Understanding the often intricate gender dynamics and norms within households can be challenging, yet it is crucially important (see Chapters 8 and 13). Properly targeting dietary intake assessments to nutritionally vulnerable individuals within households must be a priority goal.

Simple food group-based diet diversity scores may mask important variation in diets. Thus it is important to look at other measures of diet quality as nutrition outcome indicators. Diet diversity indicators focus on the contribution of food groups to diets and as such do not consider how variation within food group consumption can contribute to diet quality, nor do they fully capture the contribution of diverse species to the diet. Several studies have assessed the association of intragroup variety with energy and nutrient intake and health outcomes. Foote et al. (2004), for example, demonstrated that intragroup variety for some food groups (i.e., dairy, grain, fruit, and vegetables) showed strong associations with nutrient adequacy of single nutrients. Dietary species richness (DSR) is an indicator that measures biodiversity (as a count of the number of different species consumed) in the diet and is associated with individual micronutrient intake (Lachat et al. 2018). The indicator allocates a score for each unique species consumed in the daily diet, even when multiple food sources from a single species are present (e.g., if cow meat, cow milk, cow yoghurt, cow cheese, and chicken meat are consumed, then the DSR is 2). Presenting the species richness per food group consumed has the potential to allow for more in-depth understanding of how diversity within food groups can contribute to diet quality and nutrition. This indicator also offers an opportunity to be used with landscape or farm species richness indicators (i.e., counts of the number of unique species in a landscape) to measure and compare biodiversity using a single metric (species richness) in diet and agriculture. However, it remains important to consider the nutritional contribution and significance of multiple uses of the same species. For example, both the root and the leaves of cassava (*Manihot esculenta*) are consumed and provide distinct nutritional contributions to the diet.

The modernization and transitioning of food systems is a common global trend (see Chapter 6 and 8), and has been accompanied by an increase in diet-related noncommunicable diseases (e.g., diabetes, overweight, and obesity) in both urban and rural environments (HLPE 2017). There is extensive evidence related to the relationship between the increased consumption of ultraprocessed foods (i.e., foods that are often high in saturated fats, salts, and sugars and low in micronutrients and fiber) and an increased risk of noncommunicable diseases (Forouzanfar et al. 2015; Louzada et al. 2015). The NOVA classification system was developed by Monteiro et al. (2016) to assist with identifying ultraprocessed foods; however, the broad definition of ultraprocessed foods can make it difficult to operationalize (Gibney et al. 2017). Efforts to measure diet quality could also consider the dietary balance between fresh, minimally processed foods and ultraprocessed foods by presenting the percent of daily energy consumed that comes from ultra-processed foods.

Finally, nutrient adequacy indicators can provide rich insights into the details of micronutrient consumption and how biodiverse foods contribute to daily nutrient intakes. Recommended daily intakes (RDIs) of different nutrients have been established by different international bodies (e.g., the World Health Organization and Institute of Medicine), and many countries also have established their own RDIs. Including indicators that allow for comparison of population intakes compared to RDIs can provide detailed information on specific micronutrient gaps. The technical skills and costs (financial and time) required to collect and analyze these data are often reasons why these indicators are not included.

Each of the diet quality indicators have strengths and weaknesses (summarized in Table 10.1). Such a suite of diet quality indicators would ideally be selected for research studies to ensure different dimensions of diet quality are included that capture different elements of biodiversity.

Policy Recommendations

The recognition that hunger, food security, nutrition, and sustainable agriculture are deeply interrelated is clearly stated in the second sustainable development goal of the United Nations, yet food systems are currently not necessarily delivering healthy and sustainable diets (HLPE 2017). Universal, specific priorities for policy action aimed at producing more diverse foods and improving diet quality are clearly spelled out in the Foresight Report of GLOPAN (2016) and include recommendations such as "making fruits, vegetables, pulses, nuts, and seeds much more available, more affordable, and safe for all consumers." Integrated policy approaches and actions are required across the environment, agriculture, and health sectors to promote greater diversity and to better mainstream agrobiodiversity into relevant sustainable development goal indicators for improved tracking of the multiple long-term ecosystem services food-based approaches can deliver for human well-being (Hunter et al. 2015).

To drive change at global and country levels and to inform policy, however, significant research gaps need to be addressed to better link agrobiodiversity,

Table 10.1 Summary of commonly used indicators to measure diet quality and their relevance to agricultural biodiversity.

Diet Diversity and Quality Indicators	Strengths	Weaknesses
Household diet diversity score Food consumption score	Quick to administer Limited technical skills required Widely used Collected with qualitative diet recall	Food security (access) indicator, not nutrition Does not consider intra-household dynamics Does not capture biodiversity May not consider biodiverse foods if limited food list used
Individual diet diversity score for women of reproductive age and infants aged 6–23 months	Quick to administer Limited technical skills required Widely used Collected with qualitative diet recall Validated as a proxy for micronutrient adequacy from the diet	Limited to food group diversity Does not capture biodiversity
Dietary species richness	Quick to administer Captures diversity in the diet Collected with qualitative diet recall	Requires additional technical skills to differentiate one species from another
Percent of energy from ultraprocessed foods	Captures one component of dietary balance	Requires quantitative dietary recall and specific technical skills Additional financial and time cost required Difficult to categorize ultraprocessed foods
Micronutrient intake	Provides specific detail about the nutrient intake of the diet Can capture contribution of biodiversity in the diet if not using a limited food list	Requires quantitative dietary recall and specific technical skills Additional financial and time cost required
Prevalence of micronutrient inadequacy	Considers intrapersonal variation in daily diet	Requires repeat dietary recalls on a subsample of individuals
Mean micronutrient density adequacy ratio	Only requires one dietary recall	Does not consider intrapersonal variation in diet

agriculture, nutrition as well as to document the role of agricultural biodiversity in improving nutrition (Hunter et al. 2016). Perhaps unexpectedly, areas that are rich in agrobiodiversity also often suffer from high rates of micronutrient deficiencies. The nutritional content of many traditional foods are often undocumented and as such, there is the need for more dietary intake and food composition data, particularly of traditional foods, which are often more affordable, available, and accessible to vulnerable groups. Food composition data is currently only available for a minor portion of the world's edible biodiversity, and national governments often lack the resources and capacity to collect information about what people actually consume (GLOPAN 2016). The International Network of Food Data Systems (INFOODS)—a forum for the international harmonization of food composition activities aimed at improving the quality, availability, reliability, and use of food composition data—has made efforts to collate food composition data on biodiverse foods (FAO/INFOODS 2013). Further efforts, however, are needed for countries to acknowledge the importance of biodiverse foods for inclusion in national food composition tables. Improving the evidence base on the importance of agrobiodiverse foods in diets, through use of appropriate dietary assessment tools and indicators, can better inform policy makers on how to identify foods to include in national programs including *in situ* conservation to improve nutrition (Bioversity Intl. 2017; Hunter et al. 2015; Hunter et al. 2016).

Research is needed on different policy options that can create incentives for people to diversify agricultural production systems for better nutrition and quality diets. This may include incentives for food companies to integrate neglected and underutilized species into national biodiversity conservation strategies and targets, as well as identifying opportunities for public procurement programs and public institutions including schools and hospitals to utilize nutritious biodiverse foods in feeding programs (Bioversity Intl. 2017; Kennedy et al. 2017a).

One pioneering example is provided by the Biodiversity for Food and Nutrition (BFN) project in Brazil, which is using nutrition information on biodiverse foods to guide its policies for food and nutrition security. Here, promising results in reshaping the food system by using agricultural biodiversity are being obtained through the BFN project, which is funded by the Global Environment Facility. Making use of Brazil's multisectoral plans to address malnutrition, six national ministries and partner organizations analyzed gaps between development and biodiversity plans, identifying new partnerships and making available new budgets to assess the nutrient content of 70 promising species of Brazilian flora. The government of Brazil has given the BFN initiative full support, agreeing to use nutrition information generated by the project to inform their food and nutrition security policies, particularly the two national policy instruments with the greatest potential for nutrition impact: The Food Acquisition Program (PAA) and the National School Feeding Program (PNAE). The two policies, which regulate food procurement and distribution to school children and vulnerable segments of the population, also provide economic incentives to family farming for the sustainable production of biodiversity, creating positive downstream benefits (Beltrame et al. 2016). Another important step toward mainstreaming biodiversity for enhanced food and nutrition security in Brazil was the signing in 2016 of Ordinance 163 that officially defines and recognizes "Brazilian Sociobiodiversity Native Food Species of Nutritional Importance." Aside from facilitating the procurement of sociobiodiversity species by national school feeding programs and the incentives for family farmers to continue to grow and market these species, the ordinance is helping to better monitor and track the consumption of biodiversity within the PAA and PNAE. Increased purchases of sociobiodiversity products by the national food procurement programs have already been reported but still remain negligible compared to the bulk of total foods purchased (Beltrame et al. 2016).

Investments in research and development of innovative, light-weight end-user technologies, such as mobile apps, that facilitate collection and analysis of quantitative dietary recalls to capture information on food agrobiodiversity will facilitate the uptake of these tools into large-scale agricultural projects where nutrition-assessment capacity is limited. This will assist in building the evidence base around how medium- and high-agrobiodiverse diets contribute the necessary micronutrients, which can be used to promote diverse local and Indigenous foods that have the potential to improve diet quality.

Other evidence gaps include investigations into the profitability and sustainability of promoting food systems that put agricultural biodiversity at their core, as well as cost-benefit analyses of investing in nutrition and expected nutrition benefits of interventions that use agricultural biodiversity.

11

Exploring Pathways to Link Agrobiodiversity and Human Health

Victoria Reyes-García and Petra Benyei

Abstract

Growing evidence indicates that the overall reduction of biodiversity in agricultural systems is concomitant to dietary simplification and related health effects, yet our understanding of the complex relationship between agrobiodiversity and health is still poor. This chapter explores pathways that could mediate this relationship at the local level. It begins by revisiting the definition of agrobiodiversity to disentangle its social components. In addition, the concept of health is broadened from the physical perspective. Pathways are then explored to link agrobiodiversity with physical health (diet, nutrition, and beyond) and mental health, including considerations of how food culture and traditional agrobiodiversity management knowledge contribute to identity and self-esteem. Discussion follows on the social aspects related to the production and consumption of agrobiodiversity that promote health and well-being. In conclusion, the chapter contextualizes how issues addressed at the local level fit within a broader political context.

Introduction

The twentieth century witnessed a drastic reduction of biodiversity in many agricultural production systems (i.e., agroforestry, home gardens, shifting agriculture), a change concomitant with the industrialization of agriculture and the expansion of monocultures that affected individual productive units (i.e., farms) as well as the broad agricultural landscape (Thrupp 2000). Such changes in the agricultural system, while not novel, intensified during the last century as the direct result of policy measures aimed at "feeding the world," in line with a productivist approach to agriculture, which prioritized crop yield increase over crop diversity maintenance (Chapter 6). Moreover, such changes were generally imposed on farmers, following a top-down approach to agricultural decision making and, in many areas of the developing world, promoting the introduction of cash crops, with consequent losses in food sovereignty (Friedmann and McMichael 1989).

Changes in the agricultural production system appear to have had unexpected effects on human health. For example, agricultural simplification seems to be connected to dietary simplification (i.e., the global trend in increasing the consumption of cereals parallels the global trend in decreasing the intake of pulses, legumes, and traditional grains or wild green leafy vegetables) (Khoury et al. 2014). These changes, in turn, have been associated with an increase in malnutrition (i.e., stunting, anemia, micronutrient deficiencies, overweight) and related health effects (Frison et al. 2006, 2011; Powell et al. 2015; Vincenti et al. 2008). Thus, agreement has grown that loss of agrobiodiversity matters, not only in itself, but because of its effects on human health (Johns and Eyzaguirre 2006; Kahane et al. 2013).

Despite growing evidence of linkages between an overall reduction of biodiversity in agricultural production systems and dietary simplification, our scientific understanding of the complex relationship between agrobiodiversity and health is still poor. While it is highly plausible that there is a causal link leading from the reduction of agrobiodiversity to health problems through diet simplification, empirical evidence showing this connection is scarce (Powell et al. 2015). Providing evidence on the direction of causality is important because, without it, we cannot rule out that both phenomena are only correlated (i.e., they occur at the same time) or that they are both caused by a third factor, such as abandonment of traditional livelihoods. In addition, beyond nutrition, there could be other pathways through which agrobiodiversity could relate to health. For example, some research has shown that diversified landscape elements, such as gardens, are associated with better human well-being (Litt et al. 2011; Milligan et al. 2004). So, a potential pathway between agrobiodiversity and health—albeit one that is largely unexplored—could be through the satisfaction of cultural ecosystem services associated with diversified landscapes (Calvet-Mir et al. 2012b).

In this chapter, we discuss the potential pathways through which agrobiodiversity can be related to human health. Our exploration focuses on the local level: pathways that might relate to effects on individuals, households, communities, or local landscapes (i.e., landscapes managed by closely linked communities). We have chosen this level of analysis because many decisions regarding agricultural production and dietary or other health-related choices occur at the local level. Afterward, we discuss how the local scale is connected to processes at larger scales through agricultural and health policies. Figure 11.1 provides a graphical representation of the hypothesized connections between agrobiodiversity and health. As many ideas in this chapter are new, they are not yet supported by empirical research. Our intent has not been to provide a review of existing findings but rather to spur future research on the linkages between agrobiodiversity and human health.

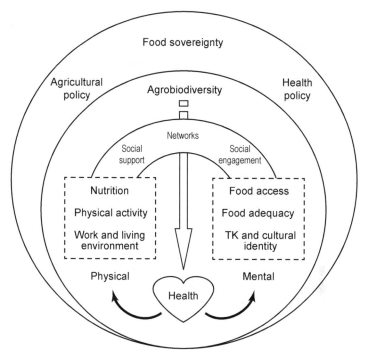

Figure 11.1   Potential pathways linking agrobiodiversity and health. TK: traditional knowledge.

Revisiting the Definitions of Agrobiodiversity and Health

The focus of this chapter requires us to revisit the definitions of agrobiodiversity and health carefully. The Food and Agricultural Organization (FAO 1999a:5) defines agrobiodiversity as

> the variety and variability of animals, plants, and microorganisms that are used directly or indirectly for food and agriculture production, including crops, livestock, forestry, and fisheries. It comprises the diversity of genetic resources (varieties and breeds) and species used as food, fodder, fiber, fuel, and pharmaceuticals. It also includes the diversity of nonharvested species that support agricultural production (soil microorganisms, predators, pollinators), and those in the wider environment that support agroecosystems (agricultural, pastoral, forest, and aquatic) as well as the diversity of the agroecosystems.

This definition rests on the fundamental idea that agrobiodiversity entails human management of natural resources (i.e., species, land, water, insects, and biota) to produce food and to satisfy other human needs. Consequently, the study of agrobiodiversity requires assessing not only the diversity of species in the system, but how they are managed and how both species diversity and management practices are embedded within cultural structures, institutions, and social relations that allow for the production, distribution, and consumption of food species as well as the transmission of knowledge regarding the properties, the material and symbolic uses, and culturally adequate ways to cultivate, use, and consume such agrobiodiversity (Powell et al. 2015). Moreover, the study of local agrobiodiversity also requires examining the influences of social institutions and cultural factors operating at larger scales (e.g., national and international policy decisions) that frame the management of systems enhancing or constraining agrobiodiversity (Johns et al. 2013). For a more inclusive definition of agrobiodiversity, see Zimmerer et al. (Chapter 1).

Taking a holistic approach, the World Health Organization (WHO) has defined health as "the state of complete physical, mental, and social well-being and not merely the absence of disease or infirmity" (WHO 1946). In its attempt to contextualize health, WHO also highlights that factors such as the social (i.e., being connected to others), the economic (i.e., income and employment), and the physical (i.e., access to safe water and clean air) environments are as important as a person's individual characteristics (i.e., genetics) and behaviors (i.e., diet or exercise) in determining health. Under the umbrella of WHO's holistic approach to health, researchers have analyzed, for example, the physical and mental health consequences of social aspects, such as income inequality (Diener and Seligman 2004; Wilkinson 2000), contact with nature (Milligan et al. 2004; Shillington 2008), belonging to meaningful social networks (Perkins et al. 2015), or adherence to specific cultural models (Dressler and Bindon 2000).

One of the main insights from this research is that health is best understood as a multidimensional state that is socially mediated and manifested through physical and mental well-being. Defining health as a multidimensional concept allows for a broader exploration of the potential relations between agrobiodiversity and health. If health has many dimensions, the pathways to health are likely multiple and probably intermingled in a complex web that challenges the "quick technical fix" approach to protecting or improving health—one that calls for a more comprehensive exploration of the complex relations between agrobiodiversity and the multiple dimensions of health.

Pathways between Agrobiodiversity and Physical Health

Most research examining the links between agrobiodiversity and health has focused on how agrobiodiversity relates to physical health through diet and nutrition (see Chapters 9 and 10). In connecting overall agricultural simplification and reduced dietary diversity, a major finding has been that even under adequate caloric intake, the reduction of dietary diversity might lead to "hidden hunger" or micronutrient deficiencies (Kahane et al. 2013), with pervasive consequences in physical health (i.e., immunostimulation or the worsening of preexisting health conditions) (Johns et al. 2013; Vincenti et al. 2008). Dietary changes have also been associated with an increase in malnutrition (i.e., stunting, anemia, undernourishment, obesity) and related health effects (Frison et al. 2006b, 2011; Powell et al. 2015; Vincenti et al. 2008).

In this context, and partly as a reaction to mainstream efforts to address micronutrient deficiencies through fortification (e.g., iodine in salt), supplements (i.e., high doses of vitamin A) or biofortification (i.e., increasing staple crops' micronutrient content), some researchers have argued that enhancing agrobiodiversity in systems could be an adequate approach to prevent "hidden hunger" and related health effects (Frison et al. 2011; Powell et al. 2015; Ruel 2003). The increase of species and varietal diversity in a system's agrobiodiversity is assumed to increase dietary diversity, which would not only provide adequate micronutrients but also boost the ingestion of foods containing phytochemicals with discrete bioactivities toward human biochemistry and metabolism, or nutraceutical foods (Carlos et al. 2007; Dillard and German 2000). Research suggests that the ingestion of a diversity of phytochemicals contained in cultivated and noncultivated plants—characteristics of local culinary traditions from agrobiodiverse-rich systems—enhances physical health (Heinrich et al. 2005; Pieroni et al. 2005), although variations should be expected across case studies.

While the benefits of a diversified diet continue to be productively examined, we still lack the empirical research to link agrobiodiversity and dietary diversity (and therefore nutrition and health) at the local level. According to recent reviews (e.g., Powell et al. 2015), relatively few empirical studies have tested the links between on-farm agricultural diversity and diversified household dietary choices (e.g., Jones et al. 2014), and not all of them report a positive association. Furthermore, the effect of intraspecific diversity has not yet been thoroughly explored. The weak evidence that has been found to link agrobiodiversity and dietary diversity has theoretical and methodological explanations.

At a theoretical level, many confounding factors (other than on-farm agrobiodiversity) shape dietary choices of households. For example, even most self-sufficient households participate in markets, either by supplying labor or by acquiring agricultural inputs or foods, a situation that certainly might affect producers' dietary choices in both directions (i.e., facilitating access to both nutritious and nonnutritious foods and beverages).

At a methodological level, authors have argued that research on the topic needs to be methodologically stronger, including long-term data and rigorous monitoring through impact evaluations (Jaenicke and Virchow 2013). The field also needs to develop proxies of dietary diversity that can be better matched with proxies of agrobiodiversity; at the moment they both use different taxonomies. Dietary diversity is often measured through intake of species in a set of food groups (i.e., at the species level). Agrobiodiversity, however, includes other levels above (i.e., landscape) and below (i.e., varietal diversity) the species level. The development of a taxonomy that allows correlations to be drawn is a prerequisite to exploring potential links between the various levels at which agrobiodiversity and health can be measured.

Beyond nutrition, there are many other potential pathways through which agrobiodiversity might relate to physical health. For instance, the maintenance of high yields in simplified agricultural systems generally requires the use of external inputs with proven detrimental effects on the environment (e.g., air, water, soil, biodiversity) and physical health of individuals (Lang and Heasman 2004). The use of pesticides constitutes a major health problem, causing death, acute and chronic neurotoxicity, lung damage, chemical burns, infant methemoglobinemia (caused by ingestion of nitrates in drinking water), various cancers, immunological abnormalities, as well as adverse reproductive and developmental effects (Eddleston et al. 2002; Weisenburger 1993). Because agricultural systems which aim to enhance agrobiodiversity generally do not heavily rely on such external inputs as pesticides (Altieri et al. 2012), they might directly contribute to physical health by reducing producers' exposure to agrochemicals. Such an impact, moreover, could extend beyond the farmer's level and reach the whole society, as has been shown in research associating the use of antibiotics in livestock farming and the increase in antibiotic resistance among bacterial pathogens (Mathew et al. 2007; Shea 2003).

Agrobiodiversity-rich systems might also contribute indirectly to physical health because of their overall positive environmental impact. It has been argued that the physical environment in which a person lives is an important determinant of health: access to clean water, sanitation, and diverse productive ecosystems has a positive impact on health and well-being (Pinstrup-Andersen 2009; Schmidhuber and Tubiello 2007). It has also been shown that industrial agriculture exerts an enormous environmental impact: emissions of greenhouse gases contributing to climate change, land-use change leading to deforestation, salinization of soils due to over irrigation, and water and land pollution from nitrogen and phosphorous fertilizers (Tilman 1999; see also Chapters 7 and 8). So, one could argue that agricultural systems that have lower environmental impacts, such as sustainable agrobiodiverse systems (Altieri et al. 2012), might positively impact physical health through the promotion of healthier environments. However, just as it is true that agrobiodiversity-rich systems are not always managed in an environmentally sound manner, monocultures do not always leave highly detrimental environmental footprints (i.e., cultivated pastures). This makes it reasonable to argue that more research is needed to elucidate the conditions under which positive health effects of environmentally managed agroecosystems can be expected.

Finally, another potential pathway that links agrobiodiversity to physical health is physical activity. The management of diversified landscapes—which include a diversity of crops (i.e., minor grains, pulses, fruits, vegetables, and root and tuber crops in addition to common staple crops) with different management techniques and various requirements at different points of time—might require greater physical effort than the management of more simplified agricultural systems, which are easier to mechanize (although this might not always be the case). Higher physical effort maintained throughout the year might, in turn, affect physical health. Research into the health impacts of gardening, for instance, has reported benefits to physical health, especially among the elderly (Litt et al. 2011). Since this study, however, assessed the impact of leisure gardening in a sample of people new to gardening, results are not easily generalizable to the continuous management of larger agroecological units. Nevertheless, future research could replicate the approach to compare farmers managing farms with different levels of agrobiodiversity. Again, we note that the relationship can go both ways: manually managing farms with high agrobiodiversity could promote a healthy level of physical activity, but such activity could also create a work overload with negative health effects. Moreover, health effects from an increased workload could vary across groups (e.g., women, children, men). Research is needed to identify the conditions under which positive health effects of physical activity associated with the management of agrobiodiverse systems can be expected.

Pathways between Agrobiodiversity and Mental Health

WHO defines mental health as the state of well-being in which an "individual realizes his or her own potential, can cope with the normal stresses of life, can work productively and fruitfully, and is able to make a contribution to her or his community." Anxiety, depression, and stress are common mental health disorders. Are there any links between agrobiodiversity and mental health? The answer is: we mostly do not know. Research on this the topic is largely nonexistent, perhaps because of important methodological challenges not only in measuring a potential link between agrobiodiversity and mental health, but also in attributing a causal link to the relation. Although empirical research on the topic is largely nonexistent, we propose that there are potential links worth exploring. Some of the links explored below might not be exclusive to agrobiodiversity-rich systems; they could also be explored in many other agricultural systems (e.g., urban gardens). Thus, empirical research is needed to determine whether the proposed links exist and manifest differently in agricultural systems with varying levels of agrobiodiversity.

Perhaps the most important link between agrobiodiversity and mental health relates to feelings associated with access (or lack of access) to sufficient food, to food that is considered culturally adequate, and to culturally acceptable productive landscapes. Using the FAO's concept of food security as the continuous and affordable access to nutritious, culturally appropriate foods (FAO 1996), researchers working in urban and rural settings have established a causal relation between food insecurity and mental morbidity, including anxiety and depression. This research shows that people who experience food insecurity, especially pregnant women or women with young children, are more likely to report depressive episodes or anxiety disorders (Hadley and Patil 2006; Whitaker et al. 2006). As it has been argued, if agrobiodiversity-rich systems are more able to ensure food security at the household level by providing continuous access to local foods, then agrobiodiversity-rich systems could help protect mental health by reducing the stress caused by situations of food insecurity.

Furthermore, there are reasons to think that the lack of access to food considered culturally adequate can affect mental health in the same way that lack of access to sufficient food does. Dietary choices are deeply embedded within cultures and ways of living; they are a function of different socioeconomic processes that range from culture to class, income, age, or profession (Greenberg 2003; Kuhnlein et al. 1996; Noack and Pouw 2015). In this sense, it has been argued that communities that value their traditional food systems—foods people know and have access to from their local environment through farming or wild harvesting (Kuhnlein et al. 2009)—are better able to conserve local food specialties and the associated agrobiodiversity (Chapter 12). Moreover, these communities exhibit a lower prevalence of diet-related chronic diseases (Johns et al. 2013). Do they also show less prevalence of mental health problems associated with food insecurity? We just do not know.

Conversely, the stigmatization of dietary choices (e.g., traditional foods being considered "backward" or low in status) might generate negative feelings (i.e., shame) among those who identify with these foods (Cruz-Garcia and Howard 2013). Research shows that stigmatization affects dietary choice and that people sometimes abandon the production and consumption of stigmatized foods, even when they continue to be locally available (Bharucha and Pretty 2010; Reyes-García et al. 2015). We know that such behavior has adverse effects on nutrition, as happened with the reduction of pearl millet consumption in African countries (Johns et al. 2013). Does stigmatization also affect mental health? Again, the answer is: we do not know. Certainly, more research is needed.

The argument can be extended to include considerations related not only to the food system but also to the cultural knowledge associated with it. To a high degree, agrobiodiversity-rich systems are knowledge intensive: to maintain agrobiodiversity, complex information on species selection, combination, and management is required (Altieri et al. 2012). Consequently, the management of agrobiodiversity-rich systems is often dependent on traditional ecological knowledge, or the cumulative and evolving body of knowledge, practices, and beliefs held by communities about their relations with the ecosystems in which they are embedded (Berkes et al. 2000; Kuhnlein et al. 2009, 2013).

Traditional ecological knowledge, however, is not only essential for the creation and maintenance of biodiversity-rich cultural landscapes, it is also a cornerstone of local cultural identities (Barthel et al. 2013). Existing research shows that the loss of cultural identity is a very important determinant of mental health both among migrants (Mossakowski 2003; Schwartz et al. 2010) and Indigenous Peoples (Kirmayer et al. 2000; Kral et al. 2011). Does the loss of agrobiodiversity-rich systems, and the associated loss of traditional knowledge and management practices, affect the mental health of knowledge holders? Do changes in knowledge systems, dietary choices, and agrobiodiversity associated with processes of human migration (Chapter 8) relate to mental health? Future research is needed to address these issues.

We emphasize the importance of adopting a gender perspective in pursuing such lines of enquiry. Since men and women assume different roles in various agricultural production systems, they might be affected in different ways by changes. The gendered distribution of agricultural work related to the management of agroecosystems might impact the physical health of women and men disparately. The same argument could be constructed around mental health. For instance, many agrobiodiversity-rich systems are highly dependent on women's role as seed custodians (Howard 2006; Zimmerer 2003b; Zimmerer et al. 2015). The loss of agrobiodiversity-containing systems might have a differentiated gendered effect: women might suffer more, in terms of mental health, through the loss of self-esteem associated with being marginalized in decisions relative to production and income use (Ravera et al. 2019). Effects of agrobiodiversity changes on physical and mental health might be context specific, but adopting a gender perspective might elucidate patterns of differentiated effects within a single case study.

Agrobiodiversity-Based Social Networks as Pathways to Physical and Mental Health

Social network research typically characterizes the web of social relations around an individual, including those with whom a person relates and how (Smith and Christakis 2008). Despite the growing evidence that the composition and structure of social networks affect an individual's physical and mental health (Smith and Christakis 2008; Valente 2010), and despite the growing body of knowledge which highlights the importance of social networks in understanding agrobiodiversity management (Calvet-Mir et al. 2012a; Reyes-Garcia et al. 2013; Ricciardi 2015), the role of social networks in explaining the association between agrobiodiversity and health remains completely unexplored. The main argument is that social networks related to the production and consumption of agrobiodiversity (i.e., agrobiodiversity-based social networks) create pathways which propagate attitudes, behaviors, and emotions, as well as financial, physical, informational, labor, and social resources (Perkins et al. 2015) which, in turn, could affect health. Based on the work of Berkman and Glass (2000), one could hypothesize that there are four main pathways through which agrobiodiversity-based social networks might mediate an association between agrobiodiversity and health:

- Access to resources (e.g., seeds, information; see Chapter 13)
- Social influence (e.g., spread of nutrition related behaviors)
- Social engagement
- Provision of social support (both perceived and actual)

Agrobiodiversity-based social networks may be particularly vital to health by being a *source of resources* (e.g., seeds, stems, associated knowledge), which could be critical in times of food stress or insecurity (Calvet-Mir et al. 2012a; Reyes-Garcia et al. 2013; Ricciardi 2015). Research among farmers in the Catalan Pyrenees reveals that networks of seed exchange act as human corridors to facilitate the flow of local landraces and associated knowledge (Calvet-Mir et al. 2012a). Moreover, farmers who hold positions of centrality in agrobiodiversity-based social networks are also more likely to maintain local landraces and associated knowledge (Kawa et al. 2013; Reyes-Garcia et al. 2013), acting as seed banks in case of need (Coomes et al. 2015). Therefore, because they provide access to material resources and information, agrobiodiversity-based social networks could be critical to ensure agrobiodiversity and germplasm conservation, and consequently to enhance food security (Chapters 13 and 14).

Agrobiodiversity-based social networks could also mediate the relation between agrobiodiversity and physical and mental health through *social influence*, or the spread of ideas and behaviors within and between communities (Valente 2010). Haselmair et al. (2014) have shown the importance of social networks in spreading food-related behaviors among migrants, and Zimmerer (2010) found how preferences for traditional, local dietary items spread among people who support local foods, organic agriculture, fair trade, and multifunctional agriculture, thereby reinforcing biological diversity in agriculture. Such findings suggest that social influence could mediate the spread of ideas and behaviors related to agrobiodiversity, which in turn could ultimately relate to physical and mental health outcomes through, for instance, the adoption of a more diverse and culturally accepted diet.

Another potential pathway through which agrobiodiversity might relate to health through agrobiodiversity-based social networks is *social engagement*, or one's degree of participation in a community or society (Chapter 12). The maintenance of landscape agrobiodiversity is often done through social structures and institutions that require social interactions. For example, farmers need to be embedded in a social network to exchange seeds and agricultural products: those exchanges might also confer a source of personal status and satisfaction (Hardon-Baars 2000). Similarly, the management of common resources (i.e., water, forest, pastures) requires coordination between people and has given rise to many common management systems (Ostrom 1990). Such institutions provide ample opportunities for social engagement and social activities that ultimately might promote physical and mental health (Berkman et al. 2000).

Finally, *social support*—the perception that one is cared for, has assistance available from other people, and is part of a supportive social network—is one of the most well-documented psychosocial factors influencing physical and mental health (Berkman et al. 2000; Uchino 2009). We argue that participation in social networks of seed exchanges or other agrobiodiversity-based social networks (sometimes fostered by information communication technologies) might provide a range of supportive resources, including emotional (e.g., nurturance), tangible (e.g., seeds), informational (e.g., advice), or companionship (e.g., sense of belonging), which might ultimately relate to both physical and mental health. We know of no research that has directly addressed the linkages considered in this section and thus recommend their inclusion in future research programs.

Again, while we know of no research directly addressing those links, we consider that the different topics outlined in this section should be part of the agenda linking agrobiodiversity and health in a comprehensive way.

Strengthening the Link between Agrobiodiversity and Health: The Role of Local Decision Making and Food Sovereignty

Having explored links through which agrobiodiverse-rich systems could relate to the physical and mental health of individuals, households, communities, and local landscapes, we wish to broaden the discussion by asking:

- Is agrobiodiversity really a local choice?
- What are the elements that influence the presence or absence of agrobiodiversity-rich food systems (and its potential effects on health) at the local level?

We begin by placing the reduction of biodiversity in a historical context: in many agricultural production systems, reduction occurred as the direct result of policy measures and development paradigms that were aligned with productivist approaches, which reinforced top-down agricultural decision making (see Bonanno et al. 1995; Friedmann and McMichael 1989; McMichael 2009). Together with the spread of a neoliberal globalized food system, such measures resulted not only in agrobiodiversity loss, but in a generalized loss of local power to control and hold authority in decisions related to the food system (Otero 2012; Wolff 2004). Although some localized agricultural systems remain diversified and controlled by communities (e.g., Baker 2008; Chase Smith et al. 2013), the trend of the past century has been toward intensification and centralization of food systems (Friedmann and McMichael 1989).

In Europe, for example, after World War II, a series of centralized policy measures (included in the new Common Agricultural Policy) were designed to enhance the productivity of farms by coupling economic support with agricultural production regardless of economic, social, and ecological impacts (Gatto et al. 2013). In a nutshell, the new agricultural paradigm promoted high-yield crop selection, crop homogenization, and the standardization of agricultural products while removing decision-making power from the farmers through the concession of subsidies to specific crops (Wolff 2004).

For some time now, social movements created by farmers (e.g., La Via Campesina) and consumers (e.g., Slow Food) have started to claim the need to restore some level of *food sovereignty*, or the downscaling of decision making related to agricultural production and consumption to local arenas (Chapters 12 and 13). Such claims contest the neoliberal globalized food system while stressing the need to enhance local knowledge systems through farmer-to-farmer networks and the promotion of agroecological innovations and ideas (Altieri and Manuel Toledo 2011; Chapter 4). Thus, as they strive for local food sovereignty leading to food systems that are more supportive of agrobiodiversity and that can reverse the simplification and industrialization processes, many local initiatives have withdrawn from general food policy and economic trends. For instance, the growing social movement for food sovereignty in Latin America and beyond (e.g., La Via Campesina) provides an integrated approach to local agricultural decision making and farmers' empowerment that has strong links to agrobiodiverse farming and that consequently might enhance some of the local pathways to health previously discussed. In urban scenarios representing the consumer side, initiatives such as food consumption groups can also be seen as enhancing these linkages, since these groups promote shorter food chains, higher farmer bargaining power, and alternative (organic) food production systems (Chapters 8 and 15). Both of these emerging initiatives, which could also be seen as shifters of the symbolic contexts in which local decisions take place, mobilize traditional ecological knowledge and enhance agrobiodiversity, factors considered by some as drivers of agroecological transition processes (López-García and Guzmán-Casado 2013).

In summary, despite the success of many of these civil society initiatives (Renting and Wiskerke 2010), we must bear in mind that local choices—specifically those that impact health (physical and mental) and food systems (production and consumption)—do not happen in a void: local decision making interacts and is affected by policies (global, national, regional) and economic trends. Thus, if public health policies continue to be viewed in isolation, we will miss the potential role for sustainable food systems in the health of individuals and communities and fail to strengthen linkages to agrobiodiversity.

Acknowledgments

Research leading to this chapter received funding through the Spanish Ministry of Economy and Competitiveness, both through the project CSO2014-59704-P and through P. Benyei's pre-doctoral grant (BES-2015-072155). A preliminary version of this text was used at the Ernst Strüngmann Forum on *Agrobiodiversity*; we thank the participants for their comments. We also thank L. Calvet-Mir, G. Cruz-Garcia, D. López-García, M.

Rivera, and T. A. Zwart for comments to a previous version of the manuscript. Reyes-García thanks the Dryland Cereals Research Group at ICRISAT-Patancheru for providing office facilities. This work contributes to the "María de Maeztu Unit of Excellence" (MdM-2015-0552).

Governance, Including Policy, Cultural, and Economic Frameworks

12

Indigenous Agrobiodiversity and Governance

Gabriel Nemogá

Indigenous Peoples have the right to self-determination. By virtue of that right they freely determine their political status and freely pursue their economic, social, and cultural development. —Article 3, United Nations Declaration on the Rights of Indigenous Peoples (2008)

Abstract

This chapter addresses agrobiodiversity as experienced by Indigenous Peoples in the context of current policy and legislation, and aims to broaden understanding of how current agrobiodiversity governance impacts Indigenous Peoples worldwide. To date, the ability of Indigenous Peoples to determine agrobiodiversity governance has not been fully recognized, and thus is not covered explicitly by international policy and legislation. This chapter develops a biocultural perspective to expand the inclusion of Indigenous Peoples' practices regarding agrobiodiversity—one that takes into account their worldviews and rights. It reviews the epistemological and political barriers that inhibit recognition of the distinct characteristics and relevancy of Indigenous agrobiodiversity; these barriers must be overcome before a research agenda can be advanced that truly contributes to the development of a consistent policy for agrobiodiversity use and *in situ* conservation. It also analyzes policy and legal instruments related to Indigenous agrobiodiversity in both international and national contexts. At the global level, self-determination for Indigenous Peoples has been recognized. Still, work remains to ensure that the role and contributions (past and present) of Indigenous Peoples to agrobiodiversity are recognized globally and nationally. Proper recognition and protection are necessary for the development of more robust approaches to a broad definition of agrobiodiversity governance, which will contribute to overcoming worldwide hunger and malnutrition.

Introduction

Researchers and nongovernmental organizations (NGOs) have increasingly documented and highlighted the valuable contributions of Indigenous farming practices to the potential of sustainability involving agrobiodiversity, as well as to the use and conservation of genetic resources pools (see also Chapters 6 and 8). This trend stands in stark contrast to the modernization and development programs of the past and their continuation in many places to the present day. Such programs have mistakenly assumed Indigenous Peoples' agrobiodiversity knowledge and cultivation practices to be misguided, often treating them as "backward" (e.g., the recurring treatment of Indigenous swidden farming and agrobiodiversity as old fashioned and unsustainable). At the same time, agroecological and ethnoecological evidence have demonstrated the valuable insights of Indigenous knowledge in diverse areas such as environment, medicine, health, cosmetics, and nutrition (Berlin and Berlin 2015; Brush et al. 1981; Dutfield 2010; Hayden 2003; Magalhães et al. 2011; Pati et al. 2014; Posey et al. 1984; ten Kate and Laird 1999). Once Indigenous knowledge was seen as providing useful ecological information (e.g., hints of the presence of valuable biochemical compounds in plants), applied researchers and bioprospecting missions targeted this knowledge and associated resources for collection, scientific validation, and commercial exploitation. Still, to date, Indigenous Peoples and their contributions to agrobiodiversity have not been recognized for their full value, even though substantial (albeit partial) changes have been registered in governance initiatives, such as the Convention on Biological Diversity (CBD), the International Treaty for Plant Genetic Resources in Forestry and Agriculture (ITPGRFA), and the Nagoya Protocol.

The biocultural framework offers a culturally appropriate approach to document, recognize, and guarantee the rights of Indigenous Peoples on their practices, innovations, and traditional knowledge (Nemogá 2016) and is used in this chapter to rethink agrobiodiversity governance. The distinctive characteristic of the biocultural framework is the recognition and respect for the worldviews of Indigenous Peoples as they interact with their ecosystems, which often include food-producing landscapes with high agrobiodiversity. Programs and local conservation initiatives in Andean ecosystems, inspired by this framework, have developed the notion of Indigenous biocultural heritage, which acknowledges ancestral practices and customs (IIED 2016; Swiderska 2006; Swiderska et al. 2009). The concept of biocultural heritage has the potential to orient practical actions for *in situ* conservation of agrobiodiversity (Argumedo and Pimbert 2006), although significant challenges for implementation remain, (e.g., Indigenous governance and community inclusivity in complex initiatives involving a wide range of social actors).

In undertaking the research and documentation of agrobiodiversity in the lives of Indigenous Peoples, the biocultural approach stresses the need to recognize the customary laws and land rights of Indigenous Peoples and local communities as well as the community worldviews that give meaning to community practices and relations with the environment (Nemogá 2016). This approach has been used to document cultural diversity as well as to orient biological conservation initiatives (Gadgil et al. 1993; Gavin et al. 2015; Gorenflo et al. 2012; Infield 2001; Maffi 2005). It needs, however, to be advanced further to influence policy and legislation of biodiversity on diverse levels. Here, I focus on overcoming the insufficiencies currently associated with the majority of international and national policy and legislation frameworks (see also Chapter 13), which threaten agrobiodiversity of Indigenous Peoples with erosion and misappropriation. These difficulties are partially explained by epistemological barriers and political interests that deserve critical attention.

Indigenous Agrobiodiversity: Challenges and Conflicts

Indigenous Extensive Family and Community: The Epistemological Barrier

Posey (1999b) pointed out epistemological barriers that hinder academic researchers and development agents from attaining true understanding and respect for the ways of living and knowledge of Indigenous Peoples. Perspectives could vary between two different types of relationships and understandings about nature that affect the agrobiodiversity field. The first encompasses large-scale production and urban market-oriented contexts that rely on a techno-economic approach that separates humans and nature. This separation is instrumental to postulate neutrality and objectivity in the knowledge creation process. For the researcher, detachment from the object of study is a key requirement: nature, as the object, is assumed to be characterized, fragmented, and analyzed into component parts so as to permit control over their functioning and to use them productively. This perspective underlines a rational understanding and controlling of nature for economic exploitation; in large-scale agriculture, for example, seeds are reduced to a raw input for agribusiness.

The second embodies a nature–human unit that embraces intimate interrelationships between Indigenous People and their seeds, other living organisms, and landforms. In terms of an Indigenous worldview, Posey (1999b:4) states:

> Knowledge of the environment depends not only on the relationship between humans and nature, but also between the visible world and the invisible spirit world.

Humans are intrinsically intertwined with nature, and the practice of mutual relations of coexistence generates knowledge about plants, animals, and other local components of nature, such as climate (see Chapter 7; Nemogá 2016; Pierotti 2011). From Indigenous Peoples' perspectives, respect, intimacy, and ceremonies characterize their experiential interactions with seeds and the natural world, rather than detachment and the quest for rational control of nature (Machaca 2016).

Similar characterizations of Indigenous worldviews and ways of living are recurrently reported in different parts of the world (Berkes 2012; Lloyd et al. 2012; Plenderleith 1999). For Indigenous People in the Andean region, for instance, the notion of *Kawsay* in Quechua "integrates and fuses the natural realm (*pacha*) and the human–social world, in contrast to the separateness of nature and culture in the classical tradition of modern Western thought, including many strands of environmental studies and resource management" (Zimmerer 2012:603). The sensorial interaction of nature guiding Indigenous Peoples' agricultural practices is found in other geographies. Salas (2005:17) describes how Indigenous communities on the Thailand–Myanmar border rely on their sensorial appreciation of the external world, rather than on their rationality when working their agrobiodiversity:

> The classification, selection, conservation, and reproduction of seeds relies on senses: touching and smelling to see whether seeds are healthy, distinguishing the colors of varieties, spotting the particular temperature and quality of the soils where the seed will grow, interpreting signs in the behavior of the birds or the weather.

From the perspective of Indigenous worldviews, intuition and senses rather than reliance on purely instrumental technocratic modes of knowledge take priority when relating to plants and animals. In Western nomenclature, this is a relational ontology or metaphysical holism that describes human–nature relationships as an extensive family or community, including other humans as well as animals, plants, landforms, water bodies, rocks, and deities and agrobiodiversity such as seeds (see Chapter 13; Keller 2009). This approach coincides to a certain extent with the central normative principle of deep ecology regarding the intrinsic value of nature: every action that demeans, disturbs, or affects the integrity of nature also affects humans, because the human species is not conceived as a discrete component, separate from nature (Leopold 1966; Naess 1973).

Indigenous Ways of Living and Land Rights: The Political Conflict

For political reasons, the protection and conservation of Indigenous agrobiodiversity has been limited, as has full recognition of its legitimate creators. Recognizing and defining Indigenous Peoples is politically loaded work. *Indigenous, Native, Aboriginal, Indian,* and *First Nations* are terms with unavoidable political content, because they generally entail collective ancestral rights to land and resources that challenge the political and economic interests of dominant classes within and among countries. The particularities of some countries—where state policies deny the very existence of Indigenous Peoples (e.g., Cuba, Laos) or where Indigenous People govern themselves in autonomous regions (e.g., the Inuit Peoples in Nunavut, Canada)—add tensions to any attempt at a universal definition. In 1993, disagreement within the United Nations Working Group on Indigenous Populations prevented a general consensual definition of Indigenous as they worked toward the UN Declaration on the Rights of Indigenous People (UNDRIP) (Daes 1996; UN 2008).

Working criteria for identifying Indigenous Peoples were introduced in 1989. Article 3 of the 1989 International Labour Organization (ILO) Convention No. 169 established objective and subjective factors for identifying tribal and Indigenous Peoples in independent states when specifying the scope of the Convention. Under Article 1b, it defines Indigenous Peoples as "peoples in independent countries who are regarded as Indigenous on account of their descent from the populations which inhabited the country, or a geographical region to which the country belongs, at the time of conquest or colonization or the establishment of present state boundaries and who, irrespective of their legal status, retain some or all of their own social, economic, cultural, and political institutions." Ancestral origin from precolonial inhabitants, occupancy of geographic homelands, and social, cultural, and political continuity are factors that can be verified objectively. In addition, Article 1.2 of the ILO 169 includes self-identification as a subjective factor that is fundamental to determine the application of ILO Convention 169. Overall, this Convention provides strategic criteria about Indigenous Peoples and aims to prevent states from denying Indigenous identities within their borders. The element of self-identification is a step forward in supporting the rebuilding of Indigenous nation processes that otherwise would be obstructed by a strict definition. At the same time, the Convention offers objective criteria to prevent false and utilitarian claims about indigeneity by other minority groups within national borders and internationally.

Other international institutions have developed operative definitions. The World Bank recognizes that Indigenous Peoples maintain intrinsic relations with their lands and depend on the use of natural resources (World Bank 1991, 2005); however, its operational definitions include the existence of Indigenous language as an identifier, despite the fact that many Indigenous Peoples have lost their language. Their definitions also require actual territorial links to lands and natural resources, though exception is noted in case of forced severance. CBD's efforts to distinguish clearly local communities and Indigenous Peoples did not succeed in adopting an enforceable distinction (UN 2014a, b). For its part, the Food and Agriculture Organization (FAO) developed multiple criteria to guide its work with Indigenous Peoples, including priority in time of territorial occupation, the perpetuation of cultural distinctiveness (without restricting it to a single cultural aspect such as the language), self-identification, and a past or current experience of subjugation, dispossession, or marginalization (FAO 2010a). Indigenous academics have proposed alternative conceptualizations via the notion of peoplehood as a framework for self-identification (Corntassel 2003). Peoplehood—a concept developed by anthropologists and Indigenous scholars—underpins the connections of Indigenous Peoples with land, spirituality, and language. Further, Corntassel (2003) proposed a flexible definition that centers on self-determination and highlights peoples' connections with sacred history, ceremonial cycles, language (spoken or not), and ancestral homelands. Overall, the ILO Convention 169 definition is the most widely accepted in international fora.

The political consensus that only Indigenous People should be entitled to decide on Indigenous membership stands in contrast to national policies and practices. Although countries are moving toward self-identification criteria for census purposes (CEPAL 2014), they endorse different criteria for recognizing Indigenous identities, as demonstrated in Latin America. In Mexico, Indigenous identities were strongly associated with speaking a native language (INEGI 2000, 2003), whereas in Colombia the government exercises the power to recognize the Indigenous status of communities. When there is doubt about the Indigenous identity of a collectivity, Colombian law orders ethnological studies to determine the legitimacy of Indigenous identity (Decree 2164 of 1995, Article 2.1).

Legal and administrative manipulation of Indigenous definitions by states produces an unexpected rise or fall in the Indigenous population. Consequently, Indigenous and mestizo (i.e., mixed ancestry) populations could shrink or expand as a result of administrative measures. In Bolivia, for example, the Indigenous population was set at 62% in 2001 but fell to 40% in the National Census on Housing and Population of 2012, after eight years of Evo Morales's government (INE 2013). Explanations for the 2012 outcome included changes in the survey design, which forced Bolivians to identify with one of the 36 Indigenous nationalities or reject their Indigenous ancestry by choosing none; the mestizo category was not included for self-identification (CEPAL 2014).

Mestizo populations are the result of mixed racial backgrounds and have been targeted by dominant classes for the liberal nation-building projects in Latin America. Marchi (2018) describes how Indigenous identities were denied during the post-agrarian reform of 1917, forcing the adoption of mestizo as a national ethnic identity in Mexico. In a general sense, mestizos have combined European and Indigenous ancestry.[1] In a racist hierarchic structure, mixed people of African and Indigenous ancestry are not considered mestizo. In the majority of the rural agrarian communities in Latin America, Indigenous cultural ancestry is apparent in community social customs, family relations, and agricultural practices, which include cultivation of native seeds inherited from Indigenous ancestors. Despite phenotypical characters, Indigenous surnames, and other objective factors, some mestizos do not self-identify as Indigenous; instead, they willingly reject their Indigenous lineage and self-identify with a monoethnic national culture, such as Bolivian, Colombian, Mexican, Peruvian (CEPAL 2014; De La Cadena 2000; Hernández 2001; Villarreal 2014). Openly

---

[1] Contrary to Latin America, descendants from Indigenous Peoples and European settlers constitute the Métis nation in Canada. One of the aboriginal peoples recognized under the Constitution Act of 1982, the Métis National Council defines a Métis as "a person who self-identifies as Métis, is distinct from other Aboriginal peoples, is of historic Métis Nation Ancestry and who is accepted by the Métis Nation."

or subtly, mestizos deny their Indigenous genetic and cultural background because of the social and racial prejudice associated with the Indigenous being identified as poor and backward people. Mestizos live and want to live more aligned with the predominant social, economic, and cultural patterns promoted by the liberal state. The extent of self-identification as Indigenous can be an indicator of success or failure of the assimilationist policy for de-Indianization through old and new Christianization enterprises.

What is relevant in this discussion is that ethnicity influences Indigenous agrobiodiversity in diverse ways. Mestizo family farmers have been found to be more inclined to adapt agrotechnological packages (e.g., high-yield seeds, fertilizers, and financial credits) and are generally more connected to market exchanges by selling their labor power or their produce outside their communities. In Chiapas, for example, Brush and Perales (2007) found that mestizo families were more integrated into the market than Indigenous Mayan families. Additionally, the ethnic affiliation of Indigenous and non-Indigenous families was more explanatory of maize race distribution than environmental factors because "significant differences exist between the two ethnic groups in the distribution of maize races, types, colors, and seed systems, and the ethnic differences are significant regardless of environment" (Brush and Perales 2007:219).

Morphological and agronomic differences in maize were previously correlated with different communities as well as linguistic groups and still continue to demonstrate these associations among certain Indigenous Peoples. For example, two distinct maize races have been found to correspond to two different Mayan groups: the Tzotzil and Tzeltal (Perales and Hernández 2005).

In summary, Indigenous Peoples should not be conflated with mestizo farmer communities. Estimates of the extent of the Indigenous population in Latin America include 826 Indigenous Peoples (i.e., collectives once termed Indigenous or ethnic groups), totaling 44.8 million individuals and accounting for about 8.3% of the total population in 2010 (Mato 2016). That year, estimates regarding the representation of the Indigenous population in each country varied substantially, from 0.5% in Brazil to 62.2% (2001 data) in Bolivia (CEPAL 2014). Notably, a national census does not always accurately include Indigenous Peoples in urban settings. Thus, these figures cannot be taken as absolute, since they reflect diverse and shifting administrative criteria in each country. However, the existence of at least 826 recognized Indigenous Peoples in this part of the world highlights the wealth of knowledge, practices, innovations, and plant diversity that need to be protected and conserved. For the purposes of discussing Indigenous agrobiodiversity, the definition in ILO Convention 169, including objective and subjective criteria, serves as an important reference for the usage in this chapter. Alternative definitions from Indigenous scholars have not reached universal consensus and will continue to be challenged in international fora as lacking legal support.

With this backdrop, let us now discuss a definition of agrobiodiversity that captures Indigenous worldviews.

Defining Indigenous Agrobiodiversity

Many definitions of agrobiodiversity are limited to a biological perspective. To overcome the inherent limitations of a strictly biological approach, I argue that agrobiodiversity must also be framed in terms of Indigenous epistemologies:

- Humans must be understood as part of an extended relational community of animals, plants, and spiritual entities.
- The connection between biological and cultural diversity is essential.

Agriculture can provide "balance for well-being through relationships not only with people but also with nature and deities" (Posey 1999b:5). Indigenous agriculture also brings a different sense of connection to the territory, plants, and animals, as Salas (2005:20) illustrated:

> So territory and this Indigenous [swidden] practice are inextricably linked. This is what the Karen mean when they say that swidden agriculture is a way of life.

Moreover, what external observers see as "pristine nature," "wild," or "primitive" in Indigenous landscapes is actually mediated by human action. "Wild" and "wilderness" are inappropriate terms for plant and animals in Indigenous territories; these terms "imply that these landscapes and resources are the results of 'nature' and as such have no owners—they are the 'common heritage' of all humankind" (Posey 1999b:8). By extension, the external observer assigns no value to Indigenous knowledge associated with biodiversity and assumes that it is free for collection and exploitation.

In this context, Indigenous agrobiodiversity can be defined as the diversity of plants, animals, insects, microorganisms, landform, and deities and their interactions with peoples who self-identify as descendants of those who inhabited their territories since precolonial times and who retain some of their social, economic, cultural, and political institutions. As such, it is not restricted to plant material, generally termed traditional (folk) crop varieties. Agrobiodiversity is embedded within a biocultural framework that includes Indigenous epistemologies and customary law as they emerge from ancestral productive practices of peoples that preserve and nurture meaningful and sacred interrelationships with nature (Nemogá 2016).

Disregarding Indigenous Peoples in Agrobiodiversity

Although researchers have documented crop genetic diversity preserved in Indigenous territories by native peoples' practices (e.g., Bellon 1991; Brush 1995; Perales and Hernández 2005; Zimmerer 1997), the use of the distinctive term *Indigenous* (in the sense of ILO Convention 169) to indicate ancestral people in their territory is barely found in the descriptions and studies of agrobiodiversity in the American hemisphere. An ad hoc de-Indianization of the Indigenous population occurs when the term *farmers* (or campesino or peasants) is broadly applied to native people who keep alive the core of ancestral agricultural practices in Andean or Amazonian territories. For example, Canahua-Murillo (2016) describes the project Ingenious Systems of World Agricultural Patrimony in Puno, Peru, as aiming to understand the evolution and adaptation to the environment of the rural Andean societies. Although the communities under study are the descendants of prehispanic Aymara and Quechua peoples, Cahahua's description speaks of peasant communities rather than Indigenous communities. This de-Indianization is apparent in rural landscapes, as evidenced by the Peruvian agricultural census of 2012: in three out of four communities surveyed, the use of the Indigenous languages Quechua or Aymara was common, half of the communities had Indigenous traditional authorities, and three of every five communities obtained their land via ancestral inheritance (CEPAL 2014). In Andean agrobiodiversity studies, language, belief systems, and ancestral relations to the territory were not underlined as ethnic identifiers, but generalized as features of an Andean population hardly different from mestizo Spanish-speaking communities, or "campesino communities."

Paradoxically, even researchers working within a framework for cultural reaffirmation describe native communities as peasants, Andean, or Andean–Amazon peasants rather than as Indigenous (Ishizawa 2010, 2016; Machaca 2016). One illustration of how the term campesino is preferred is the Andean Project of Campesino Technologies (PRATEC in Spanish), whose vision is expressed as the recuperation of campesino technology, not Indigenous technology. This is noteworthy because PRATEC's initiative and vision center is the "revitalization of collective ceremonies associated to the recuperation of Andean wisdom for the cultivation of agrobiodiversity" (Machaca 2016:353). PRATEC's cultural affirmation emphasizes a general Andean culture, rather than the Indigenous cultures that domesticated and preserved the diversity of crops and languages in the Andean–Amazonian region (Shepherd 2010:632).

The blurred description of Indigenous Peoples contributes to the neglect of the role of Indigenous Peoples' agrobiodiversity. This is the case when Indigenous identities are described under a general descriptor instead of distinctive Indigenous Peoples. Activist organizations and scholars choose to use the term "Andean identity" despite focusing on Indigenous technologies, belief systems, knowledge, and the agricultural practices of native peoples. There are also abundant studies where the subjects of study are specifically mestizo communities, though the authors' preference for naming communities as mestizo or Indigenous could be contested (De La Cadena 2000). As suggested by Abbott (2005), the conflation of ethnicity with indigeneity—and I would add with general identifiers like *Indigenous farmers, campesinos,* or *peasants*—"simplifies the multifaceted origins of landrace varieties in the Americas, limiting our options for conservation programs" (Abbott 2005:199). Such fusion obscures the distinctive Indigenous culture and Indigenous Peoples' contribution to major crops for food and agriculture.

Postmodern trends add to vanishing Indigenous categories. Through a postmodern lens, Indigenous agriculture and cosmologies are seen as remnants associated with nonreal campesinos (or as a romantic Indigenous characterization). The category of Indigenous Peoples is "socially constructed, not innately given" (Sawyer and Terence-Gomez 2013:9). Postmodern scholars have built a consensus where culture is an academic no-no (Shepherd 2010:629); they emphasize Indigenous identity as a fluid, changing, hybrid cultural assemblage (Sawyer and Terence-Gomez 2013). One result is the hybrid mestizo–Indigenous notion that engulfs Indigenous identities and deconstructs Indigenous ways of living, thus obscuring their role in the generation and preservation of agrobiodiversity.

The neglect of Indigenous agrobiodiversity is manifest when the recognition of traditional knowledge underlines traditional farmers' rights without specific attention given to Indigenous Peoples' rights to their plant varieties and knowledge; a substantial change is required in policy on conservation of agrobiodiversity to recognize Indigenous ways of living, their cosmology, knowledge, and belief systems, and the interconnectedness with their territory and agrobiodiversity. As will be described, up to now, policies and legislation on protection and conservation of agrobiodiversity have excluded a distinctive protection of Indigenous Peoples' agrobiodiversity.

Challenges Posed by Policy and Legislation on Indigenous Agrobiodiversity

International Level

Viewed historically, Indigenous collective and inalienable rights to land, territory, resources, and cultural integrity were ignored within the human rights framework. Although post–World War II human rights instruments postulated equality and freedom for all human beings, this conception of human rights, based on liberal principles, did not establish special protection for Indigenous Peoples. The Universal Declaration of Human Rights (1948), the adoption of the International Convention on the Elimination of All Forms of Racial Discrimination (1965), and the International Covenant on Civil and Political Rights (1966) all focused on individual civil and political rights: rights to autonomy, dignity, physical integrity, freedom, and security, among others, were applied to individuals but not to Indigenous Peoples as collective subjects.

In 1957, as the Green Revolution was at its zenith with its emphasis on the adoption of high-yielding varieties, agrochemicals, and mechanization for agricultural development (see also Chapter 6), ILO Convention 107 advocated the full integration of Indigenous Peoples to mainstream agricultural production. Later, in 1989, when world consensus was coalescing around the triumph of liberal democracy and the end of history, the ILO approved a new Convention (No. 169) that recognized Indigenous and tribal peoples' collective rights in independent countries. In its preamble, the Convention acknowledged "the distinctive contributions of Indigenous and tribal people to the cultural diversity and social and ecological harmony of humankind." The ILO Convention 169 entered into force in 1991 and, as of 2017, has been ratified by only 22 countries (ILO 2017). It explicitly identified the need to overcome the assimilationist orientation of ILO Convention 107.

Thus, ILO Convention 169 was the first international legal instrument that incorporated collective rights on lands, resources, cultural identity, and the duty to consult with Indigenous Peoples on projects that affect their territories or livelihood. Although this Convention did not refer to agrobiodiversity specifically, it commits its parties to recognize and respect Indigenous Peoples' path to development consistent with their social, economic, cultural, and political institutions and irrespective of their legal status. This protection is developed in other articles referring to their lands, own way of living, and system of beliefs. Article 2 established that states' actions shall promote "the full realization of the social, economic, and cultural rights of these peoples with respect for their social and cultural identity, their customs and traditions, and their institutions." With regard to control of their economic, social, and cultural development, article 7 includes the right "to decide their own priorities for the process of development as it affects their lives, beliefs, institutions, and spiritual well-being and the lands they occupy or otherwise use." In addition, article 13 stipulates that governments respect the culture and spiritual values of Indigenous Peoples' relationships with their lands or territories and clarifies that Indigenous territories cover the total environment of the areas which they occupy or otherwise use.

A decisive turning point in international law evolved in the 1990s regarding plant genetic material and biodiversity in general, when the FAO adopted Resolution 3 (in 1991), acknowledging sovereign rights of countries to their plant genetic resources. In 1992, through the CBD, the entire international community (except the United States) approved the principle of sovereign rights of countries of origin instead of the common heritage of humankind on biodiversity. Recognition of the sovereignty rights of countries of origin was a reaction to the inapplicability of the 1983 FAO International Undertaking on Plant Genetic Resources for Food and Agriculture, which conceded that all plant genetic resources were the common heritage of humankind, including plant varieties protected by intellectual property rights. However, countries that provided the bulk of plant genetic resources were unable to counteract the tendency led by some European countries and the United States to expand intellectual property rights on plant germplasm. Intellectual property rights had an early development in the United States with the Plant Patent Act of 1930 (for asexual reproductive plants) and the Plant Variety Protection Act of 1970 (for sexually reproductive plants). Some European countries developed a special regime for plant breeders' rights formalized through the International Union for Protection of New Varieties of Plants (UPOV) Convention of 1961, modified in 1978 and 1999, that is now in large part global in reach with 79 countries now participating.

The CBD is a complex agreement with provisions influenced by governments, industry, and environmental organizations. Indigenous Peoples were not directly represented in the making of this international convention, but some of its provisions suggest states' action toward effective protection, respectful use, and suitable conservation of traditional knowledge associated with biodiversity, including agrobiodiversity. In its preamble, the CBD recognized the close interconnection between the traditional lifestyles of Indigenous and local communities based on their biological resources. Additionally, under the approach for *in situ* conservation, the CBD foresees that states shall "respect, preserve and maintain knowledge, innovations, and practices of Indigenous and local communities embodying traditional lifestyles relevant for the conservation and sustainable use of biological diversity." The signatory states are also expected to promote the wider application of traditional knowledge with the approval and involvement of the legitimate holders and to encourage the equitable sharing of benefits that arise. Unfortunately, the fulfillment of the parties' responsibilities was subjected to national legislation without an enforcement mechanism. This lack of enforcement is found in most of the provisions, clarifying that states' commitments are due "as far as possible and as appropriate." Consequently, mostly minor advances have occurred. Though partial in scope, some measures for a more effective maintenance and preservation of Indigenous knowledge and associated resources have been attained in some countries (e.g., Peru and India).

Based on their sovereign rights, several countries individually or in groups have established access regimes to genetic resources. Currently, there are more than 50 access regimes with different levels of application. The Andean community, comprised of four countries today, issued the Decision 391 on Access to Genetic Resources in 1996. As of 2016, the Colombian environmental authority has signed more than 200 contracts on access to genetic resources, with only four designated for commercial application. Other countries in the Andean community have granted a significantly lower number of contracts. Decision 391 was complemented by Decision 486 in 2000 on intellectual property rights, which established the disclosure of origin of genetic material or traditional knowledge involved in patentable invention when such material or knowledge is obtained from Andean countries. Nevertheless, the disclosure requirement of Decisions 391 and 486 have had negligible application in the region and is not enforceable outside Andean jurisdiction.

As a development of the third CBD objective (i.e., the "fair and equitable sharing of benefits") the CBD Conference of the Parties established, through Decision VII/19, the ABS Working Group in coordination with Working Group on Article 8(j) to develop a special international agreement (UN 2004). As a result, the parties of the CBD signed the Nagoya Protocol (NP) in 2010 that entered into force in 2014. By September 2017, 105 countries have ratified the NP (UN 2017b). This protocol regulates the observation of the two key issues under the bilateral approach of the CBD: (a) the prior informed consent and mutually agreed terms for access to genetic resources and (b) the sharing of benefits that derive from its use.

The NP negotiation process included the discussion of fair and equitable sharing of benefits derived from genetic resources and from innovations, practices, and traditional knowledge. As a result, the NP contains several provisions that protect traditional knowledge (Articles 5, 7, and 12) (Cabrera-Medaglia 2013). In this sense, the NP calls for states to adopt measures to enforce the acquisition of prior informed consent from communities, and that measures are developed to ensure that access to traditional knowledge does not take place without mutual agreement on the terms (Article 16, NP). The NP includes the identification and acknowledgment of customary law in regards to the process of access to genetic resources and traditional knowledge (Article 12, NP). This recognition refers to principles and norms that regulate community life and relations with outside society, which are transferred through generations by Indigenous Peoples. Customary law embraces all aspects of community life: traditional authorities and sanctions, use and management of natural resources, rights and responsibilities of land, spiritual practices and beliefs, as well as traditional medicine practices. The NP needs to be interpreted in line with principles established in additional legal sources, such as the ILO Convention 169 and the UN Declaration on the Rights of Indigenous Peoples of 2007 (analyzed further below).

Plant genetic resources and associated Indigenous knowledge relevant to food and agriculture are within the scope of the NP, as they have not been explicitly covered with special international FAO instruments. However, Articles 4.2 and 4.4 of the NP establish the necessary coordination with relevant international agreements, as long as these agreements support the CBD and NP objectives. Thus, the NP acknowledges the preeminence of specialized instruments regarding specific genetic resources. The most important of these is the International Treaty on Plant Genetic Resources for Food and Agriculture (ITPGRFA).

ITPGRFA was adopted in 2001 and entered into force in 2004. It includes provisions for benefit sharing derived from the utilization of plant genetic material. Instead of bilateral negotiations between users and providers as in the CBD, with the requirement for prior informed consent and bilateral agreements, ITPGRFA established a multilateral approach, including a facilitated mechanism for all parties to have access to genetic resources with legal certainty.

Access to the listed plant genetic material in Annex 1 is contemplated through a standardized material transfer agreement that should observe provisions of the ITPGRFA; for instance, Article 12.3 a, d, and g as well as Article 13.2d(ii). Annex 1 includes 35 major crop species and 29 forage crops. The ITPGRFA includes diverse mechanisms for the fair and equitable distribution of benefits arising from the use of plant genetic resources through a multilateral system (Article 13).

The general category of farmers in the ITPGRFA covers an individual farmer or a group of farmers. The general notion of farmers and the discussions about their rights does not, however, include the distinct cultural characteristics of Indigenous Peoples. ITPGRFA refers to Indigenous and local communities (Article 5.1.d) and Indigenous communities and farmers (Article 9.1) in regard to the promotion of *in situ* conservation of wild crop relatives and wild plants for food production and farmers' rights subject to national legislation.

The multilateral system in ITPGRFA regulates access to plant species listed in Annex 1 that are under the administration and control of the contracting parties. Access to those materials is exclusively for food and agricultural purposes; other purposes would fall under general CBD provisions and eventually under NP rules. Additionally, plant material in Indigenous territories and under the control of Indigenous Peoples is not within the scope of the ITPGRFA unless Indigenous communities voluntarily decide to submit them. A case that illustrates this option is the proposed inclusion of potato collections under the ITPGRFA multilateral system by Indigenous communities from the Parque de la Papa in Cusco, Peru (Graddy 2013).

Because of the ITPGRFA's specific role, the recognition and compensation for contributions to Indigenous knowledge, innovations, and practices associated with agrobiodiversity remain subject to national legislation and the bilateral system under the CBD and the NP. NGOs and research institutions may contribute funding directly to smallholder farmers' projects and initiatives. Within the FAO forum, the search for equitable distribution of benefits has been framed in terms of farmers' rights to counterbalance intellectual property rights on plant genetic material. Resolution 5 of 1989 on Farmers' Rights, for example, concentrated on past, present, and future contributions of generations of farmers to conservation, improvement, and availability of plant genetic resources. This Resolution did not, however, make reference to historical contributions, which continue up to the present, of useful plant material and associated knowledge made by Indigenous Peoples. In the 1996 Technical Consultation on the Implementation Framework for Farmers' Rights in Madras, India, the proposed definitions for farmers did not identify Indigenous Peoples but rather farming communities. The recommendations of this Technical Consultation admitted that farmers' rights did not contain the full range of Indigenous Peoples' rights, but that many Indigenous Peoples were in farming communities and must therefore be beneficiaries of farmers' rights (Swaminathan 1996). NGOs present at this meeting identified farming communities side by side with Indigenous communities but pointed out that Indigenous communities were the central factor in the debate over farmers' rights and that their rights should thus be protected (Mooney 1996).

Twenty years after Resolution 5 of 1989 on Farmers' Rights, an FAO policy document explicitly recognized the past and present adaptiveness and resilience of Indigenous agricultural practices as well as the contributions to "domestication, conservation, and adaptation of genetic resources and agricultural biodiversity at all scales (gene, species, ecosystem, and landscape)" (FAO 2010a:7). Thus, the concept and recognition of Indigenous Peoples as defined in Convention 169 was absent from FAO international instruments until the adoption of the FAO policy on Indigenous and tribal peoples in 2010 (FAO 2010a). This recognition in a policy document shows a formal advance when compared to the Resolution 5 of 1989. However, protection for Indigenous agrobiodiversity has to be sought within the general human rights realm. It was precisely the 2007 UN Declaration on the Rights of Indigenous Peoples (UNDRIP), Article 41, that prompted the adoption of 2010 FAO policy on Indigenous and tribal peoples.

After more than twenty years of work, the United Nations General Assembly finally adopted UNDRIP. The Declaration acknowledges that

> Indigenous Peoples have suffered from historic injustices as a result of, *inter alia*, their colonization and dispossession of their lands, territories, and resources, thus preventing them from exercising, in particular, their right to development in accordance with their own needs and interests.

An important effect of colonization was the intended or unintended omission to properly acknowledge the contributions made by Indigenous Peoples in all areas of human endowment, such as medicine, food, and environment. In this sense, the UNDRIP preamble states that

> respect for Indigenous knowledge, cultures, and traditional practices contributes to the sustainable and equitable development and proper management of the environment.

As a comprehensive international instrument, although not enforceable, UNDRIP recognizes critical issues for Indigenous Peoples (e.g., the right to self-determination included as the epigraph to this chapter). This right is highly relevant because it clearly identifies Indigenous Peoples as subjects who are entitled to collective rights under international law. Rather than a call for secession, this represents the aspirations of Indigenous Peoples: full recognition of the right to remain distinct and to determine autonomously their own economic, social, and cultural path to development within contemporary states and the global context. More specifically, Articles 20.1, 23, and 32 reiterate the right of Indigenous Peoples to decide and control their own development. Article 32.1

states that "Indigenous Peoples have the right to determine and develop priorities and strategies for the development or use of their lands or territories and other resources." This right covers all aspects and resources connected with Indigenous agrobiodiversity.

UNDRIP is the first international legal instrument that explicitly recognizes the rights of Indigenous People to their seeds and associated knowledge. Article 31 states the right of Indigenous People to "maintain, control, protect, and develop their cultural heritage, traditional knowledge…as well as the manifestations of their sciences, technologies, and cultures, including… genetic resources, seeds, medicine, knowledge of properties of fauna and flora." However, the practical application of these rights faces an insurmountable obstacle, shared in common with the CBD and NP: any effective development in policy or legislation shall be decided by each state "as far as possible and as appropriate." Despite diverse national constitutions embracing Indigenous Peoples' rights, practical application continues to be limited (Mikkelsen 2014). In the following section, I will illustrate this point.

Finally, the Intergovernmental Committee on Intellectual Property and Genetic Resources, Traditional Knowledge and Folklore (IGC) of the World Intellectual Property Organization (WIPO) has made very few advances toward an international regime after almost two decades of activity. Its primary contribution has been to provide a wide range of studies and documentation, along with drafting a body of articles. In December, 2016, during its thirty-second session, IGC reviewed the "The Protection of Traditional Knowledge: Draft Articles Rev. 2," which was based on an earlier document from the 31st session, held in September, 2016. To date, IGC has been unable to advance a text for an international instrument to protect traditional knowledge within the intellectual property framework. Indeed, it is uncertain whether such a framework could really protect the collective intellectual rights of Indigenous Peoples in line with their worldviews and consuetudinary law. Critical issues remain contentious. For example, Indigenous spokespersons oppose the extension of principles regarding the common heritage of humankind and the public domain on traditional knowledge once it has been published (Hardison 2016). Additionally, consensus does not exist regarding beneficiaries; one position identifies Indigenous Peoples and local communities as the exclusive beneficiaries while another supports the inclusion of states or nations and even NGOs as beneficiaries. After the 32nd session, delegates from the United States, South Korea, the European Union, and Canada concluded that a long discussion lies ahead before a common understanding is achieved (ICTSD 2016).

National Level

National policies on Indigenous Peoples and rural development have overlooked Indigenous agrobiodiversity. In this section, I illustrate this point with examples from Colombia and Peru. By the middle of the twentieth century, in countries like Colombia, Indigenous Peoples were still legally classified as savage, semisavage, or civilized, depending on their level of assimilation, particularly to Christian doctrine.

In Colombia, for instance, the national government designed specific policies to integrate the Indigenous population into agrarian development through the Office for Indigenous Issues, situated within the Ministry of Agriculture and Ranching Industry. In 1958, Law 81 promoted the modernization of agricultural production in Indigenous communities in total disregard to Indigenous subsistence agriculture. Indigenous *resguardos* (collective lands) were targeted to promote agrarian cooperatives or to break them up when the expansion of large private estates was favored. Finally, in 1967 Law 31 ratified the ILO Convention 107 for the "protection and integration of Indigenous and tribal populations."

Created in 1960, the Colombian Institute for Agrarian Reform (*Instituto Colombiano de la Reforma Agraria*, INCORA) was charged with modernizing agricultural production in campesino populations using Green Revolution practices. In 1967, the Colombian government also created the National Peasant Association of Colombia (*Asociación National de Usarios Campesinos de Colombia*, ANUC) as a social movement to promote agrarian reform and to appease rural protest. Initially, Indigenous communities were mobilized with the ANUC, but the union with campesinos did not last too long due to Indigenous worldviews on land tenure. Whereas the campesino movement's struggle focused on access to land for agricultural production, Indigenous Peoples looked to recuperate ancestral territory taken by force in previous decades by private landholders and the Catholic Church. In 1970, Indigenous Peoples from Cauca broke the partnership with ANUC and established the first grassroots Indigenous organization: the Regional Indigenous Council of Cauca (*Consejo Regional Indígena del Cauca*, CRIC). Recuperation of territory went hand in hand with the strengthening of their cultural identity through the revitalization of their language and control of education.

In Peru, similar integrationist policies were more successful. Indigenous identities were systematically excluded from the modern nation-building project. The agrarian reform focused on Andean peasants, rather than Indigenous Peoples, as part of the social policy engineered by the Velasco government (1969–1974) to transform *indios* into market-oriented *campesinos* (Shepherd 2010:631–632).

The de-essentializing gaze, which emphasizes a homogeneous social base, is also found in other parts of the world. In the Southeast Asian country of Laos, for example, the 230 identified ethnolinguistic groups that comprise around 70% of the population, and occupy almost 80% of the territory, are not officially recognized by the government, which prefers the notion of "multiethnic peoples" rather than "Indigenous Peoples" (Dze 2005:31).

Although Colombian and Peruvian governments have increasingly recognized the contributions made by Indigenous Peoples to biodiversity conservation at the international level over the last 25 years, the effective protection of traditional knowledge at the national level has been precarious, although Peru introduced specialized legislation. National policy and legislation often reflect a strong private interest, with the support of international capital investors, for promoting monoculture farming techniques and expanding intellectual property regimes on plant germplasm. The access regimes developed under CBD have not evolved toward effective protection of either Indigenous knowledge or agrobiodiversity. In the Andean region, for example, Decision 391 included a transitory provision demanding states to undertake harmonization studies and to establish a positive protection regime. Yet it has been over twenty years since the enactment of the Andean common access regime, and states have not made any significant effort to develop such protection. In Colombia, the opposite is true: the legal framework facilitates undue appropriation of plant genetic material of Indigenous communities by plant breeders or agrobiotech companies.

The current plant breeders' rights regime in the Andean countries, including Colombia, stems from Decision 345 of 1993. The Colombian Law 243 of 1995 ratified the 1978 UPOV, but the Colombian government operates in practice with Decision 345, a regime more like UPOV 1991. The 1978 UPOV version does not cover all plant species and excludes patents on plants. The Colombian government tried to formally ratify the 1991 UPOV version in 2012. However, in 2012 the Constitutional Court declared the unconstitutionality of the Law 1518 of 2012, which approved the 1991 UPOV convention. In its decision, the Constitutional Court considered that the content of the 1991 UPOV Convention could affect the intimate and indissoluble relationship of Indigenous Peoples with their territory and its natural resources. The Court found that this law was approved without fulfilling the duty to consult Indigenous populations and declared it unconstitutional (Constitutional Court Republic of Colombia 2012).

Decision 345 of 1993 clarified that in order to create a new plant variety, the application of scientific knowledge is necessary to improve the plant genetic pool (Article 4). Plant breeders at public universities develop plant varieties that they deliver sometimes free to small farmers. This provision, however, excludes plant varieties obtained through traditional innovations, practices, and traditional knowledge because these are regarded as nonscientific methods. At the same time, the rights of Indigenous Peoples and local communities to landraces, knowledge, innovations, and practices remain without effective protection. The plant material of Indigenous communities is freely available for researchers and professional plant breeders for developing new plant varieties. Seed phytosanitary laws reinforce the plant breeders' regime by imposing limitations to a farmer's right to save seed. The Resolutions 970 of 2010 and 3168 of 2015 of the Colombian Institute of Agriculture imposed restrictions on seed quantities and land extension where farmers could replant the saved seeds. This limitation, via administrative measures, violates Article 26 of the Andean Decision 345, which provides farmers with the right to store and sow for their own use, or to sell as a raw material or food the product of the cultivation of the saved seeds. This case illustrates the complete disregard of a national government for the agrobiodiversity of Indigenous Peoples, while at the same time guaranteeing strong intellectual property rights regimes that favor monocultures in large-scale industrial agriculture.

Very few states have established mechanisms to protect Indigenous knowledge and agrobiodiversity. In Latin America, Peru introduced the "Regime for Protecting the Collective Knowledge of Indigenous Peoples related to Biological Resources" by Law 27811. This is a registration system for preventing the granting of patents on genetic resources and Indigenous knowledge associated with plants. The system is a defensive mechanism against patents on Peruvian plants with medicinal properties or cosmetic uses (Nemogá 2013; Ruiz 2011). In India, the "Protection of Plant Varieties and Farmers Rights' Act" issued in 2001 and enforced since 2007, grants direct protection to farmers of their landraces and wild relatives (this is the term formally used in the Indian legislation). This Act recognizes farmers' contribution to the conservation, improvement, and availability of plant genetic resources that plant breeders develop into new plant varieties. Farmers can protect their plant material under the concept of a farmers' variety. A farmer's variety is described as the one that "has been traditionally cultivated and evolved by the farmers in their fields, or is a wild relative or landrace of a variety about which the farmers possess the common knowledge." The Indian Act also includes the category of an extant variety referring to varieties that are in the public domain. The Indian Act contemplates two other categories, the new variety and the essentially derived variety. These two satisfy the needs of professional breeders and companies. While the registration of varieties under the two first categories is not necessarily subject to the criteria of novelty, distinctiveness, uniformity, and stability, the last two requirements need to be fulfilled for successful registration (Ramanna 2003:15–18).

In their applications, the breeders have to reveal the use of plant genetic material provided or taken from tribal communities or rural families and used to develop the new variety. This is a practical and necessary step for making effective the provisions on the sharing of benefits included in the act. By 2013, only 22 out of 748 registers corresponded to farmers' varieties within this system. Koonan (2014) attributes this low registration to the precarious educational background and limited economic situation of poor farmers. The relatively low registration of farmers' varieties in the Indian case could indicate that farmers' rights in *sui generis* systems are more declaratory than practical. In other words, framing farmers' rights under the approach of intellectual property rights is substantially limited in its ability to recognize and compensate farmers for their past and present contributions.

Final Thoughts on the Recognition and Protection
of Indigenous Agrobiodiversity

It will take a substantial effort to shift from conventional understandings and farmers' rights interpretation to full recognition of Indigenous agrobiodiversity governance. Nonetheless, contemporary agriculture would be unthinkable without the diversity of landraces, genetic material, and knowledge that Indigenous Peoples provided in the past and still provide today for food and agriculture. The overall value of Indigenous agrobiodiversity and its origins and ongoing management by Indigenous Peoples are key for innovations in global agrifood systems. This is a challenge in global institutional contexts that assume large-scale industrial agriculture is the main way to overcome hunger and provide adequate nutrition. In a biocultural diversity framework, all biodiversity is valuable in and of itself; likewise, all manifestations of Indigenous use and knowledge of agrobiodiversity have an intrinsic value, and represent the rich variety of human adaptations to diverse environments, including climate change (see also Chapter 7).

The biocultural approach questions the dominant epistemology and research paradigms, thus opening space to include Indigenous worldviews and knowledge systems. It promotes intercultural dialogue and innovative pathways needed to recognize and protect Indigenous agrobiodiversity. Agroecologists and other practitioners of sustainable agriculture should participate in the intercultural dialogue. The biocultural approach can contribute to achieving global recognition of the multiplicity of cases that provide practical demonstrations of the knowledge, traditions, and Indigenous worldview of peoples interacting with their agricultural landscapes. Many of the successful histories are not internationally funded projects led by NGOs, but represent the everyday life of Indigenous Peoples led by their traditional authorities, elders, and leaders (women and men) who practice the teachings of their ancestors. The biocultural approach successfully supported Indigenous claims in Santiago Lachiguiri, Oaxaca, México. Contrary to the federal biodiversity conservation policy, ancestral agriculture (itinerant or swidden agriculture) was reinstated by the Santiago Lachiguiri's General Assembly as a fundamental component for the survival of the Indigenous community as a distinctive people in January of 2009. The community worldview about their relationship with the forest and their ancestral agricultural practices were recognized and protected to guarantee the balance between food production and forest conservation (Marchi 2018). Moreover, biocultural heritage is a pivotal concept of the biocultural approach and has been adopted in different contexts for conducting research *with* and *for* Indigenous communities, and to promote transformative actions with a bottom-up strategy (Nemogá 2018; Toledo et al. 2010). In 2010, the global sourcebook published by Woodley and Maffi (2012) summarized relevant grassroots initiatives on Indigenous agrobiodiversity taking place in Kaski (Nepal), Yunnan (China), Kenya, and Colombia.

Powerful economic and political forces have permeated Indigenous world-views and Indigenous ways of practicing agriculture. Indeed, the cultural heritage and identity of many communities have been severely disrupted through the actions of governments, churches, and NGOs. Moreover, Indigenous communities are increasingly immersed—willingly or not—within the global capitalist economy, and their use values are increasingly transformed into commodities for external consumption. At the same time, Indigenous Peoples' open systems for sharing ancestral knowledge and varieties, and for preserving the collective diversity of seeds, are being eroded (IIED 2017). Indigenous Peoples cannot be expected to live in isolation from social and technological changes, subsisting on productive practices and strategies frozen in time. However, the lack of full recognition of Indigenous agrobiodiversity and Indigenous rights on their ancestral territories, cultural identity, and resources will accelerate current trends of cultural erosion and misappropriation.

In this chapter, I have argued that international and national policy and legislation on agrobiodiversity do not effectively address Indigenous agrobiodiversity. Human rights, by contrast, have been more proactive in creating ways to recognize Indigenous agrobiodiversity. Public law has been instrumental for guaranteeing capital investment in agriculture and extractive industries. Economic law, particularly intellectual property law, is essentially designed to protect individual and corporate private interest in market competition. The examples of special laws in Peru and India, with regard to the protection of plants and associated knowledge of Indigenous Peoples via defensive mechanisms within the intellectual property framework, are far from an optimal solution. Though a defensive mechanism is one step forward, it does not fit the UNDRIP recognition of Indigenous Peoples' right "to maintain, control, protect and develop their intellectual property over such cultural heritage, traditional knowledge, and traditional cultural expressions" (Article 31 UNDRIP).

Though human rights instruments have not directly addressed the protection of Indigenous agrobiodiversity, their application via interpretation in international and national court decisions shows some advances. An example is the application of the American Convention on Human Rights and the International Convention on the Elimination of All Forms of Racial Discrimination to guarantee Indigenous territorial rights. The case was brought before the Inter-American Court on Human Rights by the Inter-American Commission on Human Rights in 1998 because the Nicaraguan government did not satisfactorily compensate the Awas Tingni community despite the mandate of its Supreme Court. In 2016, the Inter-American Court granted protection to the collective rights on ancestral lands of the Kaliña and Lokono Peoples against Suriname's protected areas policy. At the national level, the Colombian Constitutional Court has issued numerous sentences protecting the rights of Indigenous and tribal peoples guaranteed by the ILO Convention 169 and the UNDRIP. A milestone in constitutional jurisprudence was the recognition of Indigenous Peoples as subjects entitled to fundamental collective rights such as the right to the duty to be consulted, the right to their cultural identity, and the right to their territory. Though not directly related to Indigenous agrobiodiversity, these Courts' rulings have supported the traditional subsistence agricultural, hunting, and fishing practices of Indigenous Peoples, and protected their land, culture, and subsistence rights. In countries that have ratified the ILO Convention 169, flexible and extended interpretation of its provisions pressed under political mobilization could help to guarantee and preserve Indigenous agrobiodiversity.

To put into perspective efforts regarding the present and future of Indigenous agrobiodiversity, let us look to FAO policy, which clarifies relationships and acknowledges the diverse cultural systems that sustain natural resources (FAO 2010a:34):

> The inextricable relationship between cultural and biological diversity must therefore be respected, cultivated and promoted, and the rights of Indigenous Peoples over their traditional knowledge and practices must be recognized and, when necessary, protected.

After this statement, however, the very same policy suggests that "access to markets, financial resources, and stable sources of production..." provide the main path to solve poverty and food insecurity that affects Indigenous Peoples. It is not clear how this commitment is compatible with the right to Indigenous self-determination. Market-oriented tools in Indigenous lands can erode cultural values of solidarity and reciprocity, thereby disrupting the social institutions and practices that have maintained the collective pool of biogenetic resources.

The FAO policy on Indigenous and tribal peoples is also overly narrow in formulating research as one of the mechanisms for its implementation. Rather than envisioning how to transform the research paradigm so that it recognizes the inextricable relationships between cultural and biological diversity, it adopts a traditional approach for undertaking studies on the livelihoods of Indigenous Peoples. In addition, the FAO policy fails to overcome the dominant research paradigm that promotes conducting research *on* Indigenous Peoples, rather than *with* and *for* Indigenous Peoples. As envisioned within a biocultural diversity framework, science and technology seldom embrace the values of local knowledge and traditions and very rarely employ the language of rights and control over knowledge and resources (Nemogá 2016).

Other core principles delineated in the FAO policy on Indigenous and tribal peoples (FAO 2010a) should be emphasized as they include self-determination, and cultural and collective rights. Indigenous Peoples' own concerns and priorities must be actively supported when identifying projects and programs that affect their livelihood. Their distinctive cultures should not be threatened by open or subtle assimilation measures, and their collective rights to land, territories, natural resources, and knowledge systems should be respected.

This paper focused on Indigenous agrobiodiversity within the large agrobiodiversity governance field to underline the role of Indigenous Peoples' contributions. While governance in this area is central, we cannot dismiss the work of Indigenous and local communities to preserve and enrich agrobiodiversity at the grassroots level. This chapter on Indigenous agrobiodiversity opens the field for fruitful innovative research *with* and *for* Indigenous Peoples. Some suggestions in this direction are:

- Document cases of Indigenous agrobiodiversity practices and evaluate the impact of market-oriented strategies on their preservation as distinct peoples and cultures.
- Establish the relationship between Indigenous agrobiodiversity practices, the protection of their ancestral lands, and the right to self-determination.
- Explore the correlation between cultural, linguistic, and crop diversity.
- Assess the contribution of Indigenous agrobiodiversity to ameliorate poverty, hunger, sustainability, and *in situ* conservation.
- Strengthen legitimate participation of Indigenous Peoples in agrobiodiversity governance at the national and international level.
- Explore the best mechanisms to guarantee Indigenous self-determination, collective land rights, and preservation of their languages and cultures in the context of Indigenous agrobiodiversity conservation and governance.

Acknowledgments

I thank the Ernst Strüngmann Forum for inviting me to participate in its 24th Forum on "Agrobiodiversity: Establishing an Integrative Scientific Framework for Sustainability," in Frankfurt, Germany. Additionally, I acknowledge the organizing committee's invitation to write this chapter, and the peer reviewers' comments and editorial work. Finally, I acknowledge the support from the Social Sciences and Humanities Research Council (SSHRC) Partnership Grants IDG 2016 and the University of Winnipeg while working on this contribution.

13

Seeding Relations

Placemaking through Ecological, Social, and Political Networks as a Basis for Agrobiodiversity Governance

Guntra A. Aistara

Abstract

Agrobiodiversity governance is often guided by estimates of countable and measurable objects, from the number and diversity of heirloom seeds or landraces grown in a certain location, to the frequency of seed exchange among actors and rates of disappearance of varieties. Such variables provide important information about conservation status at different scales but do not necessarily capture the dynamic social roles and relationships of seeds and agrobiodiversity to local cultures and communities. This chapter explores (a) the cultural roles of seeds in agrobiodiversity governance as a set of interwoven processes that are mediated by, and which in turn mediate, relationships between people, their practices, and knowledge systems; (b) networks with other people and other species; (c) attachments to cultural landscapes and histories or places; and (d) the broader politics of agriculture and rural development. It argues that relational processes are a necessary part of an analytical framework and crucial for understanding the role that social networks play, at multiple scales, in agrobiodiversity governance, including creating, managing, preserving, or "losing" diversity in the long term.

Introduction

Although farmers have been directly involved in the selection and saving of new varieties of plant and animal species since the beginning of agriculture, the roles of the farmer and local community within seed systems have changed commensurate with the rise of industrial agriculture, specialized breeding programs, and gene banks (Brush 1999; Kloppenburg 1988; van Dooren 2008). As a concept, *agrobiodiversity*—"the variety and variability of animals, plants and microorganisms that are used directly or indirectly for food and agriculture"

(FAO 2004)—was introduced only recently, and it directs attention to the number of species, seeds, breeds, and varieties rather than to the relations and networks through which they emerge, are managed, and maintained. These dynamic multiscalar and multispecies relationships have important historical, cultural, social, and ecological dimensions that must be understood if we are to better preserve diversity and enable effective governance.

In this chapter, I explore how farmers and communities create and maintain agrobiodiversity through their place-based seed practices and multispecies socioecological networks in different cultural contexts, analyzing complex relationships that sustain these practices. Each local seed network is embedded within long-term historical and ecological conditions as well as contemporary social and political processes, all of which determine practices and meanings that surround agrobiodiversity. I draw upon cases from different parts of the world to illustrate the types of relationships that both mediate and are mediated by the management and governance of agrobiodiversity across scales. The selected examples permit us to consider important questions in regard to seed networks and, more broadly, agrobiodiversity governance:

- How do historical trends, ecological conditions, and cultural norms influence current practices and networks of seed saving and exchange as well as symbolic meanings of seeds and agrobiodiversity?
- How are place-based seed networks situated within broader surrounding agricultural and rural policy landscapes across scales?
- How do farmer groups and communities negotiate access to resources and engage with legal codes and international treaties that influence their relationships to land, seeds, other species in the landscape, and other political actors?

Insights into these areas of concern can help us progress beyond counting levels of existing or lost agrobiodiversity toward an understanding of the cultural and affective motivations that underlie the production and maintenance of agrobiodiversity as well as the power dynamics within local communities and across scales that facilitate or hinder the governance of agrobiodiversity over time.

I begin with a discussion of farmer practices and knowledge systems that sustain agrobiodiversity, followed by a look at the social networks that maintain these practices and influence genetic diversity. Thereafter, I review power dynamics and property politics that structure possibilities for groups to perpetuate or reinvent these practices and social networks, followed by a discussion of how these practices, social networks, and broader politics constitute a form of placemaking that produces cultural and ecological landscapes. I propose that these interlocking practices, social networks, and power dynamics are key elements in agrobiodiversity governance. Particular attention is given to the cultural and political dimensions of these relationships, which to date are only marginally covered in much of the literature on agrobiodiversity governance, as they are crucial for understanding farmer motivations and possibilities to maintain agrobiodiversity in the future.

A Note on Concepts

As documented elsewhere (see Chapter 5; Maxted and Dulloo 2016), much work has gone into the collection and cataloguing of agrobiodiversity, its biogeographical distribution, and conservation status, yet more remains to be done. Nevertheless, agrobiodiversity involves more than just crop–livestock diversity per se and cannot be reduced to the number of varieties; it encompasses the dynamic processes through which varieties are gained and lost (Thomas et al. 2011). Furthermore, different knowledge systems produce different classification systems for agrobiodiversity and must be evaluated on their own terms rather than compared simply to scientific knowledge to check accuracy (Agrawal and Gibson 1999; Nazarea 2006). Both local and scientific knowledge systems require symmetrical and context-dependent evaluation (Latour 2005).

Attention to the dynamic processes and knowledge systems from which agrobiodiversity has emerged requires a *relational* understanding that extends past a genealogical explanation of diversity. In genealogical models, Ingold (2000:138) observed that "diversity is the measure of difference…that presumes a world already divided into discrete, unit-entities—'things-in-themselves'—which may then be grouped into classes of progressively higher order on the basis of perceived likeness." In contrast, a relational approach places diversity in a dynamic context. What makes things the same or different is "the shared experience of inhabiting particular places and following particular paths in an environment….The relational model, in short, *renders difference not as diversity but as positionality*" (Ingold 2000:148–149).

Different societies may use different combinations of genealogical and relational models to understand and classify agrobiodiversity (Berlin 1992; Zimmerer 2001). Furthermore, what is valued in local classification systems may not align with Western scientific taxonomic models (Caillon and Degeorges 2007). A relational approach allows us to investigate the multispecies and multiscalar networks as well as the socioecological knowledge systems that surround seeds and their domestication and exchange, and explore their significance for agrobiodiversity governance.

In this chapter, I use the term "seeds" to refer also to other types of propagating material (including roots, tubers, and cuttings of vegetatively reproduced plants) as a component of agrobiodiversity. Seeds and varieties continuously coevolve with humans, sometimes through intimate and affective relationships. Seed exchange includes transfer of seeds among farmers that need not be reciprocal and may or may not involve the transfer of other goods in return as a form of barter. In addition to seed networks among farmers, seeds may be acquired through markets, extension services, or other commercial sources. For the purpose of this discussion, I will not focus on networks related to the preservation of animal breeds, but this, too, constitutes an important aspect to include in further considerations of agrobiodiversity governance.

Practices of seed saving and exchange both depend upon and create social networks around seeds. Within these, different types of farmers (e.g., subsistence, surplus, and commercial) have specific concerns in accordance with which they use or replace different varieties of plants (Bellon 1996). Diverse social connections, such as kinship structures, ethnolinguistic differentiation, coresidency patterns, and other factors influence exchange (Labeyrie et al. 2016; Leclerc and d'Eeckenbrugge 2012). In my analysis of social networks, I follow scholars of actor–network theory and science and technology studies who insist that nonhuman actors (e.g., plants, seeds, pollinators, and soil) also mediate human social relations and thus may be seen as exhibiting a form of agency (Haraway 2008; Latour 2005). Following Tim Ingold (2011), connections among actors in a network (which he calls a "meshwork") are made through specific practices (such as seed exchange), just as a landscape or place is made by "dwelling" in it (Ingold 2000). Thus, the landscape itself can also be understood as a process, of which we can only perceive a momentary glimpse (Hirsch 1995; Ingold 2000), and places may also be "global" (Massey 2005). Seeds and agrobiodiversity are crucial components of these landscapes cocreated by farmers in conjunction with other species.

Thus, farmer seed networks function as a means of placemaking; that is, a "set of social, political, and material processes by which people iteratively create and recreate the experienced geographies in which they live" (Pierce et al. 2011:54). How these multiple relationships interact at different levels and across scales is very important for understanding the motivations to use and protect agrobiodiversity amid changing socioeconomic and environmental circumstances. I follow scholars such as Doreen Massey (2005), Arturo Escobar, and Wendy Harcourt (2002) to capture the multiple cultural, social, and ecological relationships that farmers craft to their seeds and environments as a means of placemaking. Through such practices, farmers' cultural memories, current economic situations, and ecological futures become embodied in their landscapes in what I have called "networked diversities of place" (Aistara 2018). Let us now explore how farmers position themselves in relation to seeds through their practices, social networks, and broader political contestations, and what this implies for agrobiodiversity governance.

Practicing Agrobiodiversity

For generations, agrobiodiversity has been created and maintained by farmers according to a dynamic interplay between culturally and ecologically embedded practices and knowledge systems. Farmers select plants with desirable characteristics and save seeds to plant the next cropping season. Sometimes they purposely cross plants with distinct traits; they exchange seeds with relatives, friends, and neighbors, and acquire new seeds from other sources outside of their immediate social networks (e.g., through trade networks). Farmers may also practice staggered planting or make use of microclimates to take advantage of diversity in space and time across the farm, in an effort to manage pests and diseases as well as to provide insurance against loss of one or more varieties (Altieri 1999; Cleveland et al. 1994).

Conceptualizing farms as agroecosystems helps us understand how farmer practices create ecological relationships through the maintenance of agrobiodiversity and expands our view of diversity from one simply focused on more species or varieties (Jackson et al. 2007). Instead, farmer seed practices are embedded in their cultivation systems. Agroecological, traditional, and Indigenous agricultural models are often more beneficial for maintaining agrobiodiversity than industrialized agricultural systems (Altieri 1999). On-farm diversity assists in the management of pests, diseases, and soil fertility; it also helps a farmer avoid the need to use external inputs, conserve ecosystem structure, and maintain nutritional diets (Bianchi et al. 2006; Brussaard et al. 2007; Thrupp 2000). Particularly under marginal conditions, farmer seed practices help plants adapt to local conditions over time (Coomes 2010; Perreault 2005). Such practices serve not only to provide food, fuel, and fodder but also support ecological functions and help regulate the climate (Jackson et al. 2007). In addition, agroecology and agroecosystems enable a better understanding of how farmer practices contribute to creating and managing diverse landscapes that allow for more interactions between human-managed and wild landscapes (Altieri 1999; Jackson et al. 2007; Zimmerer 2010).

Farmer practices have tangible effects on agrobiodiversity, although further research is needed to expand our understanding of these processes (Alvarez et al. 2005). Farmer choices and trade-offs affect plant population structure and diversity through seed selection, how plants are distributed spatially during planting, and what proportion of seed is saved and sown from one season to the next (Alvarez et al. 2005). To analyze how farmers decide which varieties to save, incorporate, or discard from their crop repertoire, Bellon (1996) proposed a conceptual framework which predicted that surplus-oriented farmers who produce both for subsistence and the market would exhibit the highest levels of diversity because they must balance a great number of concerns that can only be met by choosing to cultivate various plants. When they choose to include a new crop in their repertoire, it may or may not replace other varieties, depending on which needs the variety is able to satisfy (Bellon 1996). Indeed, farmers often maintain both modern varieties and landraces; one-on-one replacement does not necessarily occur (Brush et al. 1992).

Despite widespread fears of diversity loss, farmers in many regions of the world continue to manage a high level of crop diversity (e.g., Perales and Golicher 2014; Roy et al. 2016). In Peru, for instance, de Haan (2009) studied farmer-driven cultivation practices (as distinguished from externally driven projects aimed at *in situ* conservation) in potato diversity and found that farmers from eight villages in the Huancavelica area preserved high levels of species, morphological, and molecular diversity. The highest rates of diversity were found within, rather than between, genetically isolated farmer populations, and there was no evidence of genetic erosion. Farmers managed multiple fields with a high diversity of cultivars in each to manage gene–environment interactions for yield stability rather than as a specific adaptation to environmental niches.

Indeed, conservation of agrobiodiversity may not be the explicit goal of farmer practices: it can be an outcome or beneficial side effect of more immediate goals, such as food security and risk management (Zeven 1999). The regeneration of agrobiodiversity by farmers is also driven by a variety of economic, cultural, and social reasons and is, to a certain extent, the result of unconscious social pressures on selection. For instance, dishes that are typical of a certain culture, prestige, and identity or family tradition are typical cultural drivers. However, several studies in Mexico have found that belonging to different ethnolinguistic groups may create barriers for movement of seeds and gene flow (Perales et al. 2005).

The interplay of conscious and unconscious effects on diversity raises important questions about the role of farmer knowledge systems in preserving agrobiodiversity. Almekinders and Louwaars (2002) noted a wide variation in farmer knowledge about seed practices. Traditional knowledge is not a static set of information passed down from one generation to the next; it is actively learned in local contexts (Ingold and Kurttila 2000). Farmers develop their knowledge systems through experimentation and evaluation of different management practices, including species and varieties, according to particular needs or demands. For example, Caillon and Lanouguère-Bruneau (2004) found that farmers in Vanuatu cultivate 96 different varieties of their staple crop taro on the west coast of Vanua Lava. Men possess specific cultivation skills but guard their knowledge with secrecy, as it is central to their identity and a source of competition. On average farmers grow about twenty different varieties; the top six make up over 80% of the cultivated area and the forty rarest ones constitute less than 1% of taros raised by each farmer. Each taro is valued for different characteristics. The five most popular types had the best yield or agroecosystem adaptation, whereas a range of others were grown for different reasons related to social identity. Some were cultivated for particular culinary properties and their use in boiled, roasted, or raw foods. Many other varieties were maintained, however, even though they lacked desirable culinary attributes, because they had been discovered by an ancestor or were featured in a founding myth, thus providing the basis for intergenerational narratives. Some of the oldest cultivars were central to male initiation rituals or specific magic ceremonies.

Local knowledge, therefore, may or may not map onto Western scientific knowledge and classification systems. As documented by Caillon and

Degeorges (2007), the diversity of taro varieties in Vanuatu serves as a living index of ancestors for whom they were named. Their continued cultivation is important to maintain the cultural memories of the people. For this reason, cultivation of taro is more highly valued than, for example, coconut palms, which have higher genetic diversity but are culturally associated with white colonizers and thus not as highly valued by local people. This example illustrates how biological and cultural values associated with diversity may not always coincide, and that the cultural motivation to continue certain practices may be more powerful than biological characteristics. It also demonstrates that local knowledge about agrobiodiversity can be "sticky" (von Hippel 1994); that is, information is context specific and may be difficult to transfer from place to place because it loses meaning when abstracted from the context within which it emerged.

As Stone and Glover (2017) have shown, social embeddedness of local agrobiodiversity knowledge has given modern varieties an advantage to spread globally, because agricultural knowledge during the Green Revolution was abstracted from local knowledge systems and social networks by design. Thus, rapid intensification of agricultural systems and advancement of specialized plant breeding in the twentieth century in industrialized countries changed the role of farmers in creating and maintaining new varieties (Kloppenburg 1988). Industrialization and governmental policies increased pressure on farmers to abandon "traditional" agricultural practices related to the selection and saving of seeds, which were perceived as backward (Escobar 1995; Stone and Glover 2017). With its focus on monocultures of high-yielding varieties reliant on external inputs, industrialized agriculture is usually much less diverse and provides fewer ecosystem functions. At the same time, since the beginning of the twenty-first century, the number of breeding companies has diminished, often as a consequence of privatization, and investment in public breeding has declined (Chapter 6).

There is, however, an ever greater intermingling of knowledge and seed systems. The introduction of modern high-yield varieties does not always exert negative effects on diversity, as modern varieties are often incorporated into existing seed networks (Bellon and Brush 1994). Instead of precipitating a loss in previous forms of knowledge regarding seed saving, hybrid forms of knowledge have evolved along with hybrid plants (Coomes 2010). For instance, many farmers create creolized hybrids of old and new varieties (Salazar et al. 2007), thus making the desired characteristics of modern varieties available to poor farmers (Bellon and Risopoulos 2001). Farmers continue to experiment and use what is most advantageous at a particular time, and their knowledge base grows at times or becomes more limited. For instance, maize diversity has been reduced in Mexico through the introduction of modern varieties, particularly in lowland elevations, making *in situ* conservation even more important in highland environments with specific environmental adaptations and higher sensitivity to mid- or low-level elevation conditions (Bellon and Brush 1994;

Louette et al. 1997; Mercer et al. 2008). This raises questions about how seed networks can facilitate such adaptations.

We cannot, therefore, assume that "modern and scientific knowledge" will simply replace traditional farmer knowledge systems or that farmer knowledge systems themselves are unscientific. Instead, it is imperative that we investigate how farmer knowledge is valued, how such knowledge networks are transformed, and how they are likely to change in the future (Perreault 2005). An increasing number of farmers in nonindustrialized contexts cultivate crops both for subsistence purposes and for the market; they also rely on a variety of wage labor and other nonfarm activities (Zimmerer 2010), and the changing migration and livelihood options of people must be taken into account in our understandings of seed networks and agrobiodiversity governance (Zimmerer 2014; Zimmerer et al. 2015). In some cases, farmers with greater access to markets (i.e., those who migrate or live closer to cities) also cultivate more varieties and have higher rates of diversity (Perreault 2005).

To summarize, farmer practices of seed saving and selection have resulted in high levels of varietal diversity for different purposes (e.g., livelihood, culinary, ritual), thus creating and maintaining agrobiodiversity. Knowledge systems have emerged out of and are embedded in local contexts. Industrialized agriculture risks disembedding such knowledge (Stone and Glover 2017), and has allowed for the quicker spread of modern seed varieties, which may not fulfill all of the same needs as landraces. Farmers have, in turn, appropriated and integrated modern varieties and in many cases "creolized" them, embedding them in their local contexts. Therefore, farmer knowledge and cultural rationales for maintaining diversity need to be understood as dynamic processes within the place-based contexts where they emerge.

Networking Agrobiodiversity

Farmer practices are impacted by the wider social relations out of which they emerge and which they help create in the form of seed networks. Leclerc and d'Eeckenbrugge (2012) have shown that we need a better understanding of how social organization affects crop diversity beyond individual farmer decisions or the interaction between genetic and environmental factors. Seed movement, and ultimately seed diversity, is influenced by rules for marriage, residence, inheritance, and other aspects of social life. In many agricultural societies, farmer seed exchange networks among members of kin groups or neighbors provide not only access to seed; they are a demonstration of trust, reciprocity, and solidarity as well as an extension of cultural values (Coomes 2010; Nazarea 2005a; Thomas et al. 2011).

As Thomas et al. (2011:338) note, "seed exchange cannot be reduced to only its biological or economic dimensions" but must take into account its important social and cultural role. Many seed exchanges still surround marriage rituals with prescribed inheritance patterns. For example, Delêtre et al. (2011) found that seed exchange patterns in Gabon differed in matrilineal versus (virilocal) patrilineal societies. In matrilineal societies, seeds are passed from mother to daughter, whereas in patrilineal ones they are transferred mother-in-law to daughter-in-law. The education of the daughter-in-law in the cultivation of the inherited varieties also serves to initiate her into her new familial responsibilities. Thus, seed diversity and social reproduction are mutually constituted. In many cases, these site-specific social relations have also been shown to directly influence the genetic diversity of seeds (Delêtre et al. 2011).

The scale and range of actors included in farmer seed exchange networks, however, varies greatly. Zimmerer (2003a) found that Andean potato and ulluco seed networks are socially differentiated at multiple scales, but that this is often not taken into account in *in situ* conservation and participatory plant breeding projects, which tend to focus on narrower agroecological ranges than those useful to farmers. In Guatemala and Mesomerica, van Etten (2006) noted that seed exchange historically took place among preexisting social contacts, though not necessarily only locally. In Oaxaca, Mexico, exchange was more likely between people who already knew each other than through a broader form of collective social action (Badstue et al. 2006).

Seed exchange networks extend beyond family and kin ties; class, social differences, and social tensions can also influence exchange relations. For example, in India, networks of reciprocity challenge caste boundaries (Pionetti 2006). Yet seed exchange networks may also create and perpetuate local power dynamics and hierarchies. In Ethiopia, for example, some farmers would rather purchase seeds in a more anonymous market setting to avoid complicated social entanglements (McGuire 2008). Monetary wealth and wealth in seeds are not always correlated: status in seed exchange networks and relations to other farmers also fluctuates with age, gender, and intrahousehold relations (Coomes 2010; Perreault 2005). Furthermore, in addition to farmers, seed networks involve myriad other actors: local breeders, researchers, extension agents, traders, consumers, and others (McGuire and Sperling 2016).

At times, seed exchange networks may also facilitate new social relations, particularly when farmers and organizations create new networks. There has been a revival of seed exchange networks among gardeners and farmers interested in conserving agrobiodiversity in industrialized countries (Balázs and Aistara 2018; Da Vià 2012). Although in many industrialized countries *in situ* conservation of agrobiodiversity is considered nearly extinct, having been replaced by high-yielding and industrialized agriculture, specialized seed saver networks began to form with the awareness of the ecological consequences of industrialized agriculture and fear of genetic erosion. In the United States and Australia, such networks began forming in the 1970s (Balázs et al. 2015). In Western Europe, many seed saver networks emerged during the 1990s and 2000s to protest the rise of genetically modified organisms and defend a model of small-scale diversified agriculture based on the production of local and quality products valorized within short supply chains (Da Vià 2012). These new networks are not necessarily organized only in local communities, but also in national networks that are spread over large territories. The governance of seeds brings such people together into new social networks, coordinated by formalized seed exchange fairs and festivals and often run by nongovernmental organizations or farmer organizations, sometimes with external funding (Balázs and Aistara 2018). Biodiversity seed fairs have also become successful new events in many centers of crop origin (Shen et al. 2017; Tapia 2000). While these new networks often focus on preserving already named heirloom varieties, little research exists on how such national-level networks affect conservation and the continued evolution of plant varieties (Thomas et al. 2012).

New seed networks can challenge assumptions about scientific knowledge and foster new types of collaboration. In France, scientists found that their models of breeding innovation were not applicable to new peasant seed networks, and that they had to rethink selection and production at the population level (Demeulenaere 2014). This dynamic management approach fostered collaboration between scientists and other stakeholders, both in terms of research and political mobilization (Chable et al. 2014; Goldringer et al. 2001). Indeed, there are increasing efforts to create collaborative networks that go beyond farmers to link producers and consumers, farmers and scientists, social and natural scientists, as well as *in situ* and *ex situ* conservation (Demeulenaere 2014; Jackson et al. 2007). Within such collaborations, one must also be attentive to the power dynamics and processes that structure cooperation. As Graddy (2014) has shown, efforts such as those to repatriate potatoes from *ex situ* collections for *in situ* cultivation in the Potato Park in the Peruvian Andes have the potential to reverse or intensify previous contentious power dynamics between *in situ* and *ex situ* conservation. Seed networks are also an important part of ensuring food security and food sovereignty in post-conflict areas, where local and adapted seeds can provide a means of returning to previous land and building resilience against climate change, and often work better in the long-run than hastily developed seed aid programs (McGuire and Sperling 2013; Zimmerer 2017a; see also Chapter 8).

Finally, seed networks are also a way of governing relations with nonhumans. In Costa Rica, for example, organic farmers used seed exchange as a means of creating new types of relationality with other farmers when previous kin-based exchange systems broke down. They created new networks not only with other organic farmers but also with the seeds, pollinators, and other species in the landscape with which they collaborated (Aistara 2011). These affective relations between humans and nonhumans serve as a foundation for the creation of agrobiodiversity and can be studied through multispecies ethnography (Mueller 2014; Tsing 2012). Farmers involved in social networks often see themselves as cooperating and "becoming with" other farmers as well as with other species (Aistara 2011, 2018; Demeulenaere 2014; Nazarea 2005a; van

Dooren 2008). These connections with other species are part of what embed such networks into cultural and ecological landscapes.

To date, the genetic impacts of seed exchange networks are not well understood. Social and seed exchange networks may directly or indirectly influence the genetic diversity of plant populations. Empaire and Peroni (2007) found that manioc diversity in Brazil was the result of an interaction between farmer knowledge, practices, and their social networks. In sorghum cultivation in Cameroon, Alvarez et al. (2005) observed that a small number of more mature farmers serve as seed sources whereas younger farmers serve as sinks, but this status changes as farmers mature. Changes in seed exchange networks could thus alter the migration–drift equilibrium and diminish genetic diversity (Alvarez et al. 2005). Orozco-Ramírez et al. (2016) have shown that, in some cases, ethnolinguistic differences can hinder seed exchange and influence genetic diversity more than altitude. They suggest that further study of seed networks is necessary to detail the links between social and genetic patterns in diversity.

In summary, it is important to avoid oversimplification, romanticization, or unnecessary dichotomization of seed networks and other complex social relations that are intertwined with much of agrobiodiversity governance, such as markets or migration (see Chapters 8 and 15). Coomes et al. (2015) discredit four common, longstanding misconceptions about seed networks: farmer seed networks are not necessarily inefficient; they are not closed and conservative; they are not necessarily egalitarian; and they are not likely to disappear. Instead, seed exchange networks are constantly being reformed and, in the process, they undergo important transformations which must be studied carefully. This will necessitate a range of methodological approaches from various disciplines, such as network science, microbiomics, or others (Pautasso et al. 2013; Poudel et al. 2016).

Power and Politics in Agrobiodiversity

Accounts of farmer practices and seed networks must be contextualized within discussions of prevailing political economic systems, current legislation, and contests over resources that may either facilitate or hinder the preservation of agrobiodiversity in particular places. These are enmeshed within national, regional, and global governance structures that carry their own power dynamics. The international legal system to protect breeders' rights has followed closely on the heels of the technological specialization of breeding, increasingly criminalizing the saving and exchange of seeds by farmers through patents, plant variety protection under the Union for the Protection of New Varieties of Plants (UPOV) convention, and associated national laws that are spread throughout the world via free-trade agreements and other political means (GRAIN 1999). The most important legal regimes are the UPOV Treaty of 1961, the Convention on Biological Diversity (CBD) of 1992, and the International Treaty on Plant

Genetic Resources for Food and Agriculture (ITPGRFA) of 2001. The UPOV
regime, which was developed in 1961, but revised in 1978 and 1991, grants
breeders intellectual property rights (IPRs) on new varieties. While the "breeders' exemption" allows scientists to use protected varieties for research purposes to develop new varieties, this same right is not allowed for amateur
farmer breeders; furthermore, more recent versions of the treaty restrict farmer
seed exchange and increasingly limit the "farmers' privilege" (Aistara 2011,
2012, 2018). Countries are often required to harmonize their national legislation with the UPOV Treaty when joining Free Trade Agreements, previous
national agreements notwithstanding (Aistara 2011, 2012, 2018). The World
Trade Organization's 1995 TRIPS Agreement also requires joining members
to adopt some form of plant variety protection. These agreements thus in some
ways limit state sovereignty to govern their genetic resources, even though
the CBD granted states sovereign rights over genetic resources in 1992. The
CBD requires "fair and equitable" access and benefit sharing and "prior informed consent" in the commercialization of genetic resources, but defining
the country of origin (let alone particular communities or individuals) for crops
or varieties is complicated to impossible (Winge 2016). The ITPGRFA came
into force in 2004, creating farmers' rights as a counterbalance to breeders'
rights, and a multilateral system for access and benefit sharing for particular
crops. Rabitz (2017:629) argues that while this system has facilitated access,
in the form of 3.3 million transfers of plant genetic resources as of February
2017, "no corresponding payments, either mandatory or voluntary, have so
far been made." He argues this is due to important differences in institutional
design and the types of incentives for benefit sharing within the multilateral
framework of the ITPGRFA and the bilateral approach of the CBD's Nagoya
Protocol, which entered into force in 2014 (Rabitz 2017). Guidelines for access
and benefit sharing in the Nagoya Protocol do not contradict the ITPGRFA,
but they also do not develop specific rules for plant genetic resources for food
and agriculture (Chiarolla et al. 2012). As a result, progress on access and
benefit sharing has been slow because most countries have not developed the
national legislation required for implementation, few successful models exist,
and technological advances in gene editing are outpacing legal mechanisms
to guarantee fair and equitable access and benefit sharing (Girard and Frison
2018; Roa et al. 2016). Although IPRs are meant to stimulate innovation, there
is increasing concern that they may do the opposite, and fail to support food
security, adaptation, and resilience (Halpert and Chappell 2017). Some seed
companies engage in voluntary benefit-sharing projects or arrangements with
particular communities as a means of demonstrating corporate social responsibility, but numerous conflicts still surround the implementation of seed legislation (requiring registration and testing), intellectual property rights, and access
and benefit-sharing agreements for plant genetic resources.

    The role of laws and international treaties in agrobiodiversity governance is
discussed in more detail by Visser et al. (Chapter 14). Here I wish to emphasize

their potential to alter relationships farmers have established through seed practices and social networks (Aistara 2018). Intellectual property rights limit the number of people eligible to benefit from plant innovation, and thus can potentially sever or alter relationships among kin, friends, and neighbors, in what Strathern (1996) has called "cutting the network." At the farm level, seed laws and intellectual property rights threaten to transform farmers' relationships to their seeds through bureaucratization, which requires the registration, certification, or testing of all seeds and varieties. At the national level, seed laws may alter farmers' relationships to states and markets. First, they may inhibit farmers' rights to save and work with their seeds. In the worst case, legislation can criminalize farmer seed-saving practices or appropriate seeds by placing variety protection on seeds that have been selected by farmers for generations (Aistara 2011; Graddy 2014; Kloppenburg 1988). This also devalues their knowledge. There is a long history of devaluing peasant knowledge and ways of life, beginning with colonialism and extending through modern development paradigms (Escobar 1995). Farmers who protect agrobiodiversity *in situ* provide valuable ecosystem services for the future and should thus be honored for their work and perseverance throughout these contested histories. Seed laws and international treaties often prioritize new contractual bonds between parties and states over relationships between famers and their seeds and social networks (Aistara 2018). In some cases, however, national governments may make more amenable legislation or hybrid models, often as a result of pressure from farmers' and citizens' groups (Aistara 2014a; Andersen and Winge 2013; Santilli 2012).

At the international level, geopolitical power asymmetries historically structured the appropriation of seeds and knowledge from the Global South by the Global North (Kloppenburg 1988). While some positive examples of protecting farmers' rights or creating access and benefit-sharing regimes are beginning to be documented (Andersen and Winge 2013), on a global level, the geopolitical power dynamics regulating free-trade agreements reinforce hierarchies between the Global North and Global South, between former colonizers and colonies, between resource-rich and resource-poor countries. Seed politics must also be situated within broader rural development policies and politics, which seem to perpetuate rural poverty, despite endless attempts to eradicate it, and to dictate differentiated access of different groups to productive resources (Brush 1999; Dove 1996). As Michael Dove (1996) has observed, it is unlikely that the same systems of intellectual property rights regulations that threaten Indigenous rights and biodiversity will also save them.

Nevertheless, changes in laws that affect biodiversity have been contested in multiple ways, which may have had the effect of stimulating more active seed networks and political action in both industrialized and nonindustrialized countries. Large-scale street protests and the contestation of UPOV-related legislation (called the Monsanto laws) in many Latin American countries have temporarily halted or even reversed the adoption of such laws (Gutiérrez

Escobar and Fitting 2016). There have also been numerous efforts to create new laws that protect Indigenous and farmers' rights, to establish special niches in legislation. For example, several U.S. states changed legislation to prevent seed exchange after newly formed seed libraries were found to be illegal (Balázs et al. 2015). Campaigns in Europe by seed saver groups halted a comprehensive reform of EU seed legislation, which would have eliminated leeway for member states to interpret the European directives in specific ways (Balázs et al. 2015). A brief "tomato rebellion" in Latvia caused a change in national legislation to allow memories and tastes from bygone eras to be preserved alongside EU regulations in a new legal category of "collectors' varieties" (Aistara 2014a). These examples show that seed networks have the potential to defend agrobiodiversity through political mobilization. Finally, defending agrobiodiversity does not necessarily need to take the form of contestation and protest. It can also be achieved through alternative socioeconomic models of relationships between farmers, breeders, and consumers, such as with the creation of the Open Source Seed Initiative (OSSI), analogous to the open software movement (Kloppenburg 2014). This stimulation of political resistance and creative alternatives reflects the deep relationships with practices, people, and places that seeds and agrobiodiversity foster.

Placing Agrobiodiversity

The examples above can help us see how in many cultural contexts, agrobiodiversity is an important part of placemaking. Graddy (2014) has noted that paying attention to place brings into focus critical spatial dimensions of agrobiodiversity governance, and how it fits into the social reproduction of place, seeds, food, knowledge, and memory. Because seeds and foods travel across contexts, placemaking is part of constructing cultural landscapes not only at the local level but across scales (Khoury et al. 2016; Chapter 8).

Relationships between agrobiodiversity and placemaking can be historically traced. Zimmerer (2014) has observed that the concept of cultural landscapes has great analytical potential, but it has not been widely applied in understanding agrobiodiversity. Zimmerer (2015a) has explored how particular cultural landscapes in the Andes were constructed through colonial practices and continue to affect agrobiodiversity governance today, arguing that relational placemaking must be incorporated into future planning for landscape connectivity. African slaves cultivating rice in the Americas in the sixteenth century reinforced their cultural identity, transformed the landscape, transferred an Indigenous knowledge system to a new continent, and increased their negotiating power as they were able to control their own subsistence (Carney 2001; Carney and Rosomoff 2009).

Local knowledge and seed practices are intimately tied to social and cultural identities and relations to place through taste and culinary traditions. Nazarea

(2006) argues that the sensory embodiment of local knowledge, along with emotional memories, are what give such knowledge its power. These memories are tied to particular places and inform practice. For example, displaced persons often try to recreate an "out of place sense of place" through their gardens and kitchens, thus reconstructing shared memories of the places they left behind to share that knowledge with the next generation. Such memories offset the disappearance of older varieties and tastes brought about by the industrialization of agriculture and may become the source of counternarratives (Nazarea 2006). Memories are also often tied to particular cultural symbols. In the Ecuadorian Amazon, Perreault (2005) observed how tending diverse swidden gardens and preparing traditional foods from these gardens were part of what defined being Kichwa, which they celebrated through cultural rituals and hoped their children would continue.

Taste, as a sensory experience, can also be an important driver of memory and motivation for conservation. In Latvia I have shown that the tastes of tomato varieties cultivated during the Soviet era, now illegal because they are not listed in the European Common catalogue, served as motivation to protect the varieties, to protest EU legislation, and to critique current EU policies (Aistara 2014a).

Examples from Nazarea, Graddy, Zimmerer, Carney, and others remind us that the revitalization of place is in direct reaction to previous denigration and marginalization of the very same knowledge systems, people, and places. Early taxonomic projects decontextualized plants and privileged certain types of knowledge over others. As Foucault (1994/1996) noted, plants entered into collections were reduced to "nontemporal rectangles," stripping them of all but their individual names. More recent taxonomic projects try to valorize ancestral knowledge, memories, and local practices (de Haan and Salazar 2006; de Haan and Villanueva 2015) or facilitate a multispecies sense of "care of the species" (Hartigan 2017). Thus, viewing seed practices, networks, and politics as forms of placemaking is a means of recuperating and recontextualizing knowledge and memories associated with particular plants and places. As Graddy (2014) shows, these projects involve a relearning of cultivation as well as culinary and medicinal traditions; they are also ways of creating novel, innovative economies. This is as much about reinventing places in the midst of changed and changing circumstances as it is about preserving old varieties. Because agrobiodiversity is intimately linked with histories, memories, and places, its preservation or maintenance is a deeply political issue. Placemaking is a fundamentally political project, involving political contestation and recreating political subjectivities.

Conclusion: Networked Relational Diversities

My purpose in this chapter has been to highlight how the practices, knowledge, and social networks through which farmers manage seeds are anchored within

cultural memories and future imaginaries of place. These places are further embedded in interlinked ecological, social, and political processes across scales.
Thus, I propose that agrobiodiversity governance be studied as a set of nested
and networked relationships of people to their seeds, practices, and knowledge
systems; to other people and species in their landscapes; to the broader politics
of rural development; and to the cultural imaginary of place and landscape.

Place-based networks for agrobiodiversity governance have multiple
outcomes:

- *Ecologically*, they promote functional relationships at the farm level,
  ensure gene flow and diversity, control pests and diseases, promote the
  creation of mosaic landscapes, and facilitate resilience to withstand
  shocks and adapt to marginal environments and to climate change.
- *Economically*, they provide access to seeds, subsistence, food security,
  income, and a means to diversify economies, create market niches, provide insurance against unforeseen events, and at times even facilitate
  economic innovations such as the Potato Park, OSSI, or others (Bellon
  et al. 2016; Kahane et al. 2013).
- *Socially*, they facilitate sharing of information and knowledge, social
  networks among kin, neighbors, friends, or other potential sources of
  seeds, and promote social innovation through, for example, the creation
  of new seed exchange networks in industrialized countries (Balázs and
  Aistara 2018).
- *Culturally*, because seeds, plants, and tastes embody cultural memories of people, practices, cuisines, places, and times, seed networks facilitate the cultivation and reinventions of these pasts as futures and as
  cultural landscapes.
- *Politically*, they promote the mobilization of political subjectivities of
  resistance and push for legal changes to protect farmers' rights and mobilization for more sustainable food systems.
- *Scientifically*, they promote interaction and links between farmers and
  scientists and between *in situ* and *ex situ* conservation.

Utilizing a placemaking lens allows us to study these nested social, cultural,
and ecological relations and their associated power dynamics and explore the
various processes through which farmers create, protect, or perpetuate diversity on their farms as well as the historical and cultural rationale for doing so.
Relationships are formed between farmers and their land, seeds, pollinators,
and other species in their landscapes; with other farmers, stakeholders, and
institutions; and within the broader political–economic contexts and legislative frameworks in which they operate. A nested and relational approach is
essential if we are to understand the drivers and potential for preserving agrobiodiversity in the future.

14

The Governance of Agrobiodiversity

Bert Visser, Stephen B. Brush, Guntra A. Aistara,
Regine Andersen, Matthias Jäger, Gabriel Nemogá,
Martina Padmanabhan, and Stephen G. Sherwood

Abstract

Agrobiodiversity relates to humans and their environments. It is the result of interactions between humans and nature, and thus is simultaneously social and biological by nature. Without humans, agrobiodiversity would not exist. Seeds, as carriers of major agrobiodiversity components, are not mere material objects that exist outside of social relations: they are also sociobiological artifacts embedded in these relations. The multifaceted, highly dynamic realities of agrobiodiversity mean that those interested in questions of governance need to understand the limitations and political implications of the complementary and sometimes contradictory instrumental and relational perspectives on seeds; that is, the understanding of seeds as a production input or as the subject of a social network, in which agrobiodiversity brings together production and social linkages. International instruments aim to provide a legal basis for mediating competing interests and methodologies. In addressing governance, the global framing of these instruments reflects the dynamics of agrobiodiversity in global socioeconomic and environmental changes. From the earliest recognition of the potential value of crop diversity, crop genetic resources were treated as public goods in the public domain. Breeding companies have opposed this treatment. Breeders sought exclusivity and reward for their creative activities in using genetic resources to create novel varieties. Governance of agrobiodiversity—defined by a set of relationships that influences the access to and conservation, exchange, and commercialization of agrobiodiversity—reflects underlying value systems. Conflicting approaches (e.g., "stewardship" vs. "ownership" approaches) toward governance based on divergent value systems and rationales can be

Group photos (top left to bottom right)   Bert Visser, Stephen Brush, Guntra Aistara, Regine Andersen, Gabriel Nemogá, Matthias Jäger, Stephen Sherwood, Martina Padmanabhan, Stephen Brush, Gabriel Nemogá, Bert Visser, Stephen Sherwood, Guntra Aistara, Matthias Jäger, Regine Andersen, Bert Visser, Martina Padmanabhan, Gabriel Nemogá, Stephen Brush, Guntra Aistara, Martina Padmanabhan

distinguished. It is important to identify the actors involved, from local to global, to understand the power dynamics that influence the interactions among these various actors and their ability to influence or control the management of agrobiodiversity. The governance of agrobiodiversity and the power dynamics involved are increasingly crucial in the context of rapidly changing farming and food systems, especially in the context of globalization, migration, and urbanization. This chapter elaborates an emergent research agenda, focusing on aspects of power relations in agrobiodiversity governance, agrobiodiversity and food systems, nutrition, taste and health, and the governance of genetic information.

The Concept and Scope of Governance

Together with agriculture, agrobiodiversity has developed over the last 10,000 years, and in localities across the globe, multiple forms of governance have coemerged with this development. Agrobiodiversity relates to humans and their environments. It is the result of interactions between humans and nature, and thus it is simultaneously social and biological. Without humans, agrobiodiversity would not exist. Agrobiodiversity and seeds, as carriers of major agrobiodiversity components, are not mere material objects that exist outside of social relations: they are also sociobiological artifacts embedded in these relations. In particular, in small-scale and traditional agriculture, many people have intimate and strongly affective relationships with their environment and the agrobiodiversity embedded within it, as part of broader social systems and cultures (Nazarea 2006). While humans select plants for agriculture and food, the resulting crops and their ensuing biological and ecological consequences on the environment help to shape humankind. This seamless sociobiological character warrants speaking of agrobiodiversity as a highly relational product (see Chapters 12 and 13 as well as Zimmerer 1997:186–205). It requires not only describing, developing and conserving species, varieties, and traits in agricultural biodiversity, but also understanding the linkages between natural artifacts and the human activity involved in maintaining, losing, and further developing them.

The multifaceted, highly dynamic realities of agrobiodiversity mean that those interested in questions of governance need to understand the limitations and political implications of the complementary and sometimes contradictory instrumental and relational perspectives on seeds; that is, the understanding of seeds as a production input or as the subject of a social network, in which agrobiodiversity brings together production and social linkages (Caillon and Degeorges 2007). These linkages unfold between seeds and users, but they also occur among users (Leclerc and d'Eeckenbrugge 2012). The relational perspective, which largely has been neglected in the social research on agriculture and food, leads to at least two insights. First, when considering the social value and "life" of seeds, the identity and qualities of the seed take on specific significance for the user and can provide an important basis for shaping her or his identity and sense of being (Padmanabhan 2007). Second, with the perception of genetic resources as centers around which social networks emerge, seeds enter the realm of ownership, power, inclusions, and exclusions (Aistara 2011).

Moreover, agrobiodiversity is not a stable phenomenon; it is the result of continuously evolving interactions between people and their environments, and thus a product of the multifaceted coevolution of human societies and their biological environments. Therefore, conservation management approaches regarding agrobiodiversity need to recognize the essential role of ongoing processes in the field of practice—be it that of farming, marketing, and circulation of goods and services as well as crop utilization and food consumption (Brush 2000; Veteto and Skarbø 2009). Extracting and isolating agrobiodiversity from its natural environment greatly reduces its scope and capacity to coevolve. When extracted from its social or environmental context, for example, or when leaving a field or plate and entering a laboratory or gene bank, only a limited number of the multifaceted and highly nuanced sociobiological qualities of a seed survive. Nevertheless, new traits are unraveled and new uses of crops are developed in breeding and research, sometimes resulting in game-changers for farmers and consumers globally.

Not only do substantial differences underlie the science and governance of agriculture and food (Sherwood et al. 2016), there can also be substantial heterogeneity between modern and traditional peoples or practitioners outside the mainstream, with important implications for how people experience, think about, and seek to govern agrobiodiversity and seeds. In particular, for Indigenous and other native peoples, seeds and other life reproductive forms are not necessarily understood or experienced as a materiality or object; they may take on other, more inclusive and integrated meanings and expressions as well as spiritualities. Such views and beliefs can form a fundamental part of their autonomy, livelihoods, and collective identity. As such, socially inclusive, responsible agrobiodiversity governance demands that conservationists and scientists as well as policy makers find ways to create space for and accommodate the unique worldviews and needs of traditional peoples (Nemogá 2016). Many other family and smallholder farmers and their rural communities that contribute to the maintenance of agrobiodiversity and plant genetic resources share similar experiences and concerns.

In many communities, agrobiodiversity forms a major part of the living environments of farmers and often plays major roles in shaping cultural identity and food systems. The seeds maintained in such farming systems travel through social exchange networks with their own internal norms (Pautasso et al. 2013) and rules and within specific cultural and geographic spaces (Zimmerer 2003a). When seeds are brought from place to place, they may serve as markers of memory, place, and family ties, as well as embody sociobiological relationships between the people who nurtured and exchanged them, be it over shorter or longer periods of time and space (Nazarea 2005a, b). The social networks that people build through and around seeds also allow for the sharing of experience and knowledge concerning natural elements, such as the soils, microbes, pollinators, and other living organisms that form part of local agroecosystems and food production. In most cases, Indigenous communities and other traditional peoples do not seek to conserve agrobiodiversity, as such, nor for its own sake. They maintain their gardens, farms, and landscapes where their seeds help to secure certain livelihoods as well as ways of living and being. Through such processes, they come to affect, effect, and determine the agrobiodiversity that surrounds them and to which they are attached (Almekinders and Louwaars 2002).

In modern industrialized food production systems, in both developed and developing countries, farmers increasingly have become detached from the agrobiodiversity that originated and surrounds their crops and livestock. Seeds and livestock may feature traits that have been developed in remote locations and been acquired in commercial markets serving largely different ecosystems and divergent geographies. In this context, seeds have become dispossessed and commoditized, and hence, they are reduced to a store-bought input, not unlike fertilizers, pesticides, and equipment.

Historically, the collection, taxonomic classification, and adaptation of seeds and plants in accordance with practices in research and industry have often come to mean the separation of these seeds and plants from the sociobiological context in which they were domesticated as well as from the knowledge systems in which they functioned, effectively rendering their nuanced context redundant and irrelevant (see Chapter 6 and Kloppenburg 1988). To appreciate the role of agrobiodiversity and its relationship with knowledge systems, farming, and food production practices, the function of seeds and crop plants in small-scale and traditional agriculture must be understood in the situated context in which they historically developed and continue to function.

This reality poses a dilemma. The multiple ways in which people relate to agrobiodiversity reveal a myriad of lifestyles, visions, cultures, and beliefs as well as the social systems that help to determine how resources are owned, exchanged, and distributed. Rather than simply reflecting different views and experiences, these nuanced relationships reflect unique histories and different ways of living (Kohn 2015). Methodologically, such considerations lead to unique questions and a different type of research (Nemogá 2016). To appreciate fully the potentialities of agrobiodiversity and the wealth of options for conservation and governance, the physical, biological, social, and cultural contexts must all be taken into account and multiple worldviews need to be managed or accommodated; hence, other and new questions need to be raised and addressed.

Researchers wishing to work with people of unique experience and value systems must not only be respectful of multiple and sometimes incompatible worldviews, they must also have the willingness and ability to represent competing worldviews as equally valid, thus strengthening the unique plurality that historically gave rise to rich patterns of agrobiodiversity as well as promoting relationship building and trust implicit in a highly multi-cultural, cosmopolitan world. This requires a critical awareness of both the process and the outcome of research and governance as well as their political utilizations.

Governance and International Policies and Institutions

International instruments aim to provide a legal basis for mediating com-peting interests and methodologies. The recognition of Indigenous Peoples, for instance, is reflected in the United Nations Declaration of the Rights of Indigenous People, the International Labour Organization's Convention 169, the Convention on Biological Diversity (CBD), its Nagoya Protocol, and the International Treaty on Plant Genetic Resources for Food and Agriculture (ITPGRFA). Full implementation of these international legal instruments, and in particular the access and benefit-sharing regimes and the Farmers' Rights provisions contained in these instruments, is required to support and assist lo-cal and Indigenous communities and farmers in maintaining and conserving the agrobiodiversity that is part of their environments and cultures. This is essential to support the farmer-driven development and conservation of global biodiversity and the utilization of its components for the purpose of food and agriculture, and for global food and nutrition security. In addressing gover-nance, the global framing of these instruments reflects the dynamics of agro-biodiversity in global socioeconomic and environmental changes (Andersen 2016, 2017; Zimmerer 2010; see also Chapters 6 and 8).

Researchers identified the cradle areas of domestication of major food crops almost a century ago, and this analysis rested in part on the geographical loca-tion of an abundance of biological diversity of crop species (Vavilov 1926a). In subsequent years, other regions of diversity were discovered elsewhere and linked either to different crops or to major crops as secondary centers of origin (Harlan 1992). Crop genetic diversity was recognized as a significant asset to the emerging science of crop breeding and to the farmers who maintained it. Indeed, the successful use of crop diversity soon led crop scientists to worry that diversity in crop species was vulnerable to loss—or "genetic erosion"—as newly created varieties replaced older and more diverse ones (Harlan and Martini 1936).

Four decades later, alarm was raised about significant loss of diversity in major crops that had been the centerpieces of the Green Revolution, in par-ticular self-pollinating wheat and rice (Harlan 1975). The evidence of the loss of diversity was based on observing the spread of high-yielding varieties in the prime production areas. Few ecological studies of crop populations in the regions of crop diversity were available when gene banks were established, but such studies have now confirmed that agrobiodiversity continues to exist (Brush 2004; Zimmerer 1997) not only in relatively marginal agricultural areas, but also in more intensive agroecosystems (Chapter 8). The threat of genetic erosion was met by the collection of genetic resources stored *ex situ* in national and international gene banks. Accordingly, these resources have been available to crop breeders in the public and private sectors. In the meantime, various studies have shown that farmers' contemporary management of agrobiodiversity is highly resilient and that predicted loss of races, landraces, or genes has not necessarily materialized (de Haan et al. 2013; Perales and Golicher 2014; Zimmerer 2013).

From the earliest recognition of the potential value of crop diversity, crop genetic resources were treated as public goods in the public domain; that is, without specifying ownership and governed by open access (Fowler and Mooney 1990). Crop breeders, especially in the private sector, have opposed this treatment (Chapter 6). Breeders sought exclusivity and reward for their creative activities in using genetic resources to create novel varieties. Governments across the planet provided breeders' rights and, in some jurisdictions, patents as a form of intellectual property rights (Berland and Lewontin 1986). A series of laws and legal decisions provided this type of protection over the course of the twentieth century, with significant milestones being the establishment of the Union for the Protection of New Varieties of Plants (UPOV) in 1961, and the 1994 Agreement on Trade-Related Aspects of Intellectual Property Rights of the World Trade Organization.

The legal system established in the 1960s, which allowed breeders to seek intellectual property while farmers' varieties were still treated as goods of the public domain, was challenged during the Food and Agriculture Organization's (FAO) negotiations of the International Undertaking on Plant Genetic Resources for Food and Agriculture, adopted in 1983 (Mooney 1983). This challenge was fortified by the fear of the rapid loss of diversity, and subsequently voiced in the negotiations that led to the adoption of the CBD in 1992 and the ITPGRFA in 2001. Most recently, the Nagoya Protocol on Access to Genetic Resources and the Fair and Equitable Sharing of Benefits Arising from Their Utilization was adopted in 2010 to ensure better implementation of the access and benefit-sharing provisions established under the CBD. Whereas the provisions of the International Undertaking on Plant Genetic Resources reflected the principle that genetic resources were part of the common heritage of humankind, this foundation was increasingly challenged during the late 1980s within FAO as well as in other international arenas. Critics of an unbalanced regime governing plant genetic resources for food and agriculture raised the following primary concerns (Bellon et al. 2005):

- The inherent inequity of contrasting property systems for farmers' varieties (open access, public domain) and breeders' varieties (various forms of intellectual property)

- The loss of crop diversity from farming systems in cradle areas of domestication and elsewhere
- The poverty and economic marginalization of the smallholder farmers who continue to act as stewards of crop genetic resources

At the international level, the CBD replaced the principle of common heritage of humankind with that of national sovereignty over genetic resources and established a bilateral approach requiring agreements between countries owning these resources and users based on prior informed consent and mutually agreed terms. The ITPGRFA conformed to this principle but used it to establish a multilateral system of access and benefit sharing that specified continued public domain (open access) management for a list of 35 major crop species and 29 forage crops. The third development followed from both the CBD and the ITPGRFA: national systems of access and benefit sharing were established for the management of genetic resources embodied in biological diversity. Both the CBD and the ITPGRFA also had the objectives of stimulating local communities and Indigenous People to conserve the genetic resources that they manage and to provide a mechanism to address, at least partially, poverty and economic marginalization by giving economic value to their genetic resources.

An important achievement was the recognition of farmers' rights related to crop genetic resources in the ITPGRFA. However, these rights were not defined in any detail, and their implementation was left to national governments without further guidance other than three proposed measures: protection of traditional knowledge, the right to participate in benefit sharing, and the right to participate in decision making at the national level. Furthermore, the rights that farmers may have to save, use, exchange, and sell farm-saved seed were addressed but without giving specific directions or guidance regarding the implementation of those rights (for examples involving agrobiodiversity and the rights of Indigenous Peoples, see Chapter 12).

In this global policy context, the institutes of the Consultative Group on International Agricultural Research (CGIAR), established in the 1970s, were able to play a major role in addressing food security and preventing global hunger by developing and promoting new crop varieties and better agronomic practices, widely known as the Green Revolution. National agricultural research systems were provided with new materials which were then adapted to national needs and circumstances and grown over very large acreages across the developing world. Nevertheless, in particular the provision of higher-yielding crop varieties was the cause of genetic erosion in major food crops and the rise of new biotic and abiotic pressures as a result of homogenization of production methods and the promotion of higher level agricultural inputs (fertilizers and pesticides), thereby exemplifying the effects of major global trends in agricultural production outlined above. Bioversity International, formerly known as the International Plant Genetic Resources Institute, was to lessen and prevent the negative effects by promoting conservation, and from the 1980s it coordinated field collection missions and the establishment of gene bank collections across the CGIAR centers.

Currently, it is widely believed that the goals embedded in the international treaties and national laws, intended to stimulate use and conservation and provide economic compensation to stewards of genetic resources, have only partially been met (see Chapters 12 and 13). The loss of biological diversity and genetic resources for food and agriculture is generally thought to continue at a fast pace (Ahuja and Jain 2015), even though evidence for this assumption remains limited to date. Likewise, the poverty and marginalization of Indigenous People and other local communities that maintain crop genetic resources remains largely unaffected by the access and benefit-sharing systems at the national level (Peschard 2014). The emerging consensus is that these regimes have generally failed to create viable and sustainable means to address the loss of biological diversity or economic marginalization of stewards of genetic resources in agriculture (Carrizosa et al. 2004). Indeed, an impetus for negotiating the Nagoya Protocol was the perceived failure of the access and benefit-sharing systems generated by the CBD (Marion Suiseeya 2014).

Multiple Expressions of Governance

Governance of agrobiodiversity is defined by a set of relationships that influences the access to and conservation, exchange, and commercialization of agrobiodiversity and its components. Governance approaches, in particular, reflect initiatives that involve technology transfer (seeds embodying technology), levels and forms of participation by different stakeholders, and their ability to self-organize. Governance triggers policy and determines the course of action. Control over, access to, and use of agrobiodiversity form major expressions of governance at all levels (from the international to the local) and are reflected in rules and practices: from local markets to international legal instruments, and from barter and exchange of local varieties and their products to the development, regulation, and distribution of genetically modified crop seeds used by traditional and small-scale production systems to large-scale intensified production systems.

With regard to the management and conservation of agrobiodiversity, formal governance aims to regulate the access to agrobiodiversity and the use of its components, expressed in stakeholder actions, international legal instruments, and national legislation and regulations, and related policies. Not only the instruments referred to in the section above, but also seed policies and laws as well as intellectual property rights regimes are expressions of formal governance directly and expressly bearing on the management of agrobiodiversity. Implementation of these seed policies and laws as well as the intellectual property rights regimes is not necessarily in line with their intentions.

Implementation may even be lacking, depending on financial resources and capacity of the countries concerned.

Beyond formal governance expressed in policy and law, informal governance is used in various forms by divergent actors. At the community level, cultural and social norms and the membership of groups holding collective cultural identities (e.g., Indigenous Peoples as well as social networks that connect different communities) determine the governance of agrobiodiversity as well as the way it is conserved and used. Informal governance has historically been, and continues to be, the main framework through which on-farm conserved diversity is reproduced and exchanged. Informal governance is also executed through markets through

- the commodification of seeds and the exclusive development of hybrid varieties binding farmers to seed companies,
- the creation of linkages between products and their origins (such as through geographic indications, e.g., basmati rice),
- the means and conditions of production (e.g., agroecological or organic products, direct purchasing, and fair trade), and
- the informal trade networks and farmers' markets that facilitate the movement of seeds.

Moreover, both formal and informal governance regimes themselves may reflect major societal developments, as in urbanization (consumers become detached from food production) and migration (food products and the crops and varieties from which these are derived are displaced to new environments and social contexts) (Chapter 8). In summary, governance operates along a continuum of formal and informal mechanisms and processes that together affect how individuals, communities, corporate groups, and governments relate to agrobiodiversity and determine the state of agrobiodiversity.

Governance reflects underlying value systems. For agrobiodiversity, conflicting approaches (e.g., "stewardship" vs. "ownership" approaches) toward governance based on divergent value systems and rationales can be distinguished (Andersen 2008, 2016). The stewardship approach represents the concept that agrobiodiversity belongs to the common heritage of humankind: genetic resources should be shared for the common good, as part of the public domain. The stewardship approach was regarded as the dominant rationale throughout the history of agriculture until the advent of intellectual property rights regimes in agriculture. By contrast, the ownership approach holds that establishing individual or collective ownership of genetic resources provides important incentives to promote breeding as well as the conservation and sustainable use of agrobiodiversity. Furthermore, ownership enables control over genetic resources covered by ownership rights for the holders of such rights, and it makes possible their trade as well as benefit sharing. One could argue that the ownership approach underlies not only intellectual property rights regimes but also the sovereign rights of nation-states reflected in the access and benefit-sharing concept embodied in the CBD and the ITPGRFA. A farmer-based approach, including proposed farmers' rights to seeds, has been formulated as a hybrid of the stewardship and ownership approaches (Brush 1991, 1992).

Whereas the stewardship approach embodies the recognition of and respect for the value of maintaining agrobiodiversity and is often grounded in collective responsibility, the ownership approach allows for extraction of agrobiodiversity and the commodification of its components, regardless of whether their use is sustainable or not. Both approaches can result in unforeseen effects. The stewardship approach may result in misappropriation of agrobiodiversity components by third parties and consequent tendencies toward strict protection and isolation, whereas the ownership approach may result in limitations to access and disincentives to share agrobiodiversity among farmers.

As intellectual property systems are costly institutions, the capacity of many developing countries that are often rich in genetic resources but poor in financial resources to develop and effectively use such systems is limited (Andersen 2008). This situation has resulted in power asymmetries, which have been met with much protest against intellectual property rights to genetic resource products from stakeholders in the Global South, along with the demands of securing control over the genetic resources through access and benefit sharing modalities.

It is clear that—as a consequence—these divergent formal governance approaches have been adopted at different geographic and temporal scales, and by different groups of actors. Whereas the access and benefit-sharing system under the CBD and the Nagoya Protocol can be said to be part of an ownership approach, the benefit-sharing system under the ITPGRFA is based on a derived rationale. Nation-states have placed some genetic resources in a multilateral system to provide access to all users, stating that access to genetic resources is the greatest of all benefits. The ITPGRFA does not establish owners of genetic resources in the multilateral system, but rather provides that benefits should flow to the farmers who conserve and sustainably use crop genetic resources. This may be seen as a reflection of the stewardship approach. Farmers' Rights, as addressed in the ITPGRFA, can be linked to both approaches depending on how they are interpreted and implemented (Andersen 2008, 2016). In reality, in many countries and policies both the ownership and the stewardship approach are recognized and respected to a variable extent, and both are enacted in new policy and legislation, even without sufficient recognition for the incongruities between the two approaches.

An unfounded assumption dominant among many scientists and policy makers is that a global shift from stewardship to ownership is bound to occur and will be irreversible, since this shift is necessary to create sufficient investments in breeding and will thus serve the goals of global food and nutrition security. A contrasting assumption prevalent among farmer organizations, scientists involved in the conservation and sustainable use of crop genetic diversity, and some civil society organizations, holds that the ownership approach will enable different actors to exclude each other from access to and use of their genetic resources. It will thereby reduce the legal space for all to safeguard food security and contribute to the conservation and sustainable use of crop genetic diversity (Andersen 2008; Salazar et al. 2007). Whereas the stewardship approach may seem most beneficial to maintain crop diversity *in situ*, the paradox of this approach is that without sufficient measures to prevent misuse, genetic resources and information from the public domain may be privatized and thus become subject to ownership. For a stewardship approach to succeed, it is therefore necessary to introduce measures to ensure that genetic resources and the knowledge associated with them remain in the public domain and cannot be misappropriated.

Initiatives based on the stewardship approach have been undertaken across the globe to document genetic resources developed and maintained by rural and Indigenous communities, as through the potato catalogues developed by the International Potato Center and local farmers from the Huancavelica and La Libertad regions in Peru (de Haan and Salazar 2006; de Haan and Villanueva 2015; Scurrah et al. 2013). The catalogues recognize farmers' contributions in maintaining the varieties and provide access to information about the potatoes and the people maintaining them.

The values underlying the stewardship and ownership approaches may be expressed as government regulations and community norms (formal governance) as well as incentives, motivations, and social recognition for particular forms of management of agrobiodiversity (informal governance). Actors in agrobiodiversity use and management might act from different worldviews and apply different rationales governing their decisions regarding the governance of agrobiodiversity. What appears as rational is very much tied to the institutional framing of any given situation and the motivations that individuals or communities hold. Some farmers will prefer their own landraces (also referred to as farmers' varieties to stress their explicit management) to higher-yielding modern varieties, knowing that modern varieties will often outperform landraces but that their own landraces offer better harvest security and yield stability in each season. Other farmers may prefer to grow their own landraces despite lesser yield because of a preference for certain tastes or textures that a given landrace displays (Birol et al. 2006; Chapter 15). These preferences are often related to culinary traditions and use, which form the core of cultural identity. Adivasi or Indigenous farmers of the Kurychia community in Kerala, India, for example, tie religious celebrations to a communal feast of landrace rice. A moral obligation to grow appreciated varieties or sentimental attachment to certain crops are among a range of motivations to encourage agrobiodiversity (Brush and Meng 1998). Framed in the rationality of the market, a diverse cropping portfolio might appear unprofitable. Nevertheless, when evaluated in light of other values beyond commodification, such a portfolio reveals a whole range of emotional, tacit, and spiritual relations, including additional embedded production "costs" that make sense to the people maintaining them (Chapter 15). Therefore, it is important to distinguish and recognize multiple motivations and to advocate for support of such choices in government policies and through other means of governance, such as payment approaches (Narloch et al. 2011a, 2013, 2017).

These distinct rationales underlying smallholder practices and commodification are also reflected in the balance between *ex situ* conservation and *in situ* management and conservation efforts, further elaborated below in the section on power dynamics.

Power Dynamics That Influence Governance

Considering how power dynamics influence the management of agrobiodiversity, it is initially important to identify the actors involved, from local to global. In this context, actors are understood as categories of persons or institutions that reflect on their interests, take positions, and start seeking influence over the processes relating to the governance of agrobiodiversity. Not all actors take a single position or seek to exert influence at all times.

Relative to the governance of agrobiodiversity, actors may be grouped into three major categories:

1. Private sector: seed companies, food processing industry, and retail concerns as well as some public research and breeding institutions having to recoup their investments from sales of their capacity and outputs that are involved in the study and use of components of agrobiodiversity and that seek influence in decision making.
2. Civil society: nongovernmental organizations, farmers' groups or organizations, organizations and groups of Indigenous People, and consumer organizations that also seek influence in decision making.
3. Public sector: politicians, government officials, and other policy makers as well as some public research institutions under direct government control and funding that respond to actors seeking influence by creating and/or implementing relevant policies and legislation.

All actors may function at local, national, regional, and global levels. Researchers, in particular, may play roles in each of these sectors. For instance, they may perform fundamental research, engage in public–private partnerships, contribute to plant breeding and seed development, carry out contract research for the private sector, advise NGOs and farmers as well as be active in government institutions (Baranski 2015b; see also Chapter 6).

While agency is traditionally understood as a human quality, some social science approaches also consider how nonhuman actors (e.g., seeds, plants, pollinators, and soil microbes) influence and mediate human relations and agrobiodiversity governance, thus displaying a kind of agency (Kirksey and

Helmreich 2010; Kohn 2015). This is related to the broader philosophical question of whether nature and culture should be considered two separate realms or as a more intertwined nature–culture domain (Descola and Pálsson 1996). In the former perspective, humans as part of culture are considered to be dominant over nature, a stance that lends itself more easily to an ownership approach toward agrobiodiversity governance. In the latter perspective, humans themselves are considered part of nature along with other nonhuman actors— a stance that lends itself more easily to stewardship approaches to agrobiodiversity governance. The perspective in which humans see themselves as a part of nature and give agency to other natural life forms is prevalent in many Indigenous, native, and traditional peoples' cultural worldviews, and therefore important to a relational understanding of the governance of agrobiodiversity. We stress that this relation to nature cannot only be found in present-day traditional cultures but in ancient Western culture as well (e.g., the similarity between the concepts Pachamama and Gaia).

There are different ways of understanding the power dynamics that influence the interactions among these various actors and their ability to influence or control the management of agrobiodiversity. One approach (Andersen 2008) distinguishes between two forms of power: structural power and ideational power.

Structural power can be defined as the ability to shape and determine the structures of political systems within which states, their political institutions as well as public, private, and civil sectors can operate, in other words to shape and determine the rules of the game (Strange 1988). The exertion of structural power in our context aims to influence or control the management of agrobiodiversity as reflected in (a) political decisions regarding the conservation, access, and use of agrobiodiversity, and in particular (b) international agreements, codified as the conservation of biodiversity, the utilization of its components, and benefit sharing resulting from its utilization. The possibility to shape and determine the structures of political systems in relation to agrobiodiversity is not limited to politicians and policy makers, but extends to all other actors in the public, private, and civil sectors. The extent to which actors can be successful in their attempts to influence depends on their capacity and competence as well as their financial means and access to power. For example, multinational companies have greater resources and can therefore pursue their interests often with greater success than traditional agricultural communities and Indigenous Peoples. The privatization of components of agrobiodiversity in the form of intellectual property rights over plant varieties that has taken place during the last few decades can be considered a result of exerting structural power. Breeding companies have influenced political decision making to create a legislative framework conducive to allowing the privatization of plant genetic resources and promoting the use of private sector varieties, in particular through the establishment of intellectual property rights legislation covering plants and plant varieties, and seed legislation governing the registration, production, and distribution of varietal seed.

Ideational power, a term coined by Rosendal (2000), describes the power to exploit knowledge and promote certain norms and ideas (rather than influence policy and law). Exercising ideational power aims to influence the development and diffusion of certain knowledge, norms, and ideas, often by actors who share a common set of views, beliefs, and knowledge. The actors may reach from advocacy coalitions (Jenkins-Smith and Sabatier 1993, 1994) and other international expert networks (Haas 1992), which share a set of policy core beliefs or a certain knowledge base, to researchers and marketing specialists in the retail sector seeking to influence human behavior (e.g., global spread of fast food and soft drinks). Ideational power may be exerted to change consumer habits and to promote certain food habits as modern or comfortable, but may also take the form of social protest, coercion or shaming, information sharing, and social learning (Checkel 1999). The main sources of ideational power are the moral authority based on argumentation (e.g., campaigns for Fair Trade and Slow Food), or the analysis of human behavior and preferences (e.g., the spread of fast food, such as McDonald's and Coca Cola). As a further example, the development of the concept of access and benefit sharing during the negotiations of the CBD that stressed the ethics underlying the concept, may be regarded as the result of exercising ideational power (Andersen 2008; De Jonge 2011).

In responding to these different manifestations of power, the state may play an important role itself by exerting structural and ideational power and balancing the power influences within its various institutions. The extent to which the power dynamics described above translate into formal political decisions and their implementation and enforcement depends on the institutional capacity of the state. In turn, the institutional capacity of the state to respond properly is dependent on knowledge and expertise, human and financial resources, and leadership (political clout as well as inclusive decision-making processes) as well as the capacity to exclude law avoidance and malpractices and balance interests in a democratic manner (Hanf and Underdal 1998; Jänicke 1995). Over the last few decades, civil society (including NGOs, farmers' organizations, Indigenous Peoples' organizations) has striven toward an alternative policy and legal framework for the conservation and management of agrobiodiversity (Aksoy 2014). For a strong civil society to have influence on the decision-making process of the state and to contribute to the introduction and implementation of proper policy and legal frameworks, a strong and responsive state is a prerequisite. This phenomenon, in which civil society needs a strong opponent, has been labeled as the civil society paradox (Walzer 1992).

Power asymmetries are neither an accident nor an oversight, but rather a product of formal and informal processes. Asymmetries result in different levels of access and control over agrobiodiversity, but they also influence and structure relationships in rural communities with regard to the use of and property rights over land, water, seeds, credit, and information. Social dimensions (e.g., age, gender, caste, ethnicity) create power asymmetries, as the intersection of these categories determines the social appropriation of control and access, including over the use of agrobiodiversity and its components. The social collective must back the exertion of property or use rights. For this reason, it is essential to consider social stratification, internal power asymmetries, and possibly diverging interests within local communities. For instance, in the southern Kerala State of India, Indigenous Kurichya women are charged with preserving seeds of rice landraces but they have no control over the decisions of when and where to plant them. So, when Kurichya women have the chance to cultivate their own private fields, they tend to prefer nonlocal seeds from extension services or other external sources to avoid conflicts about the control over the landraces belonging to their Indigenous communities (Suma and Großmann 2017). Also, concepts developed in the context of the international agreements do not automatically resonate with local communities, which need time to understand the perspectives taken and the consequences of these concepts for their own living.

Yet another approach to the analysis of power relationships is provided by the Foucauldian tradition, according to which power is not really "held" by the state nor by any other actor (Foucault 1980). Rather, power in itself is essentially fluid, in perpetual circulation through society and dispersed among its actors, manifesting itself in every social interaction between or among various actor groups, as can be deduced from relevant discourse and practices. When repeated, these discourses and practices may reinforce themselves into particular narratives, which over time become more difficult to challenge, although counterdiscourses may also form in reaction. In this approach, because power is fundamentally unstable, one must analyze not only how power is manifested in various forms of agrobiodiversity governance, but also how the seeming stability of powerful actors and their narratives might be overturned. For instance, the collection of plants from their places of origin and their incorporation and description in collections and herbaria in the eighteenth and nineteenth centuries by European colonists can be analyzed as a set of practices and discourses whereby colonial rulers, who valued scientific knowledge over local knowledge and maintenance in collections over maintenance in local communities, exerted their power and reinforced discourses of superiority of Europeans over non-Europeans (Bonneuil 2002; Foucault 1994/1996). Such exertion of power has since been challenged, ultimately resulting in the re-appreciation of on-farm management of genetic resources and associated farming and knowledge systems in the CBD and ITPGRFA, and in the establishment of the concept of access and benefit sharing. It has also been successfully challenged, for example, in recent protests against the implementation of the European Common Catalogue of plant varieties in Latvia, which had originally excluded many varieties that had been cultivated during the Soviet era, until corrective legislative change, thus demonstrating the power of counterdiscourses even in seemingly unchangeable situations (Aistara 2014a, b). Furthermore, a more dynamic understanding of power allows us to investigate and explore how power circulates within and across different scales and how it changes relationships, for example, within local communities, and between communities and states. It allows us to analyze any intervention aiming to improve agrobiodiversity governance, including through legislative changes, and the relationships and balance between *in situ* and *ex situ* initiatives. Analysis and self-reflection on potential power asymmetries even in scientists' own initiatives is crucial for improving contributions from the scientific domain to agrobiodiversity governance.

Since the 1960s, power relations have also influenced the debate on the primacy of either on-farm management or *ex situ* management of plant genetic resources (de Wit 2016, 2017). Whereas technical arguments were often used to prioritize the one over the other, governance played a major role in how these two conservation approaches relate. *Ex situ* approaches are extractive, remove the genetic resources from farmers' fields and communities, and result in the use of collections that first and foremost serve the public and private breeding sectors. In contrast, on-farm approaches keep the access to and control of plant genetic resources in the hands of farmers and allow for a continued direct use by farmers of these resources. Clearly, opposing governance approaches serve different beneficiaries. Both the CBD and the ITPGRFA recognize the importance of on-farm management of genetic resources as being complementary to *ex situ* conservation efforts. Whereas the text of the CBD regards *ex situ* conservation of genetic resources as complementary to *in situ* conservation, many actors have since focused on strengthening *ex situ* conservation efforts; *in situ* management by smallholder farmers and Indigenous or traditional communities has only gradually gained more support (Jarvis et al. 2011; Visser et al. 2019; Oxfam Novib coordinated seed system initiatives). Yet, globally the funding for research and action supporting on-farm and *in situ* conservation is marginal compared to *ex situ* conservation.

In reality, *ex situ* conservation and *in situ* management and conservation are highly interlinked and complementary. Whereas *ex situ* conservation conditions may not be attainable in farming communities, gene banks will not be able to store the vast crop diversity occurring globally in farmers' fields. Landraces and farmers' varieties collected from farmers' fields represent a major share in the global gene bank collections, from which materials are made available to formal sector breeders, but are also repatriated to farming communities, for example, after biological disaster or political and civil unrest or made available to other communities in the same or other countries that may benefit from the adoption and adaptation of such varieties against the backdrop of climate change.

The Impact of Changing Food Systems on Governance

The governance of agrobiodiversity and the power dynamics involved are increasingly crucial in the context of rapidly changing farming and food systems, where seeds are seen as an agricultural input without attention to the abovementioned relationships between people and agrobiodiversity. This is especially true in the context of globalization, migration, and urbanization, in both the developed and developing world, which has resulted in a homogenization of food patterns and food cultures (Khoury et al. 2014). Many crops and many varieties have disappeared from the diet. The increasing disconnect between food systems and seed systems, and between agriculture and food cultures, must be addressed by changes in agrobiodiversity governance in order to inspire both growers and consumers to use and value agrobiodiversity-rich and culturally inspired food products in their daily lives. The phenomenon of urbanization marks a profound change in human relationships to food. For the first time in history, the majority of people consume food without direct engagement in its production or, to a large extent, contact with its producers. In addition, the rise of urban-based interests of access to cheap and easily processed and prepared foods has had consequences outside the city for such issues as agricultural land use, the choice of crops and varieties, the diffusion of high-input technology, rural impoverishment and agrobiodiversity conservation.

The "food system" is commonly defined by the majority of its actors and in the literature as a suite of activities by which food is produced, processed, distributed, and consumed. The linear process is frequently referred to by flows from "farm to fork" or "soil to plate." Food systems are shaped by the social, economic, and environmental outcomes of this suite of actions through a complex set of private and public interests, influences, and conflicts. A simplified generic food supply chain might include the following elements. It starts with the mobilization of "inputs" for production, including land, labor, finance, seed, feed, pesticides, fertilizers, and machinery. Value accrues at the different stages along the chain, partly determined by the way that enabling conditions, such as subsidies, trade rules, transport infrastructure, and the norms of business are organized. Value creation along the food chain may result in strengthening of the food chain, but it may also create conflict between players within and between regions. Power is distributed among the various actors throughout the food system in different ways, depending on the context. In recent years, a concentration of negotiation power can be observed particularly at the retail and wholesale level, where processing results in a growing diversity of products that are manufactured from an ever-narrowing base of genetic and species diversity of the crops and animals used in producing these products.

While seeds are recognized as a major determinant for the crop in the field and the product on the consumer plate, the conditioning and facilitating role of other components of agrobiodiversity is often overlooked and undervalued. In identifying the current gaps in the governance of agrobiodiversity within the food system, greater attention and understanding must be paid to its complex nature, which extends far beyond seeds to include environmental factors, values, and knowledge systems. Often, these more hidden dimensions are major drivers of demand for agrobiodiversity. Whereas modern markets may offer potential, it is often within the domain of rural and farmer cuisine and informal markets that demand for diversity of unprocessed, fresh, and seasonal agrobiodiversity products tends to be particularly high (Skarbø 2014; Weismantel 1988).

Various examples can be found of initiatives that aim to introduce and strengthen the role of agrobiodiversity in the food chain. The Swiss foundation ProSpecieRara, for instance, collaborates with (a) breeders to preserve old varieties and develop new ones on their basis, (b) growers and seed savers to save and distribute a diversity of seeds and breeds, and (c) a supermarket chain to develop a logistical center capable of delivering products grown from those seeds and marked with the ProSpecieRara logo to draw the consumer's attention to the importance and attractiveness of biodiversity. Another example is the Andes Potato Park, established in the Andean mountains in the Cusco Province of Peru, where local farmers maintain native potato varieties and serve these to visitors in their restaurant. Many other such consumer-oriented initiatives can now be observed across the globe. The concept of a geographical indication (GI) has been showcased as an alternative market-based strategy to overcome the problem of increasingly anonymous food, and to reestablish the lost connection between rural producers and urban consumers. A GI is a distinctive sign that identifies a product from a given place whose quality, reputation, and/or characteristics are attributable to its origin (territory). The specificity of a certain category of GIs relies on the use of native plant varieties and animal breeds typical for a defined area that may be used to stimulate the protection and use of agrobiodiversity of that given place (e.g., Aprile et al. 2012). Marketing products from certain geographical origins through differentiating them from mainstream, anonymous products has the potential to begin to move toward turning abstract commodities into particular niche products, to capture added value, and to increase revenues for producers, since a segment of conscious and demanding consumers (in relation to product quality, health, methods of production, and agrobiodiversity and environmental concerns) are willing to pay a premium price, both in developed and many developing countries (Chapter 15). In addition, GIs may increase market transparency and reduce transaction costs. More participatory research is needed to identify and strengthen pathways of how to support networks of custodian farmers in developing countries through similar approaches. In many low- and middle-income countries, links between custodian farmers and markets are already being developed. Examples include the Chefs' Alliance movement in Peru, supported by the Peruvian Society of Gastronomy (APEGA), or emerging enterprises such as Pachaa in India which sell rice landraces. Furthermore, traditional production techniques can sometimes help to conserve agrobiodiversity, keeping traditional landscape features as well as avoiding land and soil degradation.

In this context, new efforts may be undertaken to increase "agrobiodiversity literacy" as a concept and to elicit and develop interest in the status of agrobiodiversity and the quality of its products, strengthening the use of designation of origin labels and food chain initiatives such as those of ProSpecieRara, APEGA, and Slow Food. These efforts may include the following activities:

- An agrobiodiversity label could be developed, where the underlying narrative builds on the GI concept but shifts its focus from place to the environment and the production system.
- Primary education may be utilized for the purpose of conferring narratives and associated values, noting the importance of sharing inspiring stories about the role of agrobiodiversity.
- Unfamiliar or forgotten flavors, tastes, and looks could be promoted through markets and restaurants (e.g., as practiced by the Slow Food movement), and neglected and underutilized crops could be brought back into the market and onto the consumer's plate.
- The connection to health issues might be further explored and used by promoting healthier or low-allergenic food.
- The role of gardeners and seed savers might be strengthened, noting that gardeners have played an essential role in preserving vegetable and fruit diversity in Europe and North America.

All these initiatives require facilitating and supportive policies and legislation. Thus, structural and ideational power can exhibit major influences to determine the playing field and its options. Actors that have played an important role in promoting agrobiodiversity and in changing the power relations include La Via Campesina (a global platform organizing small-scale producers) and the Slow Food movement that has worked through its product and farmers networks as well as the Chefs' Alliance and Terra Madre, which have focused respectively on the interconnectedness between local and short supply chains and local culture as well as a fair income for the farmer producer (see Williams et al. 2015). Youth engagement and education are equally important to maintain autonomous and vibrant links between custodians and farmer communities and their traditional food systems. This is particularly true for products that have little market demand beyond the local food system, such as landraces characterized by long cooking times, the need for elaborate processing, pungent flavors, and perishability.

Emergent Research on Governance

From the above analysis of the governance of agrobiodiversity, a number of research gaps and recommendations for future research directions have been deduced and are elaborated briefly below, without pretending that they form a coherent and exhaustive agenda.

A Note on the Conceptualization of Agrobiodiversity and its Governance

Awareness of the importance of human–biological interactions in agriculture and food production has increased as has consumers' appreciation of the quality and origin of their food and the conditions under which it has been produced. Given this development, a more inclusive conceptualization of agrobiodiversity is needed to better acknowledge the natural environment and all its components in which food production takes place as well as the human environment and food cultures that are rooted in this agrobiodiversity, and the knowledge and value systems that underpin such production processes and food cultures, including those of Indigenous Peoples. Such review could spark the development of a new narrative, paying tribute to biodiversity and place, to human diversity and culture, and promoting more pluralism and interconnections in agriculture and food systems. It also could contribute to initiatives designed to strengthen new public policy based on novel forms of consumer citizenship, contributing to a healthier, sustainable, and equitable future through more deliberate, strategic use of agrobiodiversity.

A Reflection on Methodologies Employed in Studying Agrobiodiversity Governance

For a meaningful research agenda involving interventions in agrobiodiversity management and governance, the research methodology must take into account the research topic and stakeholders. It should fit the research partnership with the men and women farmers, gardeners, and seed savers involved, from the formulation of the research problem to the reflection of one's role in the process. A research design based on participatory and action-oriented methods is a precondition for a wide array of meaningful agrobiodiversity research, whether involving the development of community seed banks or farmer seed enterprises, the improvement of local diets and nutrition, or the exercise of influence in local to international policy making. Methodologies should be developed and improved in such a way that these can be applied beyond anecdotal scale, and alliances with government as well as public and private sectors that up-scaling requires brings new government challenges that should be carefully addressed. Ideally, farmers and researchers should be involved from the beginning in the shared design of research projects, each bringing their assessments of the problem to the table. The next step of co-creation of knowledge requires the integrated use of different knowledge sources and capacities, perhaps from different disciplinary approaches or from different practitioners and stakeholder communities with divergent insights to produce an inclusive and coherent knowledge outcome. Furthermore, a co-evaluation of the research outcomes will secure the societal relevance of the analysis. It also requires self-reflection from the researcher and his or her motives of involvement as an inseparable part of the research. Because a multitude of plant, animal, and microbial species as well as associated practices and knowledge contribute to the way of life and worldviews of Indigenous Peoples, it is particularly important to analyze formal legal measures and policy mechanisms for the recognition of the special role of Indigenous communities in sustaining agrobiodiversity.

Along with the shift toward a relational approach to understanding biodiversity and studying it in a participatory manner, the following research questions emerge.

Asymmetries of Power Relations in Agrobiodiversity Governance

Who is invited to the table of policy debates and how are particular interests represented? Whose voice is heard, and how are different views and interests resolved? To what extent do initiatives to integrate agrobiodiversity into the food system recreate or challenge existing power asymmetries in the food system?

Rather than taking any intervention to preserve agrobiodiversity as inherently good, a relational approach requires that we analyze all interventions, ranging from legislative changes to *in situ* or *ex situ* initiatives and marketing campaigns, exploring the relative and shifting power dynamics between diverse actors in any given interaction, discourse, or practice. This may begin with tracing the relations among different historic actors, an analysis of the discourses and narratives through which the intervention is framed, and an evaluation of the perceptions of involved actors. Beyond counting the number of varieties or species conserved, exchanged, or marketed, researchers also must analyze resulting proliferations or shifts in power dynamics, in particular as relates to the rights and access to benefits of historically disadvantaged groups, such as Indigenous and traditional peoples. Needless to say, analysis and self-reflection on potential power asymmetries, even in a scientist's own initiative, is crucial for improving critical reflection and contributions from the scientific domain to agrobiodiversity governance, in line with the self-reflective, participatory approach to research management and governance systems prescribed above.

Agrobiodiversity and Food Systems

Food systems have undergone major transformations in past decades (Reardon et al. 2012), reflected in a trend toward capital-intensive food production, less diversity in crop species being traded, and longer supply chains to urban consumers in which basic ingredients undergo multiple transformations on their way to becoming final food products (Hawkes et al. 2012). It is of paramount importance how processes in food systems in a rapidly urbanizing world impact the governance of agrobiodiversity and where entry points to induce change can be found. More specifically, food systems research should address the key leverage points to support local and global food systems in ways that lead to enhanced use of and demand for agrobiodiversity, thereby triggering conservation. Such research should aim at field-tested recommendations of how we can balance the social and environmental footprint of food with viable family farming in vibrant rural communities with access by the urban poor to quality non-anonymous food. It may also address how engagement of consumers and civil society and advocacy groups can more effectively influence healthier and more sustainable diets and more sustainable food systems, and how people can support and facilitate initiatives lending more resilience to local and regional foods, improving and securing their place and status in the market. Such approaches would have to be complemented with research studying market-based strategies to overcome the problem of the commercialization of increasingly anonymous food and to reestablish the lost connection between rural producers and urban consumers. A related and more specific question would regard how both producers and consumers can become interested in promoting and recognizing the value of geographical indications. A policy question may address how governments can be made interested in introducing primary school curriculums that address the value of agrobiodiversity rooted in local environments and cultures.

Nutrition, Taste, and Health in Agrobiodiversity Governance

How are connections among the issues of agrobiodiversity, taste, and health understood by diverse consumer groups, including gastronomy and the food industry, and how may they be better integrated via agrobiodiversity governance? The need for studying relations generated and sustained in, among, and through agrobiodiversity, food, and human health—beyond simple nutritional indicators (e.g., also taking into account food flavor, taste, and cultural preferences)—calls for more nuanced, integral, and rigorous research on the governance dimension of these issues. Such approaches may be developed by building community inclusive agendas and developing more democratic, trustful relations with participants (rather than by visiting communities with refined research tools to collect data on additional variables). Gender roles should be part of the research agenda studying these relationships. To this end, researchers are charged with the responsibility of enabling ample involvement and participation of communities in the identification of local problems emerging from the relationship between agrobiodiversity and food, and nutrition and health issues, as well as in finding workable solutions that can contribute to an improved agrobiodiversity governance.

Governance of Information and Its Impact on Agrobiodiversity

Given the advent of genome editing and other technological developments, how will these new informational and molecular genetic capacities targeting the genomics and so-called omics of plant and animal diversity influence the power dynamics and governance of agrobiodiversity? In particular, how will these new biotechnologies exert influence on the still growing tension and deep historical controversies between the historic public ownership of genetic resources and patentable technology and private industry (Salter and Salter 2017)?

Commodification goes hand-in-hand with privatization of plant genetic resources and crop seeds. The use of intellectual property rights regimes (in particular claims of ownership over plants through patents and to some extent plant breeders' rights) as well as the focus of the private breeding sector on the development of hybrid varieties (requiring farmers to buy new seed each growing season) has led to the privatization of the access to and the use of genetic diversity. New technological developments, such as genome editing, are likely to increase the options for privatization by delinking access to genetic diversity from the physical access to a genetic resource. Instead, access to the DNA sequence information from public and private databases, and effective analysis of such information, will suffice in the future to develop new varieties. These may then be protected by intellectual property rights, even if new products have been developed using information from public databases only. This poses the need for the governance of information related to agrobiodiversity, rather than merely the governance of agrobiodiversity itself (Zimmerer and de Haan 2017), an issue that has been well recognized by policy makers addressing the implementation of the CBD and its Nagoya Protocol as well as the ITPGRFA. Without agreement on the governance of genetic information, fair and equitable benefit sharing, and hence the sense and survival of the current international agreements, the global governance basis provided by the CBD and ITPGRFA is at stake.

Agrobiodiversity lies at the heart of peoples' environments, their food production systems and livelihoods, and their diets and food experiences as well as their social and cultural identities. The governance of agrobiodiversity is in constant flux as new insights, technologies, and applications develop. It is a challenge to all stakeholders and actors to discover how they can contribute to an improved, more sustainable, plural, and equitable agrobiodiversity governance.

15

How Have Markets
Affected the Governance
of Agrobiodiversity?

Matthias Jäger, Irene van Loosen, and Alessandra Giuliani

Abstract

This chapter focuses on the role of markets (especially those of agricultural products) in agrobiodiversity governance. Over the past two decades, expansion of global agricultural product markets has, in general, furthered the simplification of agricultural and food systems, reducing the diversity within crop and animal species. Farmers who continue to conserve on-farm agrobiodiversity are providing global public goods in terms of food security and environmental sustainability insurance for the world's population, both currently and in the future. Yet because markets or other global institutions are not compensating farmers for conserving high levels of agrobiodiversity, these farmers face little private incentive to maintain on-farm conservation practices and may resort to practices that result in reduced levels of agrobiodiversity, which in turn could lead to the destruction of local food systems and general biodiversity loss. To enhance both agrobiodiversity conservation and income generation through market-based instruments, endeavors to place a value on agrobiodiversity that signal its true production cost and contributions to genetic resource usage should be further developed. It is proposed that payments for agrobiodiversity conservation schemes and niche market development through differential marketing, labels, certification schemes, and agrotourism are needed in concert to provide a robust foundation for agrobiodiversity conservation activities, building on both private sector investment and government funds. Depending on the context, these measures hold great potential for the successful marketing of agrobiodiversity and agrobiodiversity niche products through collective action. Constraints and potential unintended consequences of market-based approaches to agrobiodiversity conservation need, however, to be taken into account.

Introduction

To examine the impact of markets on the governance of agrobiodiversity, we analyze the interrelationships between agrobiodiversity, markets, and sustainability that are necessary to an integrated scientific framework of agrobiodiversity. We begin by explaining what the governance of agrobiodiversity entails and how it contributes to global food and environmental security. We then illustrate how the integration of farmers into markets generally leads to declining private incentives to conserve agrobiodiversity on the farm. We discuss several market approaches to *in situ* agrobiodiversity management that have the potential to lessen or even reverse this tendency and address factors that enable or constrain the marketing of agrobiodiversity products. In conclusion, we offer recommendations on how to support sustainable food systems and expand agrobiodiversity governance while promoting *in situ* conservation at the farm level through the marketing of agrobiodiversity products.

The Governance of Agrobiodiversity

For the purposes of this chapter, we follow the arguments of Johns et al. (2013) and Padulosi et al. (2011a): agricultural biodiversity, henceforth referred to as "agrobiodiversity," comprises cultivated plants and animals in agricultural ecosystems as well as wild foods and other products gathered by rural populations for their livelihoods through the application of traditional, locally sourced knowledge (cf. Chapter 8). A distinction is made between planned agrobiodiversity (i.e., the diversity of crops and livestock directly managed by farmers) and associated biodiversity (i.e., the biota in the agroecosystem that survive in the presence of local management and environmental conditions) (Jackson et al. 2007; Kontoleon et al. 2008).

The sustainability of global agriculture and related ecosystems is dependent on the use, enhancement, and consequent conservation of agricultural biodiversity (Bardsley 2003; Lockie and Carpenter 2010). Agrobiodiversity plays a pivotal role in enhancing farm productivity, developing resilient farming systems, generating income, providing ecosystem services, and climate regulation as well as creating food and nutrition security for the world's population (Kruijssen et al. 2009b; Padulosi et al. 2011a; Thrupp 2000).

Governance of agrobiodiversity has three components: access, use, and management. Access entails the legal entitlement, permission, or (free) admission to obtain available plant and animal genetic resources for food and agriculture (Andersen 2006). Use of these species and varieties for subsistence or sale implies having access to them. Management can take place both *in situ* (on the farm) or *ex situ* (outside natural habitats, normally in gene banks) (De Boef et al. 2012; Gauchan et al. 2005). Most agrobiodiversity is actively managed and consequently maintained *in situ* as part of smallholder family farming practices (Padulosi et al. 2011a). As certain key elements of genetic resources cannot be captured and stored outside natural habitats, dynamic *in situ* strategies that result in sustainable conservation are necessary to maintain traditional knowledge, to increase the adaptation and resilience potential of species and varieties, and ultimately to prevent global loss of plant and animal genetic resources (De Boef et al. 2012; Narloch et al. 2011a).

Farmers around the world as well as human society at large depend on agrobiodiversity for their multiple production objectives and livelihoods (Lockie and Carpenter 2010). Over many centuries, ancient agricultural settlements have made use of diverse plant and animal species and varieties to enhance productivity and adapt to new social and environmental challenges. Through on-farm diversification of crop species and varieties as well as landscape-level effects, smallholder farmers frequently aim to reduce the risk of food shortages and production fluctuations that result from abiotic shocks (such as drought), biotic stress (e.g., pest, disease outbreaks), and seasonality (Frei and Becker 2004; see also Chapter 6). Many traditional practices that were applied in the past to utilize, improve, and adapt agrobiodiversity in smallholder farming systems are still operational today in both large- and small-scale production systems; examples include the exchange of seed within and between different regions and the selection of best breeds for adaptive production (Thrupp 2000).

It is well recognized that smallholder family farmers in low- and middle-income countries have important roles to play in maintaining a dynamic and evolutionary state of agrobiodiversity conservation (Lockie and Carpenter 2010). Simultaneously, these agrobiodiversity-producing smallholders have become part-time farmers integrated within a wide range of product and labor markets (Zimmerer et al. 2015; see also Chapter 8). Indeed, smallholders conduct farming under highly varied circumstances and manage the majority of the world's rich stock of animal breeds and crop varieties. Through their governance practices, genotypes with unique and valuable traits are created and maintained for breeding and research (Frei and Becker 2004; Johns et al. 2013). The diversified agroecosystems maintained by smallholders are crucial for ensuring global food security because of their high resilience to environmental shocks, particularly in the context of climate and socioeconomic change (Frei and Becker 2004; Gonzalez 2011; see also Chapter 7). Furthermore, these diversified agroecosystems and landscapes provide a habitat for a large range of associated biota and contribute to sustainable production (Berg 2009; Bianchi et al. 2006; Frei and Becker 2004; Thrupp 2000). In summary, traditional food systems link the socioeconomic resilience of smallholder farmers with global food and nutrition security (Johns et al. 2013).

Currently, however, diversity hotspots (i.e., traditional "home gardens" and genetically diverse small-scale polycultural systems that include multiple landraces) are primarily found in environmentally heterogenic or marginal parts of Asia, Africa, as well as Central and South America, where pressure for intensification and specialization potentially conflict with diversification (Thrupp 2000; Van Dusen and Taylor 2005). The often lower levels of market infrastructure and agricultural technology in low-income countries can make farmers more reliant on local agrobiodiversity management. In richer and middle-income countries, genetic improvement to enhance the quality or quantity of food has become increasingly managed by professional plant breeders and formalized seed systems (Gauchan et al. 2005). It could be argued that there is less need for individual farmers in richer and middle-income countries to govern and invest in agrobiodiversity as a natural insurance against environmental risks (Van Dusen and Taylor 2005). However, current global trends suggest that certain farmers, agricultural and food institutions, and consumer groups are employing expansion practices that entail agrobiodiversity use and governance in richer and middle-income countries (Chapters 8 and 13).

Product Market Integration and Impacts on Agrobiodiversity Conservation

Changes in the production or marketing environment, or both operating simultaneously, can induce farmers to grow landraces and modern varieties because of their relative advantages (Gauchan et al. 2005; see also Chapter 8). The greatest threat to agrobiodiversity is nonuse that occurs as farming systems become increasingly homogenized and specialized (Lockie and Carpenter 2010). It has been suggested that 75% of the world's richness in agrobiodiversity has been lost over the course of the twentieth century (Brush et al. 2015; FAO 2010b; Gonzalez 2011; Padulosi et al. 2011a), through the introduction of genetically uniform modern varieties which have superseded local varieties. Regional studies, however, indicate that this supposed one-to-one replacement is less straightforward than commonly assumed (Brush 2004; Zimmerer 1997). Relative loss, commonly reflected as a reduction of the area dedicated to landraces, can partly be attributed to the perceived low economic potential of landraces compared to modern varieties. In practice, traditional plant and animal products in remote and marginalized areas often suffer from a lack of value additive methods, or the infrastructure and technology for transformation (Padulosi et al. 2011a), as well as missing or incomplete markets that result in high transaction costs (Van Dusen and Taylor 2005).

The word "market" can refer to several meanings, such as the physical location where the produce is exchanged (the local market) or a form of exchange based on a price mechanism. Markets offer smallholder farmers opportunities to participate and benefit from consumer demand and economic growth (Ferrand et al. 2004). Since smallholders are a heterogeneous group, the markets in which they participate differ in terms of size, location, links to other markets, power relations among market actors, and institutional environments (see also Chapter 8). Local-to-global agricultural product markets encompass periodic (e.g., daily, weekly) assemblies of buyers and sellers in a given place. Normally this takes the form of open-air markets but sometimes they are situated in more permanent, covered structures (Anderson et al. 2010). The useful description of a local agricultural market includes the elements of farmers being sellers and buyers in transactions involving commodities and seeds (and thus agrobiodiversity) (Fafchamps and Vargas (2005). These markets need to be seen as local institutions through which buyers and sellers (and those conducting barter exchanges) enact their transactions.

Some of the factors commonly related to market and economic development can have an adverse effect on the demand of farmers for agrobiodiversity on their farms by reducing the incentives to maintain it. This is due to increasing opportunity costs of maintaining diversity, the availability of new consumer products and substitutes for previously self-grown or collected products, as well as social and cultural change. Since most markets aim for homogeneity and nonseasonality, their capacity to handle diversity is limited. In addition, the availability of hired labor, inputs, or machinery may decrease the demand for diversity as will an increase in net return from agriculture due to increased income or cheaper inputs. Nonfarm sources of income, including income from smallholder migration and household remittances, can level out fluctuations in farm income and may decrease agrobiodiversity (Bellon 2004), though migration can also support higher levels under certain conditions (e.g., Andean maize in Bolivia) (Zimmerer 2014). The substitution of consumption products takes place when a crop with a high personal value (and its substitutes), for which previously the market was missing, becomes available (Van Dusen and Taylor 2005).

World population growth coupled with urbanization, supermarketization, and changing dietary preferences and consumption patterns demand high productivity from our agricultural systems (Devaux et al. 2009; Westhoek et al. 2014). Emphasis is placed on food industry requirements for crop and animal products, such as "prices, relative advantage, consumer tastes, yield, standardized production and uniformity of maturity" (Padulosi et al. 2011a:141). Varieties that cannot comply with these standards are either ignored or marginalized into niche markets (Padulosi et al. 2011a). Another major constraint of traditional agriculture, commonly aggravated by lack of access to resources, is its low productivity in comparison to intensive farming systems (Johns et al. 2013).

During the Green Revolution in the 1960s and 1970s, technology packages transformed traditional agricultural systems into large-scale commercial monocultures (Frei and Becker 2004; Gonzalez 2011; Holt-Gimenez et al. 2006). These intensive farming systems are characterized by several requirements for high-yielding varieties, such as enriched seeds, synthetic pesticides, fertilizer inputs, and large quantities of water and fossil fuel energy (Frei and Becker 2004; Holt-Gimenez et al. 2006). Despite greatly increasing the global food supply by producing much higher yields than traditional farming systems, intensive agriculture leads to a range of social and environmental problems, including the marginalization of traditional farmers and worldwide loss of genetic diversity (Bardsley 2003; Frei and Becker 2004; Gonzalez 2011; Jackson et al. 2007; Thrupp 2000).

The expansion of global markets over the past two decades—a nonuniform and complex process—has, in general, led to a simplification of agricultural and food systems and has significantly reduced crop and animal species diversity in several regions (Devaux et al. 2009; Khoury et al. 2014a; Narloch et al. 2011a; Van Dusen and Taylor 2005; Westhoek et al. 2014). As areas become increasingly integrated in regional and global markets, the opportunity costs to conserve on-farm diversity rise (Van Dusen and Taylor 2005). To sell their produce at these markets and receive economic incentives, farmers must comply with food industry requirements (Narloch et al. 2011a). Consequently, these incentives frequently lead them to disinvest in agrobiodiversity as an asset (Pascual and Perrings 2007), though countertrends also exist (discussed further below as well as in Chapter 6).

The introduction of intensive agricultural systems led farmers around the world to substitute agrobiodiversity as a form of natural income insurance for financial insurance from the market. Therefore, the adverse impacts of environmental and market conditions were no longer (or only partially) mitigated by growing a large variety of crop and animal species but rather by relying on agricultural policies such as crop yield insurance, extension services, subsidies, and other financial assistance (Baumgärtner and Quaas 2008; Lockie and Carpenter 2010). This tendency, however, leads to a market failure problem: in addition to providing on-farm benefits, agrobiodiversity provides public benefits such as a genetic reserve to cope with future change, increased control of diseases and pests, and an abundance of crop varieties, animal breeds, and food products for consumers worldwide (Pascual and Perrings 2007). In addition, agrobiodiversity provides environmental services and contributes to the restoration of degraded lands (Gruère et al. 2009b). The market valuation of agrobiodiversity, however, is considerably below the levels associated with its services of as public good (Baumgärtner and Quaas 2008). This undervaluation by the market poses a sizeable hurdle for agrobiodiversity valuation (see Chapter 6).

Market integration often times entails a simplification of agricultural landscapes, the expansion of the agricultural frontier, and increased uniformity of agricultural practices that result in disinvesting in natural capital and the governance of agrobiodiversity (Bardsley 2003; Kontoleon et al. 2008; Padulosi et al. 2011a; Thrupp 2000). Consequently, the world currently relies on a small number of crops and animals with a narrow genetic base for its global food security. For example, out of several hundred thousand known plant species, just nine species supply over 75% of global plant-derived human food (Padulosi et al. 2011a). The global area trade-off of both high levels of intraspecific and interspecific genetic diversity with monocultures endangers the global food system by increasing the likelihood of crop failure as a result of environmental shocks or outbreaks of pest and disease (Bardsley 2003; Gonzalez 2011; Holt-Gimenez et al. 2006).

Declining Private Incentives to Conserve Agrobiodiversity

Farmers who conserve *in situ* agrobiodiversity provide a global public good—one that underpins food security and environmental sustainability for the world's population, both now and in the future (Holt-Gimenez et al. 2006; Kontoleon et al. 2008). Although this conservation process provides a public ecosystem service, the actual landraces being conserved fall under national control over which countries have sovereign rights (Sullivan 2004). Landraces and local animal breeds tend to perform well in marginal production environments and often have nutritional values and harvesting cycles complementary to modern varieties (Narloch et al. 2011a). The *in situ* conservation of genetic diversity results in a range of ecosystem services: the supply of highly nutritious food with unique flavors, maintenance of local cultures and associated traditional knowledge, as well as the provision of natural insurance against extreme events and global change (Narloch et al. 2011a).

Most providers of agrobiodiversity services reside in remote areas of low-income countries (Narloch et al. 2011a), whereas agrobiodiversity services function more generally across middle-income and rich countries (Zimmerer et al. 2015). Since markets or other global institutions are not compensating farmers for conserving higher levels of agrobiodiversity, many of these farmers have little private incentive to maintain on-farm conservation practices and may resort to reducing agrobiodiversity, leading to impoverished local food systems and possible biodiversity loss (Holt-Gimenez et al. 2006; Lockie and Carpenter 2010; Perrings et al. 2009). Once connected to regional and global markets, and facing the demands of economies of scale, it is often more profitable for agriculturalists to specialize—that is, to grow only a few varieties favored by the market (Pascual and Perrings 2007)—or to resort to cultivating a combination of traditional and modern varieties for home consumption and income generation (Jackson et al. 2007).

Smallholder farmers in low- and middle-income countries tend to be disproportionally disadvantaged in terms of their inclusion into national and global economies: they have to operate within imperfect market conditions, have limited technical skills to comply with markets demands and bargaining power, and often lack access to information and the other inputs required for building competitive production systems (Devaux et al. 2009; Kruijssen et al. 2009b). In addition, many smallholder households are subject to increased shortages of labor time, frequently gendered through migration and other off- and nonfarm activities, and are often associated with agrobiodiversity loss (Zimmerer et al. 2015; Zimmerer and Vanek 2016). Consequently, their efforts to offset high transaction costs by maximizing production over the short term can lead to the erosion of local environments and natural habitats (Bardsley 2003; Holt-Gimenez et al. 2006; Thrupp 2000). Indeed, unsustainable intensification is a common pitfall when new boom crops, including potential booms of high agrobiodiversity crops, enjoy sudden demand (Hermann 2013). In addition, smallholders lack well-functioning financial insurance mechanisms that compensate for agrobiodiversity conservation, and farmers in low- and middle-income countries are increasingly at risk for adverse weather conditions, volatile global agricultural commodity markets, and the accumulation of debt (Frei and Becker 2004; Holt-Gimenez et al. 2006). The trade-off between income increase and on-farm agrobiodiversity conservation can potentially lead to a reduction of agrobiodiversity and a general degradation of natural resources through a simplification of production systems (Bardsley 2003; Kontoleon et al. 2008; Thrupp 2000).

Underlying dynamics are "the failure of markets to signal the true cost of biodiversity change in terms of ecosystem services, the failure of governance systems to regulate access to the biodiversity embedded in 'common pool' environmental assets, and the failure of communities to invest in biodiversity conservation as an ecological public good" (Perrings et al. 2009:231). A simplified conceptual framework, presented in Figure 15.1, aims at better understanding the complex relationship between markets and agrobiodiversity as well as the trade-offs between income generation and on-farm agrobiodiversity maintenance (Kruijssen et al. 2009a:416):

> One could imagine a farm household with a given level of household income and a certain level of agricultural biodiversity (inter- and intraspecific) present on its farm [in Figure 15.1, this is point 0]. Economic development takes place leading to an increase in farm household income from I to I^i [Intervention 1]... [T]his leads to a reduction in diversity from A to A^i, either because of a reduction in the variety of the original crop or because the crop portfolio has been replaced with a different (uniform) crop. The change thus leads to a shift of this particular farm household [in Figure 15.1, from 0 to I^i] along curve I [the income curve]. The theoretical trade-off that takes place is then the difference between the income increase and the agrobiodiversity lost. Quantifying the trade-off requires assigning a monetary value to the resources lost...Reducing the trade-off would require a change in the relationship between agrobiodiversity maintained on-farm and household income. This is represented in Figure 1 by the shift to curve II [in Figure 15.1, through Intervention 2 or 3]...[T]he reduction in agrobiodiversity is now reduced to A^{ii}–A and the outcome for this farmer is [in Figure 15.1, Intervention 2], leading to a reduced trade-off. The scale of the axes can vary among cases and will change the slope of the curve.

Nevertheless, farm household decisions are not solely based on profit maximization. Smallholder farmers often use this diversity to address a range of ecological niches on their farms, to reduce risk, or to provide internal demand for a variety of products, cultural values, identity, and taste preferences. In some cases, market chains of specialized products positively affected the agrobiodiversity maintained on farm as well as the livelihoods of those participating in the chain (Devaux et al. 2011; Keleman et al. 2009). For instance, the marketing development of African garden eggs demonstrated an increase in social welfare: not only did this generate income for local producers and chain actors in Ghana, it promoted the sustained use and conservation of agricultural biodiversity (Horna et al. 2007). The cases of emmer in Italy and kokum in India (Kruijssen et al. 2009a), have shown that the trade-off can potentially be reduced with proper interventions and bring substantial economic benefits to poor smallholder farmers.

Placing a Value on Agrobiodiversity

As demonstrated by Figure 15.1, to enhance both agrobiodiversity conservation and income generation through market-based instruments, value must be placed on agrobiodiversity that signals its true cost and contribution to genetic resource functions. This value can be subdivided into

- use value, which encompasses the direct contribution of agrobiodiversity to food security, nutrition, and income generation through cultivation practices,
- nonuse value, which refers to the ethical value of agrobiodiversity and its role in food culture, and
- option value, which represents the potential to realize future value by providing genetic material for innovation.

We note that use value of agrobiodiversity is closely related to its insurance value in the case of temporal and economic stresses (Padulosi et al. 2011a).

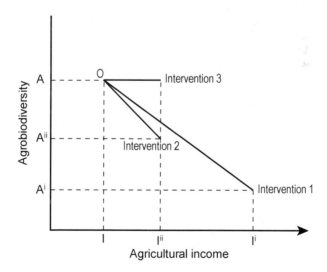

Figure 15.1 Conceptual framework of trade-off income (I) generation and *in situ* agrobiodiversity (A) maintenance (after Kruijssen et al. 2009a).

Agrobiodiversity is classified as an impure public good because of the rivalry involved in its use and the difficulty of excluding users (Kruijssen et al. 2009b; Pascual and Perrings 2007). Agrobiodiversity is prone to market failure because of its characteristics as an impure public good, with intergenerational and interregional dimensions: agrobiodiversity has both public and private economic attributes, and farmers (as a group) tend to generate less diversity than is desirable to contemporary and future societies (Kontoleon et al. 2008; Kruijssen et al. 2009b; Smale et al. 2004). In addition, there are no markets for *ex situ* ecosystem services that depend on *in situ* agrobiodiversity (Pascual and Perrings 2007). This means that public interventions are needed to increase the private value of agrobiodiversity by economically incentivizing farmers to conserve plant and animal varieties with high public value on their farms (Kruijssen et al. 2009b; Pascual and Perrings 2007). Because of its public good characteristics, agrobiodiversity will be undersupplied if left to the market (Perrings et al. 2009).

At the sociocultural level, the conservation of agrobiodiversity builds on the preservation of traditional knowledge. Economic and institutional levels require the existence of markets, infrastructure, and a supportive institutional and legal framework (Padulosi et al. 2011a). To create efficient and effective markets for agrobiodiversity conservation, several steps would be needed to capture its public good values and prevent global free-riding on the services of small-scale traditional farmers. To this end, it is crucial to reconcile private and social values with regard to agrobiodiversity (Lockie and Carpenter 2010).

First, values of agrobiodiversity need to be identified and measured in price data. To this end, tools need to be developed (Pascual and Perrings 2007:259) to help (a) assess the functional role of species in their crop- and noncrop habitats, (b) identify the biotic and abiotic components of agroecosystem structures that support the provision of ecological services at the landscape level, and (c) assess the contribution of such ecological functions to human well-being. Subsequently, mechanisms and markets should be introduced that allow the demonstrated and measured values of agrobiodiversity to be channeled between its cost bearers and beneficiaries, allowing the cost bearers to receive benefits for their roles in agrobiodiversity conservation (Kontoleon et al. 2008; Narloch et al. 2011a).

However, ensuring that the importance of agrobiodiversity is effectively valued by market mechanisms is a complex matter. Generally, there is a lack of market acknowledgement of *in situ* agrobiodiversity conservation aggravated by continuous financial support for intensive agriculture through macroeconomic policies, such as subsidies and price controls that stimulate nonbiodiverse farming practices (Lockie and Carpenter 2010; Pascual and Perrings 2007). Fortunately, worldwide recognition is growing of the value of traditional varieties, landraces, and underutilized species in the context of climate change, rural poverty, and malnutrition (Padulosi et al. 2011a). Increasingly "attention is being given to the potential role markets can play for agrobiodiversity conservation through product differentiation and increasing competitiveness in niche and novelty markets" (Kruijssen et al. 2009b:46). Over the past decades, several international governance efforts have been made to increase the value of agrobiodiversity and other types of genetic diversity. These efforts include the establishment and testing of access and benefit-sharing systems, based on the Convention on Biological Diversity (CBD) since 1992 (Reichman et al. 2016), and the introduction of participatory plant-breeding initiatives since the mid-1990s (Ceccarelli and Grando 2007). The Nagoya Protocol on Access to Genetic Resources and the Fair and Equitable Sharing of Benefits Arising from their Utilization, an international agreement enforced as of 2014, is a relatively new instrument aimed at increasing the value of genetic resources.

According to Johns et al. (2013:3437), the successful incorporation of agrobiodiversity products into modern markets requires overcoming the disadvantages of small-scale agriculture, accommodating the unique character of agrobiodiversity, taking into account the economic and social needs of smallholder communities, and avoiding or mitigating the potential negative consequences of market integration on agrobiodiversity conservation. If market chains of agrobiodiversity-specialized products (based on agrobiodiversity traits) were created, on-farm agrobiodiversity maintenance could be enhanced as could the livelihoods of the participating farmers in the chain (Kruijssen et al. 2009a). Analysis of market type, products, and situations involving agrobiodiversity products should be conducted to increase understanding of how agrobiodiversity could be used to improve livelihoods and maintain *in situ* biodiversity, keeping in mind that markets may not always be the most suitable tool to conserve agrobiodiversity (Giuliani et al. 2011). As certain types of markets and products will be more beneficial than others, it is important to analyze those circumstances under which agrobiodiversity product marketing could be most successful and sustainable, and to obtain a clear understanding of the trade-offs between increased income and maintenance of diversity in production systems (Kruijssen et al. 2009a). Similarly, we strongly recommend that future research focus on trade-offs between promoting few landraces versus multiple landraces and between different use, nonuse, and option values of agrobiodiversity.

Market Approaches to *In Situ* Agrobioversity Management

If left to market dynamics, agrobiodiversity as a public good will be under-valued (Pascual and Perrings 2007). However, by economically incentivizing farmers to change their land development decisions, innovative and flexible market approaches to *in situ* agrobiodiversity management have the potential to contribute to sustainable agrobiodiversity governance, to mitigate and adapt to climate change, and to improve livelihoods (Kruijssen et al. 2009b; Pascual and Perrings 2007). Several social movements in the Global North and South are advocating for the development of sustainable food systems that reduce greenhouse gas emissions, promote and value the use of agrobiodiversity, rely on local inputs, support the livelihoods of small farmers, and connect farmers and consumers through direct and inclusive value chains (Gonzalez 2011; see also Chapter 6 and 8; Jackson et al. 2007).

Linking agrobiodiversity products successfully to markets can play an important role in this dynamic. In fact, appropriate value chain and market instruments can induce consumers to behave in ways that are consistent with socially optimal outcomes (Perrings et al. 2009). Currently many valuable agrobiodiversity products—from exotic fruits and heirloom varieties to animal products from native breeds—are not well known among the greater public. Agrobiodiversity products could be introduced to consumers worldwide if (a) niche value chains were developed; (b) marketing, food literacy, and consumer behavior research were improved; and (c) production differentiation was better valued (De Boef et al. 2012; Padulosi et al. 2011a). Doing so might generate enough revenue for smallholders to allow agrobiodiversity conservation to pay for itself (Perrings et al. 2009). Market approaches that aim to place a direct monetary value on agrobiodiversity or contribute to the development of niche markets for agrobiodiversity products range from actual niche market development, certification schemes, and payments for agrobiodiversity conservation services (PACS) schemes to agrotourism (Lockie and Carpenter 2010; Pascual and Perrings 2007). On the supply side, some form of collective action is often indispensable to building successful value chains for agrobiodiversity products. Direct payment is possible for the direct values associated with agrobiodiversity production, consumption, and service provision, although other types of values (e.g., option or intrinsic value) are more difficult to link to market approaches.

Certification Schemes and Labels

In a context of increasing international trade and product uniformity, calls for market transparency, sustainable consumption, and the conservation of agrobiodiversity have led to the development of a range of private labels and certification schemes for food products. This, in turn, could facilitate the development of niche markets for agrobiodiversity products by allowing consumers to make informed decisions (Gonzalez 2011; Jaeger et al. 2017). As Gauchan et al. (2005:294) report:

> When consumers are willing to pay a premium for a quality that is associated uniquely with an identifiable landrace or group of landraces grown in a specific geographical area, the price differentials that result can generate an economic incentive for farmers to continue growing them.

Certification schemes are commonly implemented by using price premiums to reward sustainable producers or to exclude nonsustainable producers from certified value chains (Kontoleon et al. 2008). Examples include products that are specifically linked to regional markets, cultural knowledge, traditional processing methods, or a specific region or denomination of origin (Kontoleon et al. 2008; Lockie and Carpenter 2010). Several attributes of underutilized species could contribute to their marketability and branding strategies, such as a striking name, traditional knowledge and utilization, geographical origin, or history of a product (Will 2008). Geographical indication, protected designation of origin, and traditional specialties guaranteed—all labels with an international reputation that protects and links a product to a locality with its particular history—are useful for product promotion (Will 2008). However, we note that certification schemes and labels depend on dedicated consumers' preferences and their willingness to pay (Gauchan et al. 2005).

Payments for Agrobiodiversity Conservation Services

The conservation of agrobiodiversity is perceived as an ecosystem service that provides global benefits for which the wider community should pay (Narloch et al. 2011a). PACS are a subcategory of payments for ecosystem services (PES) schemes, which unites buyers and sellers in market-based, voluntary transactions to compensate for or trade environmental services (e.g., agrobiodiversity conservation). It also provides financial or in-kind incentives to actors that maintain or enhance the ecosystem service (Lockie and Carpenter 2010; Perrings et al. 2009). A variant of PES schemes are direct compensation payments, which are paid by government agencies to landowners for taking private land out of production and into conservation (Pascual and Perrings 2007). Less common, but upcoming are market creation methods for agrobiodiversity conservation, such as transferable development rights and auction contracts for conservation (Perrings et al. 2009).

As relatively new phenomenon, PACS increases the private benefits from *in situ* agrobiodiversity conservation of socially valuable, underutilized plant and animal species or varieties through a PES scheme, hence stimulating their conservation and utilization (Narloch et al. 2011a). We wish to stress that accurate information on the conservation status and monetary values of agrobiodiversity conservation is indispensable for the creation of effective PACS mechanisms, and that pricing access to ecosystem services could have the side effect of excluding and marginalizing local populations (Perrings et al. 2009). Furthermore, problems may arise in the identification of potential buyers and the articulation of meaningful conservation goals (Narloch et al. 2011a). PES schemes can be linked to benefit-sharing mechanisms, such as the pilot schemes being promoted by European Seed Association members on a voluntary basis.

Agrotourism

An alternative measure to promote *in situ* conservation of underutilized plant species as well as rare and useful animal breeds is agrotourism, a form of tourism that capitalizes on rural culture as a tourist attraction (Jaeger et al. 2017; Perrings et al. 2009). Through agrotourism, the use of typical regional breeds and varieties may be maintained to exhibit a landscape replete with cultural, historical, and natural characteristics. Farmer cuisine and the use of local landraces, breeds, and products are often highly valued by certain types of tourists willing to provide additional income to local farmers. Agrotourism fosters regional development and agrobiodiversity governance through schemes of equitable profit sharing. It is, however, dependent on the local context; that is, whether agrobiodiversity conservation and local economic development are consistent goals (Perrings et al. 2009).

Collective Action on the Supply Side

Without well-established value chains, underutilized agrobiodiversity products face several constraints for initial demand creation: high transaction costs for new business development, limited knowledge about technical issues for value addition, and access to capital resources. Hence, to achieve economies of scale and investment, "smallholders typically require market-based actions or incentives, some form of collective action that combines efforts of several producer households" (Johns et al. 2013:3437). Collective action can be defined as the coordinated behavior of groups toward a common interest (Kruijssen et al. 2009b). By uniting farmers and other value chain actors in the quest to deliver a stable and high-quality agrobiodiversity product, collective action can lead to a more equitable distribution of costs and benefits along the value chain, as well as enhanced market access and bargaining power, which in turn could increase market effectiveness and efficiency (Kruijssen et al. 2009b). In practice, collective action can result in producer groups, cooperatives, or other forms of clusters to organize supply.

Comparative analysis on collective action indicates that collaboration among actors could help the less advantaged actors improve their market position (Kruijssen et al. 2009b). Increased social capital can reduce risks, support sustainable production, and facilitate investments in processing technology. Nevertheless, collective action for smallholder market linkages is costly. High levels of effort and investment are required to achieve successful and sustainable collaboration among stakeholders and individuals. In addition, the environment, including the policy framework, needs to be conducive to collaboration (Giuliani et al. 2011).

Constraints for Marketing Agrobiodiversity Products

Given the private and public value of agrobiodiversity products, market access for products derived from underutilized species or landraces provides an opportunity both to enhance smallholder well-being and to contribute to *in situ* conservation. Market-based incentives are, in principle, among the most sustainable mechanisms for *in situ* conservation of agrobiodiversity because public interventions are unnecessary when they function well (Gauchan et al. 2005). Depending on the context, labels, and certification schemes, agrotourism and PES systems offer great potential for the successful marketing of agrobiodiversity and derived niche products through inclusive and integrated value chains. However, initial lack of demand, thin or small volume markets, as well as underdeveloped information systems limit opportunities for smallholder producers.

Hence, developing value chains and creating niche markets for agrobiodiversity products requires extensive start-up investment and technical assistance for adequate production, processing, packaging, and distribution processes. This could be provided by nonmarket institutions, such as government agencies and nongovernmental organizations (NGOs) (Gauchan et al. 2005; Johns et al. 2013).

Another reality for the marketing of landraces is that smallholders, who are reliant on their own production for consumption, may prefer to select taste and quality attributes that are not necessarily valued by urban consumers (Gauchan et al. 2005; see also Chapter 8). This is problematic because only those agrobiodiversity products that match consumers' tastes and preferences will be included in financially viable niche markets, resulting in the neglect of many underutilized species (Narloch et al. 2011a). Not all diversity can be marketed. To conserve varieties with uses restricted to traditional farmers, other business-led initiatives may need to be promoted (e.g., using a PES scheme). Alternatively, farmer cuisines and agricultural systems that build on diversity may need to be maintained. When global beneficiaries are not willing to pay the providers of agrobiodiversity, other agents (e.g., governments, NGOs) could act as buyers (Narloch et al. 2011a), as in the school feeding programs that have been implemented in Peru, Brazil, India, and Uganda (e.g., Beltrame et al. 2016).

Toward the Sustainable Governance of Agrobiodiversity

Synergetic valuation practices involving sustainable farming and changes in agricultural policies and institutions can potentially overcome the conflict between sustainable agrobiodiversity governance and markets (Thrupp 2000). In this context, sustainable agriculture is characterized by reducing the environmental footprint of agriculture, reliance on both farmers' knowledge and scientific innovations, and the quest to enhance and conserve agrobiodiversity (Gonzalez 2011; Jackson et al. 2007). Supporting and investing in the governance of local agrobiodiversity through sustainable agriculture renders possibilities to offset risks created by the dominant agricultural and development paradigm (Bardsley 2003).

PACS schemes and niche market development through labels, certification schemes, and agrotourism should occur in complement to provide a stronger foundation for agrobiodiversity conservation activities, building on both private sector investment and government funds (Narloch et al. 2011a). However, several possible unintended consequences of market-based approaches to agrobiodiversity conservation need to be taken into account. For instance, although market-based approaches could raise demand for agrobiodiversity products and increase income for those farmers that conserve agrobiodiversity, they may also result in specialization and homogenization (Kruijssen et al. 2009b). To generate saleable surpluses based on consumer preferences, farmers may be required to focus on the cultivation of a reduced number of plant or animal species or varieties for successful niche markets, which would lead once again to unsustainable *in situ* conservation practices and expose farmers to market supply fluctuations and adverse weather conditions (Johns et al. 2013; Kruijssen et al. 2009b). All too often, the development of niche markets has resulted in unsustainable intensification and a gradual takeover by medium- to large-scale farmers (Tobin et al. 2016). Examples include the development of potato landrace, quinoa, and maca value chains in Peru. While local markets can be important for agrobiodiversity, studies also show they may contain lower levels relative to nonmarket transactions (e.g., millet diversity in seed fairs; Smale et al. 2014). Finding the right mix of market and nonmarket incentives is a challenge and must be carefully considered if we are to achieve the viable maintenance of agrobiodiversity. Indeed, no market can absorb all of the diversity that exists. Therefore, market approaches should be complemented by activities (e.g., school curricula, diversity seed fairs) that promote livelihoods and farming systems which value diversity. Furthermore, public awareness campaigns are necessary to underline the nutritional, taste, and traditional values of products and attract consumers' preference for different species. One example of this is found in the Kolli Hills, India, where the M. S. Swaminathan Foundation has helped local farmers increase the markets for minor millets (besides that of white rice) by developing new recipes and diversifying the products (Gruère et al. 2009a).

Moreover, to manage the trade-offs that occur between market integration and maintenance of *in situ* agrobiodiversity, it is essential to consider the diverse levels of the managed agrobiodiversity (e.g., household or community/ village level). Even though there is a tendency to focus on a restricted number of varieties at the private household level, diversity at the community level can still be maintained as different households specialize in diverse crops. Hence, the outcome of a trade-off analysis (Figure 15.1) depends on the level of considered diversity as well on the type (Kruijssen et al. 2009a). Above all, it is of high importance to incorporate the interests, needs, and institutions of smallholder farmers in market-based approaches for agrobiodiversity conservation (Johns et al. 2013). Furthermore, capacity development on value addition and marketing practices for farmers and other value chain actors is crucial to create enabling environments for niche markets focused on agrobiodiversity products (Padulosi et al. 2011a).

Questions that remain to be answered focus on the public good characteristics of agrobiodiversity and its conservation. Because farmers are rational actors, there is no guarantee that, when given the choice, they will choose to conserve agrobiodiversity to benefit our global society if they do not gain a personal benefit (e.g., income or access to preference traits). Farmers do not select crops and varieties to grow based on a rationale to conserve; their decision making is informed by both market- and household-level criteria. Furthermore, from an evolutionary perspective, we must expect that some agrobiodiversity will be lost while new diversity will be added. By creating or re-governing markets in such a way that agrobiodiversity is recognized and maintained or enhanced, it may be possible to increase these benefits and maintain on-farm agrobiodiversity (Kruijssen et al. 2009a). Nevertheless, cultural, social, or political factors may support or block conservation, even when market incentives are in place to stimulate conservation (Perrings et al. 2009). For instance, when cultural change occurs and traditional knowledge erodes, accelerated by economic development, reduced demand for diversity may result (Bellon 2004). Depending on local context, the goals of local economic development and agrobiodiversity conservation may be misaligned.

To move effectively forward in the future, we need to address the following: To what extent can resource-poor farmers be expected to display sustainable agrobiodiversity governance, and how high should the benefits of *in situ* conservation be to stimulate this behavior? What is needed to develop place-specific, market-based frameworks that successfully engage farmers in agrobiodiversity conservation? To what extent do niche markets for agrobiodiversity products have the potential to reach more and larger consumer segments without compromising agrobiodiversity conservation?

Bibliography

Note: Numbers in square brackets denote the chapter in which an entry is cited.

Abbott, J. A. 2005. Counting Beans: Agrobiodiversity, Indigeneity, and Agrarian Reform. *Prof. Geogr.* **57**:198–212. [12]

Acosta-Martínez, V., C. W. Bell, B. E. L. Morris, J. Zak, and V. G. Allen. 2010. Long-Term Soil Microbial Community and Enzyme Activity Responses to an Integrated Cropping-Livestock System in a Semi-Arid Region. *Agricult. Ecosyst. Environ.* **137**:231–240. [4]

Adu-Gyamfi, P., T. Mahmood, and R. Trethowan. 2015. Can Wheat Varietal Mixtures Buffer the Impacts of Water Deficit? *Crop Past. Sci.* **66**:757–769. [4]

AFSA. 2017. Resisting Corporate Takeover of African Seed Systems and Building Farmer Managed Seed Systems for Food Sovereignty in Africa. Kampala, Uganda: Alliance for Food Sovereignty in Africa. [8]

Agrawal, A., and C. Gibson. 1999. Enchantment and Disenchantment: The Role of Community in Natural Resource Conservation. *World Dev.* **27**:629–649. [13]

Ahmed, S., U. Unachukwu, J. R. Stepp, et al. 2010. Puerh Tea Tasting in Yunnan, China: Correlation of Drinkers' Perceptions to Phytochemistry. *J. Ethnopharmacol.* **132**:176–185. [9]

Ahn, Y. Y., S. E. Ahnert, J. P. Bagrow, and A. L. Barabási. 2011. Flavor Network and the Principles of Food Pairing. *Nature Sci. Rep.* **1**:196. [2]

Ahuja, M. R., and S. M. Jain, eds. 2015. Genetic Diversity and Erosion in Plants: Indicators and Prevention. Cham: Springer. [14]

Aistara, G. 2011. Seeds of Kin, Kin of Seeds: The Commodification of Organic Seeds and Social Relations in Costa Rica and Latvia. *Ethnography* **12**:490–517. [8, 13, 14]

———. 2012. Privately Public Seeds: Competing Visions of Property, Personhood, and Democracy in Costa Rica's Entry into Cafta and the Union for Plant Variety Protection (UPOV). *J. Polit. Ecol.* **19**:127–144. [13]

———. 2014a. Actually Existing Tomatoes: Politics of Memory, Variety, and Empire in Latvian Struggles over Seeds. *Focaal* **69**:12–27. [13, 14]

———. 2014b. Latvia's Tomato Rebellion: Nested Environmental Justice and Returning Eco-Sociality in the Post-Socialist EU Countryside. *J. Balt. Stud.* **45**:105–130. [14]

———. 2018. Organic Sovereignties: Struggles over Farming in an Age of Free Trade. Seattle: Univ. of Washington Press. [13]

Aksoy, Z. 2014. Local–Global Linkages in Environmental Governance: The Case of Crop Genetic Resources. *Glob. Environ. Polit.* **14**:26–44. [14]

Alexiades, M. N., ed. 2012. Mobility and Migration in Indigenous Amazonia: Contemporary Ethnoecological Perspectives. Environmental Anthropology and Ethnobiology, vol. 11. New York: Berghahn Books. [8]

Allaby, R. G., J. L. Kitchen, and D. Q. Fuller. 2015. Surprisingly Low Limits of Selection in Plant Domestication. *Evol. Bioinform. Online* **11(Suppl. 2)**:41–51. [7]

Almekinders, C. J. M., and W. de Boef. 1999. The Challenge of Collaboration in the Management of Crop Genetic Diversity. *ILEIA News.* **4**:5–7. [6]

Almekinders, C. J. M., L. O. Fresco, and P. C. Struik. 1995. The Need to Study and Manage Variation in Agro-Ecosystems. *NJAS* **43**:127–142. [1, 6]

Almekinders, C. J. M., and N. P. Louwaars. 2002. The Importance of the Farmers' Seed Systems in a Functional National Seed Sector. *J. New Seeds* **4**:15–33. [13, 14]

Almekinders, C. J. M., L. Mertens, J. P. van Loon, and E. T. Lammerts van Bueren. 2014. Potato Breeding in the Netherlands: A Successful Participatory Model with Collaboration between Farmers and Commercial Breeders. *Food Secur.* **6**:515–524. [6]

Altieri, M. A. 1999. The Ecological Role of Biodiversity in Agroecosystems. *Agricult. Ecosyst. Environ.* **74**:19–31. [13]

Altieri, M. A., F. R. Funes-Monzote, and P. Petersen. 2012. Agroecologically Efficient Agricultural Systems for Smallholder Farmers: Contributions to Food Sovereignty. *Agronom. Sustain. Devel.* **32**:1–13. [11]

Altieri, M. A., and V. Manuel Toledo. 2011. The Agroecological Revolution in Latin America: Rescuing Nature, Ensuring Food Sovereignty and Empowering Peasants. *J. Peasant Stud.* **38**:587–612. [11]

Altieri, M. A., C. I. Nicholls, A. Henao, and M. A. Lana. 2015. Agroecology and the Design of Climate Change-Resilient Farming Systems. *Agronom. Sustain. Devel.* **35**:869–890. [6, 7]

Alvarez, N., E. Garine, C. Khasah, et al. 2005. Farmers' Practices, Metapopulation Dynamics, and Conservation of Agricultural Biodiversity on-Farm: A Case Study of Sorghum among the Duupa in Sub-Sahelian Cameroon. *Biol. Conserv.* **121**:533–543. [8, 13]

Amend, T., J. Brown, A. Kothari, A. Phillips, and S. Stolton, eds. 2008. Protected Landscapes and Agrobiodiversity Values. Protected Landscapes and Seascapes, vol. 1. Heidelberg: Kasparek Verlag. [9]

Andersen, R. 2006. Governing Agrobiodiversity: A Framework for Analysis of Aggregate Effects of International Regimes (paper presented at the IDGEC Synthesis Conf.). Bali: IDGEC. [15]

———. 2008. Governing Agrobiodiversity: Plant Genetics and Developing Countries. Farnham: Ashgate Publishing Ltd. [14]

———. 2016. Farmers' Rights: Evolution of the International Policy Debate and National Implementation. In: Farmers' Crop Varieties and Farmers' Rights: Challenges in Taxonomy and Law, ed. M. Halewood, pp. 129–152. Abingdon, UK: Earthscan. [14]

———. 2017. Who Owns Agricultural Biodiversity? Rights, Responsibilities and Roles. In: Routledge Handook of Agricultural Biodiversity, ed. D. Hunter et al. London: Routledge. [14]

Andersen, R., and T. Winge, eds. 2013. Realising Farmers' Rights to Crop Genetic Resources: Success Stories and Best Practices New York: Routledge. [13]

Anderson, C. L., L. Lipper, T. J. Dalton, et al. 2010. Project Methodology: Using Markets to Promote the Sustainable Utilization of Crop Genetic Resources. In: Seed Trade in Rural Markets, ed. L. Lipper et al., pp. 31–48. London: Earthscan. [15]

Angelsen, A., P. Jagger, R. Babigumira, et al. 2014. Environmental Income and Rural Livelihoods: A Global-Comparative Analysis. *World Dev.* **64**:12–28. [9]

Angelsen, A., and S. Wunder. 2003. Exploring the Forest-Poverty Link: Key Concepts, Issues and Research Implications, Occasional Paper No. 40. Bogor, Indonesia: Center for Intl. Forestry Research. [9]

Apffel-Marglin, F. 2002. From Fieldwork to Mutual Learning: Working with Pratec. *Environ. Val.* **11**:345–367. [2]

Aprile, M. C., V. Caputo, and R. M. Nyaga, Jr. 2012. Consumers' Valuation of Food Quality Labels: The Case of European Geographic Indication and Organic Farming Labels. *Int. J. Consum. Stud.* **36**:158–165. [14]

Araújo, M. B., A. Rozenfeld, C. Rahbek, and P. A. Marquet. 2011. Using Species Co-Occurrence Networks to Assess the Impacts of Climate Change. *Ecography* **34**:897–908. [2]

Argumedo, A. 2008. The Potato Park, Peru: Conserving Agrobiodiversity in an Andean Indigenous Biocultural Heritage Site. In: Protected Landscapes and Agrobiodiversity Values, ed. T. Amend et al., pp. 45–58. Heidelberg: International Union for the Conservation of Nature. [2, 5]

———. 2012. Decolonising Action-Research: The Potato Park Biocultural Protocol for Benefit-Sharing. In: Biodiversity and Culture: Exploring Community Protocols, Rights and Consent, ed. K. Swiderska et al., vol. 65, Participatory Learning and Action, pp. 91–100. London: International Institute for Environment and Development. [2]

Argumedo, A., and M. Pimbert. 2006. Protecting Indigenous Knowledge against Biopiracy in the Andes. London: IIED. [12]

Asseng, S., F. Ewert, P. Martre, et al. 2015. Rising Temperatures Reduce Global Wheat Production. *Nat. Clim. Chang.* **5**:143–147. [3]

Atlin, G. N., J. E. Cairns, and B. Das. 2017. Rapid Breeding and Varietal Replacement Are Critical to Adaptation of Cropping Systems in the Developing World to Climate Change. *Glob. Food Sec.* **12**:31–37. [7]

Atran, S., D. Medin, N. Ross, et al. 1999. Folk Ecology and Commons Management in the Maya Lowlands. *PNAS* **96**:7598–7603. [2]

Aubry, C., and L. Kebir. 2013. Shortening Food Supply Chains: A Means for Maintaining Agriculture Close to Urban Areas? The Case of the French Metropolitan Area of Paris. *Food Pol.* **41**:85–93. [8]

Ávila, J. V. D. C., A. S. D. Mello, M. E. Beretta, et al. 2017. Agrobiodiversity and in Situ Conservation in Quilombola Home Gardens with Different Intensities of Urbanization. *Acta Bot. Brasilica* **31(1)**:1–10. [8]

Avila-Garcia, P. 2016. Towards a Political Ecology of Water in Latin America. *Rev. Estudio. Social.* **55**:18–31. [9]

Badr, A., K. Müller, S.-P. R., et al. 2000. On the Origin and Domestication History of Barley (*Hordeum Vulgare*). *Mol. Biol. Evol.* **17**:499–510. [3]

Badstue, L. B., M. R. Bellon, J. Berthaud, et al. 2006. Examining the Role of Collective Action in an Informal Seed System: A Case Study from the Central Valleys of Oaxaca, Mexico. *Hum. Ecol.* **34**:249–273. [13]

Baker, L. E. 2004. Tending Cultural Landscapes and Food Citizenship in Toronto's Community Gardens. *Geogr. Rev.* **94**:305–325. [8]

———. 2008. Local Food Networks and Maize Agrodiversity Conservation: Two Case Studies from Mexico. *Local Environ.* **13**:235–251. [11]

Bakker, M. G., D. K. Manter, A. M. Sheflin, T. L. Weir, and J. M. Vivanco. 2012. Harnessing the Rhizosphere Microbiome through Plant Breeding and Agricultural Management. *Plant Soil* **360**:1–13. [4]

Balázs, B., and G. Aistara. 2018. The Emergence, Dynamics, and Agency of Social Innovation in Seed Exchange Networks. *Int. J. Sociol. Agricult. Food* **24**:336–353. [13]

Balázs, B., A. Smith, G. Aistara, and G. Bela. 2015. WP 4: Case Study Report: Transnational Seed Exchange Networks, Transit: EU SHH.2013.3.2-1 Grant Agreement No: 613169. [13]

Baranski, M. R. 2015a. The Wide Adaptation of Green Revolution Wheat. PhD dissertation, Arizona State Univ. [6]

———. 2015b. Wide Adaptation of Green Revolution Wheat: International Roots and the Indian Context of a New Plant Breeding Ideal, 1960-1970. *Stud. Hist. Philos. Biol. Biomed. Sci.* **50**:41–50. [6, 7, 14]

Bibliography

Barbeau, W., and K. Hilu. 1993. Protein, Calcium, Iron, and Amino-Acid Content of Selected Wild and Domesticated Cultivars of Finger Millet. *Plant Foods Hum. Nutr.* **43**:97–104. [2]

Bardsley, D. 2003. Risk Alleviation via *in Situ* Agrobiodiversity Conservation: Drawing from Experiences in Switzerland, Turkey and Nepal. *Agricult. Ecosyst. Environ.* **99**:149–157. [15]

Barker, J. S. F. 1999. Conservation of Livestock Breed Diversity. *Anim. Genet. Res. Info.* **25**:33–43. [2]

Barnaud, A., M. Deu, E. Garine, et al. 2009. A Weed–Crop Complex in Sorghum: The Dynamics of Genetic Diversity in a Traditional Farming System. *Am. J. Bot.* **96**:1869–1879. [2]

Barrett, C. B., M. R. Carter, and C. P. Timmer. 2010. A Century-Long Perspective on Agricultural Development. *Am. J. Agric. Econ.* **92**:447–468. [8]

Barrett, C. B., T. Reardon, and P. Webb. 2001. Nonfarm Income Diversification and Household Livelihood Strategies in Rural Africa: Concepts, Dynamics, and Policy Implications. *Food Pol.* **26**:315–331. [7]

Barrios, E. 2007. Soil Biota, Ecosystem Services and Land Productivity. *Ecol. Econ.* **64**:269–285. [4]

Barthel, S., C. Crumley, and U. Svedin. 2013. Biocultural Refugia: Combating the Erosion of Diversity in Landscapes of Food Production. *Ecol. Soc.* **18**:71. [11]

Bassett, T. J., and D. Gautier. 2014. Regulation by Territorialization: The Political Ecology of Conservation & Development Territories. *EchoGéo* DOI : 10.4000/echogeo.14038 [8]

Baumgärtner, S., and M. F. Quaas. 2008. Agro-Biodiversity as Natural Insurance and the Development of Financial Insurance Markets. In: Agrobiodiversity, Conservation and Economic Development, ed. A. Kontoleon et al., pp. 293–317. London: Routledge. [15]

———. 2010. Managing Increasing Environmental Risks through Agrobiodiversity and Agrienvironmental Policies. *Agric. Econ.* **41**:483–496. [7]

Bazakos, C., M. Hanemian, C. Trontin, J. M. Jiménez-Gómez, and O. Loudet. 2017. New Strategies and Tools in Quantitative Genetics: How to Go from the Phenotype to the Genotype. *Annu. Rev. Plant Biol.* **68**:435–455. [3]

Bazile, D., S.-E. Jacobsen, and A. Verniau. 2016. The Global Expansion of Quinoa: Trends and Limits. *Front. Plant Sci.* **7**:1–6. [5]

Beaumont, M. A., W. Zhang, and D. J. Balding. 2002. Approximate Bayesian Computation in Population Genetics. *Genetics* **162**:2025–2035. [3]

Bebbington, A., and J. Carney. 1990. Geography in the International Agricultural Research Centers: Theoretical and Practical Concerns. *Ann. Assoc. Am. Geogr.* **80**:34–48. [6]

Bedmar Villanueva, A., M. Halewood, and I. López Noriega. 2015. Agricultural Biodiversity in Climate Change Adaptation Planning: An Analysis of the National Adaptation Programmes of Action. CCAFS Working Paper No. 95. Copenhagen: CGIAR. [7]

Beebe, S., O. Toro, A. V. Gonzalez, M. I. Chacon, and D. G. Debouck. 1997. Wild-Weed-Crop Complexes of Common Bean (*Phaseolus Vulgaris* L., Fabaceae) in the Andes of Peru and Colombia, and Their Implications for Conservation and Breeding. *Genet. Resour. Crop Evol.* **44**:73–91. [4]

Beeching, J. R., P. Marmey, M. C. Gavalda, et al. 1993. An Assessment of Genetic Diversity within a Collection of Cassava (*Manihot Esculenta* Crantz) Germplasm Using Molecular Markers. *Ann. Bot.* **72**:515–520. [2]

Belimov, A. A., I. C. Dodd, V. I. Safronova, et al. 2015. Rhizobacteria That Produce Auxins and Contain 1-Amino-Cyclopropane-1-Carboxylic Acid Deaminase Decrease Amino Acid Concentrations in the Rhizosphere and Improve Growth and Yield of Well-Watered and Water-Limited Potato (*Solanum tuberosum*). *Ann. Appl. Biol.* **167**:11–25. [4]

Bellon, M. R. 1991. The Ethnoecology of Maize Variety Management: A Case Study from Mexico. *Hum. Ecol.* **19**:389–418. [12]

———. 1996. The Dynamics of Crop Infraspecific Diversity: A Conceptual Framework at the Farmer Level. *Econ. Bot.* **50**:26–39. [13]

———. 2004. Conceptualizing Interventions to Support On-Farm Genetic Resource Conservation. *World Dev.* **32**:159–172. [8, 15]

Bellon, M. R., and S. B. Brush. 1994. Keepers of Maize in Chiapas, Mexico. *Econ. Bot.* **48**:196–209. [13]

Bellon, M. R., E. Dulloo, J. Sardos, I. Thormann, and J. J. Burdon. 2017. *In Situ* Conservation: Harnessing Natural and Human-Derived Evolutionary Forces to Ensure Future Crop Adaptation. *Evol. Appl.* **10**:965–977. [1, 5]

Bellon, M. R., and J. Hellin. 2011. Planting Hybrids, Keeping Landraces: Agricultural Modernization and Tradition among Small-Scale Maize Farmers in Chiapas, Mexico. *World Dev.* **39**:1434–1443. [8]

Bellon, M. R., D. Hodson, B. D., et al. 2005. Targeting Agricultural Research to Benefit Poor Farmers: Relating Poverty Mapping to Maize Environments in Mexico. *Food Pol.* **30**:476–492. [14]

Bellon, M. R., D. Hodson, and J. Hellin. 2011. Assessing the Vulnerability of Traditional Maize Seed Systems in Mexico to Climate Change. *PNAS* **108**:13432–13437. [2, 7]

Bellon, M. R., G. D. Ntandou-Bouzitou, and F. Caracciolo. 2016. On-Farm Diversity and Market Participation Are Positively Associated with Dietary Diversity of Rural Mothers in Southern Benin, West Africa. *PLoS One* **11**:e0162535. [13]

Bellon, M. R., and J. Risopoulos. 2001. Small-Scale Farmers Expand the Benefits of Improved Maize Germplasm: A Case Study from Chiapas, Mexico. *World Dev.* **29**:799–811. [13]

Bellon, M. R., and J. E. Taylor. 1993. Folk Soil Taxonomy and the Partial Adoption of New Seed Varieties. *Econ. Dev. Cult. Change* **41**:763–786. [4]

Bellon, M. R., and J. van Etten. 2014. Climate Change and On-Farm Conservation of Crop Landraces in Centres of Diversity. In: Plant Genetic Resources and Climate Change, ed. M. Jackson et al., pp. 137–150. [7]

Beltrame, D. M. O., C. N. S. Oliveira, T. Borelli, et al. 2016. Diversifying Institutional Food Procurement: Opportunities and Barriers for Integrating Biodiversity for Food and Nutrition in Brazil. *Rev. Raizes* **36**:55–69. [10, 15]

Bennett, E. M., G. S. Cumming, and G. D. Peterson. 2005. A Systems Model Approach to Determining Resilience Surrogates for Case Studies. *Ecosystems* **8**:945–957. [8]

Bentley, J. W. 1991. What Is Hielo: Honduran Farmers Perceptions of Diseases of Beans and Other Crops. *Interciencia* **16**:131–137. [4]

Berg, G. 2009. Plant–Microbe Interactions Promoting Plant Growth and Health: Perspectives for Controlled Use of Microorganisms in Agriculture. *Appl. Microbiol. Biotechnol.* **84**:11–18. [15]

Berkes, F. 2012. Sacred Ecology. New York: Routledge. [12]

Berkes, F., J. Colding, and C. Folke. 2000. Rediscovery of Traditional Ecological Knowledge as Adaptive Management. *Ecol. Appl.* **10**:1251–1262. [11]

Berkman, L. F., and T. Glass. 2000. Social Integration, Social Networks, Social Support, and Health. In: Social Epidemiology, ed. L. F. Berkman and I. Kawachi, pp. 137–153. New York: Oxford University Press. [11]

Berkman, L. F., T. Glass, I. Brissette, and T. E. Seeman. 2000. From Social Integration to Health: Durkheim in the New Millenium. *Soc. Sci. Med.* **51**:843–867. [11]

Berland, J. P., and R. Lewontin. 1986. Breeders' Rights and Patenting Life Forms. *Nature* **322**:785–788. [14]

Berlin, B. 1992. Ethnobiological Classification: Principles of Categorization of Plants and Animals in Traditional Societies. Princeton Princeton Univ. Press. [13]

Berlin, E. A., and B. Berlin. 2015. Medical Ethnobiology of the Highland Maya of Chiapas, Mexico: The Gastrointestinal Diseases. Princeton: Princeton Univ. Press. [12]

Bertazzini, M., and G. Forlani. 2016. Intraspecific Variability of Floral Nectar Volume and Composition in Rapeseed (*Brassica Napus* L. Var. *Oleifera*). *Front. Plant Sci.* **7**:288. [4]

Berthet, E. T. A., C. Barnaud, N. Girard, J. Labatut, and G. Martin. 2016. How to Foster Agroecological Innovations? A Comparison of Participatory Design Methods. *J. Environ. Plann. Manage.* **59**:280–301. [7]

Berthrong, S. T., D. H. Buckley, and L. E. Drinkwater. 2013. Agricultural Management and Labile Carbon Additions Affect Soil Microbial Community Structure and Interact with Carbon and Nitrogen Cycling. *Microb. Ecol.* **66**:158–170. [4]

Berti, P. R., and A. D. Jones. 2013. Biodiversity's Contribution to Dietary Diversity: Magnitude, Meaning and Measurement. In: Diversifying Food and Diets: Using Agricultural Biodiversity to Improve Nutrition and Health, ed. J. Fanzo et al., pp. 186–206. London: Earthscan. [10]

Berti, P. R., J. Krasevec, and S. FitzGerald. 2004. A Review of the Effectiveness of Agriculture Interventions in Improving Nutrition Outcomes. *Public Health Nutr.* **7**:599–609. [10]

Bhandari, B. 2009. Summer Rainfall Variability and the Use of Rice (*Oryza Sativa* L.) Varietal Diversity for Adaptation. M. Sc. dissertation, Swedish Univ. of Agricultural Sciences, Uppsala. [4]

Bharucha, Z., and J. Pretty. 2010. The Roles and Values of Wild Foods in Agricultural Systems. *Philos. Trans. R. Soc. Lond. B Biol. Sci.* **365**:2913–2926. [11]

Bhattarai, B., R. Beilin, and R. Ford. 2015. Gender, Agrobiodiversity, and Climate Change: A Study of Adaptation Practices in the Nepal Himalayas. *World Dev.* **70**:122–132. [6, 9]

Bianchi, F. J. J. A., C. J. H. Booij, and T. T. Tscharntke. 2006. Sustainable Pest Regulation in Agricultural Landscapes: A Review on Landscape Composition, Biodiversity and Natural Pest Control. *Proc. R. Soc. Lond. B Biol. Sci.* **273**:1715–1727. [2, 13, 15]

Bianchi, F. J. J. A., V. Mikos, L. Brussaard, B. Delbaere, and M. M. Pulleman. 2013. Opportunities and Limitations for Functional Agrobiodiversity in the European Context. *Environ. Sci. Policy* **27**:223–231. [9]

Bioversity Intl. 2017. Mainstreaming Agrobiodiversity in Sustainable Food Systems: Scientific Foundations for an Agrobiodiversity Index. Rome: Bioversity Intl. [8, 10]

Birol, E., M. Smale, and Á. Gyovai. 2006. Using a Choice Experiment to Estimate Farmers' Valuation of Agrobiodiversity on Hungarian Small Farms. *Environ. Resour. Econ.* **34**:439–469. [14]

Bishopp, A., and J. P. Lynch. 2015. The Hidden Half of Crop Yields. *Nat. Plants* **1**:15117. [4]

Bjørnstad, Å., S. Tekle, and M. Göransson. 2013. "Facilitated Access" to Plant Genetic Resources: Does It Work? *Genet. Resour. Crop Evol.* **60**:1959–1965. [5]

Blanco, J., L. Pascal, L. Ramon, H. Vandenbroucke, and C. S. M. 2013. Agrobiodiversity Performance in Contrasting Island Environments: The Case of Shifting Cultivation in Vanuatu, Pacific. *Agricult. Ecosyst. Environ.* **174**:28–39. [8]

Bodin, J. 2010. Observed Changes in Mountain Vegetation of the Alps during the Xxth Century: Role of Climate and Land-Use Changes. Ph.D. thesis, Dept. of Ecology, Univ. Henri Poincaré and Univ. Hannover. [5]

Bodirsky, B. L., A. Popp, H. Lotze-Campen, et al. 2014. Reactive Nitrogen Requirements to Feed the World in 2050 and Potential to Mitigate Nitrogen Pollution. *Nat. Commun.* **5**:3858. [9]

Bonanno, A., D. Constance, and M. Hendrickson. 1995. Global Agrofood Corporations and the State: The Ferruzzi Case. *Rural Sociol.* **60**:274–296. [11]

Bonman, J. M., H. E. Bockelman, Y. Jin, R. J. Hijmans, and A. Gironella. 2007. Geographic Distribution of Stem Rust Resistance in Wheat Landraces. *Crop Sci.* **47**:1955–1963. [5]

Bonnave, M., G. Bleeckx, J. Rojas Beltrán, et al. 2014. Farmers' Unconscious Incorporation of Sexually-Produced Genotypes into the Germplasm of a Vegetatively-Propagated Crop (*Oxalis Tuberosa* Mol.). *Genet. Resour. Crop Evol.* **61**:721–740. [3]

Bonnave, M., T. Bleeckx, F. Terrazas, and P. Bertin. 2015. Effect of the Management of Seed Flows and Mode of Propagation on the Genetic Diversity in an Andean Farming System: The Case of Oca (*Oxalis Tuberosa* Mol.). *Agricult. Human Values* **33**:673–688. [6]

Bonneuil, C. 2002. The Manufacture of Species: Kew Gardens, the Empire and the Standardisation of Taxonomic Practices in Late 19th Century Botany. In: Instruments, Travel and Science: Itineraries of Precision from the 17th to the 20th Century, ed. M.-N. Bourguet et al., pp. 89–215. New York: Routledge. [14]

Boster, J. S. 1985. Selection for Perceptual Distinctiveness: Evidence from Aguaruna Cultivars of *Manihot Esculenta. Econ. Bot.* **39**:310–325. [2]

Boyd, C., and T. Slaymaker. 2000. Re-Examining the More People, Less Erosion Hypothesis: Special Case or Wider Trend? ODI Natural Resource Perspectives, vol. 63. London: Overseas Development Institute. [8]

Bradbury, E. J., A. Duputié, M. Delêtre, et al. 2013. Geographic Differences in Patterns of Genetic Differentiation among Bitter and Sweet Cassava (*Manihot Esculenta*: Euphorbiaceae). *Am. J. Bot.* **100**:857–866. [5]

Bradbury, E. J., and E. Emshwiller. 2011. The Role of Organic Acids in the Domestication of Oxalis Tuberosa: A New Model for Studying Domestication Resulting in Opposing Crop Phenotypes. *Econ. Bot.* **65**:76–84. [8]

Branca, G., L. Lipper, N. McCarthy, and M. C. Jolejole. 2013. Food Security, Climate Change, and Sustainable Land Management: A Review. *Agronom. Sustain. Devel.* **33**:635–650. [6, 7]

Bretagnolle, V., and S. Gaba. 2015. Weeds for Bees? A Review. *Agronom. Sustain. Devel.* **35**:891–909. [2, 4]

Bretting, P. K., and D. N. Duvick. 1997. Dynamic Conservation of Plant Genetic Resources. *Adv. Agron.* **61**:2–51. [3]

Broegaard, R. B., L. V. Rasmussen, N. Dawson, et al. 2017. Wild Food Collection and Nutrition under Commercial Agriculture Expansion in Agriculture-Forest Landscapes. *For. Policy Econ.* **84**:92–101. [9]

Brondizio, E. S., and E. F. Moran. 2008. Human Dimensions of Climate Change: The Vulnerability of Small Farmers in the Amazon. *Philos. Trans. R. Soc. Lond. B Biol. Sci.* **363**:1803–1809. [9]

Brookfield, H. C. 2001. Exploring Agrodiversity. New York: Columbia Univ. Press. [7, 8]

Brookfield, H. C., C. Padoch, H. Parsons, and M. Stocking. 2002. Cultivating Biodiversity: Understanding, Analysing and Using Agricultural Diversity. London: ITDG Publications. [4]

Brush, S. B. 1991. A Farmer-Based Approach to Conserving Crop Germplasm. *Econ. Bot.* **45**:153–165. [14]

———. 1992. Farmer's Rights and Genetic Conservation in Traditional Farming Systems. *World Dev.* **20**:1617–1630. [14]

———. 1995. *In Situ* Conservation of Landraces in Center of Crop Diversity. *Crop Sci.* **35**:346–354. [5, 12]

———. 1999. Bioprospecting the Public Domain. *Cult. Anthropol.* **14**:535–555. [13]

———, ed. 2000. Genes in the Field: On-Farm Conservation of Crop Diversity. Rome: International Plant Genetic Resources Institute. [1, 2, 14]

———. 2004. Farmers' Bounty: Locating Crop Diversity in the Contemporary World. New Haven: Yale Univ. Press. [2, 8, 9, 14, 15]

Brush, S. B., M. R. Bellon, R. J. Hijmans, et al. 2015. Assessing Maize Genetic Erosion. *PNAS* **112**:E1. [2, 15]

Brush, S. B., H. J. Carney, and Z. Humán. 1981. Dynamics of Andean Potato Agriculture. *Econ. Bot.* **35**:70–88. [4, 12]

Brush, S. B., and E. Meng. 1998. Farmers' Valuation and Conservation of Crop Genetic Resources. *Genet. Resour. Crop Evol.* **45**:139–150. [14]

Brush, S. B., and H. R. Perales. 2007. A Maize Landscape: Ethnicity and Agro-Biodiversity in Chiapas Mexico. *Agricult. Ecosyst. Environ.* **121**:211–221. [4, 12]

Brush, S. B., D. Tadesse, and E. Van Dusen. 2003. Crop Diversity in Peasant and Industrialized Agriculture: Mexico and California. *Soc. Nat. Resour.* **16**:123–141. [5]

Brush, S. B., J. E. Taylor, and M. R. Bellon. 1992. Technology Adoption and Biological Diversity in Andean Potato Agriculture. *J. Dev. Econ.* **39**:365–387. [2, 3, 13]

Brussaard, L., P. C. de Ruiter, and G. G. Brown. 2007. Soil Biodiversity for Agricultural Sustainability. *Agricult. Ecosyst. Environ.* **121**:233–244. [2, 13]

Buchanan, R. 1992. Wicked Problems in Design Thinking. *Design Issues* **8**:5–21. [7]

Burlingame, B., R. Charrondiere, and B. Mouille. 2009. Food Composition Is Fundamental to the Cross-Cutting Initiative on Biodiversity for Food and Nutrition. *J. Food Compost. Anal.* **22**:361–365. [10]

Burlingame, B., and S. Dernini. 2010. Sustainable Diets and Biodiversity Directions and Solutionsfor Policy, Research and Action. Proc. Intl. Scientific Symposium Biodiversity and Sustainable Diets United against Hunger. Rome: FAO. [9]

Cabrera-Medaglia, J. 2013. La Relación del Protocolo de Nagoya Con el Tratado Internacional de Recursos Fitogenéticos Para la Alimentación y la Agricultura: Opciones y Recomendaciones de Política Para Una Implementación Sinérgica a Nivel Nacional. Quito: International Union for Conservation of Nature (UICN). [12]

Cadima Fuentes, X., R. Van Treuren, R. Hoekstra, R. G. Van den Berg, and M. S. M. Sosef. 2017. Genetic Diversity of Bolivian Wild Potato Germplasm: Changes during *Ex Situ* Conservation Management and Comparisons with Resampled *in Situ* Populations. *Genet. Resour. Crop Evol.* **64**:331. [2]

Caillon, S., and P. Degeorges. 2007. Biodiversity: Negotiating the Border between Nature and Culture. *Biodivers. Conserv.* **16**:2919–2931. [13, 14]

Caillon, S., and V. Lanouguère-Bruneau. 2004. Taro Diversity in a Village of Vanua Lava Island (Vanuatu): Where, What, Who, How and Why? In: Third Taro Symposium, ed. L. Guarino et al. Nadi, Fiji Islands: Secretariat of the Pacific Community. [13]

Calvet-Mir, L., M. Calvet-Mir, J. L. Molina, and V. Reyes-Garcia. 2012a. Seeds Exchange as an Agrobiodiversity Conservation Mechanism: A Case Study in Vall Fosca, Catalan Pyrenees, Iberian Peninsula. *Ecol. Soc.* **17**:29. [11]

Calvet-Mir, L., E. Gómez-Bagetthun, and V. Reyes-García. 2012b. Beyond Food Production: Ecosystem Services Provided by Home Gardens: A Case Study in Vall Fosca, Catalan Pyrenees, Northeastern Spain. *Ecol. Econ.* **74**:153–160. [11]

Camacho-Henriquez, A., F. Kraemer, G. Galluzzi, et al. 2015. Decentralized Collaborative Plant Breeding for Utilization and Conservation of Neglected and Underutilized Crop Genetic Resources. In: Advances in Plant Breeding Strategies: Breeding, Biotechnology and Molecular Tools, ed. J. M. Al-Khayri et al., pp. 25–61. Dordrecht: Springer. [2]

Camadro, E. L. 2012. Relevance of the Genetic Structure of Natural Populations, and Sampling and Classification Approaches for Conservation and Use of Wild Crop Relatives: Potato as an Example. *Botany* **90**:1065–1072. [3, 5]

Camadro, E. L., L. E. Erazzú, J. F. Maune, and M. C. Bedogni. 2012. A Genetic Approach to the Species Problem in Wild Potato. *Plant Biol.* **14**:543–554. [5]

Campbell, B. M., S. J. Vermeulen, P. K. Aggarwal, et al. 2016. Reducing Risks to Food Security from Climate Change. *Glob. Food Sec.* **11**:34–43. [6]

Campbell, L. G., J. Luo, and K. L. Mercer. 2013. Effect of Water Availability and Genetic Diversity on Flowering Phenology, Synchrony and Reproductive Investment in Summer Squash. *J. Agric. Sci.* **151**:775–786. [4]

Canahua-Murillo, A. 2016. Revaloración del Conocimiento Tradicional en la Gestión de la Agrobiodiversidad: Proyecto Sipam in Puno. In: Biodiversidad y Propiedad Intelectual en Disputa, ed. S. Roca, pp. 287–305. Lima: Esan ediciones. [12]

Cardinale, B. J., J. E. Duffy, A. Gonzalez, et al. 2012. Biodiversity Loss and Its Impact on Humanity. *Nature* **486**:59–67. [4]

Carlos, E. J., M. T. Garcia-Conesa, and F. A. Tomas-Barberan. 2007. Nutraceuticals: Facts and Fiction. *Phytochemistry* **68**:2986–3008. [11]

Carney, J. A. 1991. Indigenous Soil and Water Management Senegambian Rice Farming Systems. *Agricult. Human Values* **8**:37–48. [8]

———. 1993. Converting the Wetlands, Engendering the Environment: The Intersection of Gender with Agrarian Change in the Gambia. *Econ. Geogr.* **69**:329–349. [6, 8]

———. 2001. Black Rice: The African Origins of Rice Cultivation in the Americas. Cambridge, MA: Harvard Univ. Press. [8, 13]

Carney, J. A., and R. N. Rosomoff. 2009. In the Shadow of Slavery: Africa's Botanical Legacy in the Atlantic World. Berkeley: Univ. of California Press. [6, 8, 13]

Carrizosa, S., S. B. Brush, B. Wright, and P. McGuire, eds. 2004. Accessing Biodiversity and Sharing the Benefits: Lessons from Implementing the Convention on Biological Diversity, Iucn Environmental Policy and Law Paper No. 54. Cambridge: The World Conservation Union. [14]

Cash, D. W., W. C. Clark, F. Alcock, et al. 2003. Knowledge Systems for Sustainable Development. *PNAS* **100**:8086–8091. [1]

Castañeda-Álvarez, N. P., S. de Haan, H. Juárez, et al. 2015. *Ex Situ* Conservation Priorities for the Wild Relatives of Potato (*Solanum* L . Section *Petota*). *PLoS One* **10**:e0122599. [2, 5]

Castañeda-Álvarez, N. P., C. K. Khoury, H. A. Achicanoy, et al. 2016. Global Conservation Priorities for Crop Wild Relatives. *Nat. Plants* **2**:16022. [2, 5, 9]

CBD. 2000. What Is Agricultural Biodiversity? https://www.cbd.int/agro/whatis.shtml (accessed Jan. 17, 2018). [1, 8]

Ceccarelli, S. 2009. Evolution, Plant Breeding and Biodiversity. *J. Agric. Environ. Int. Dev.* **103**:131–145. [2]

Ceccarelli, S., and S. Grando. 2007. Decentralized-Participatory Plant Breeding: An Example of Demand Driven Research. *Euphytica* **155**:349–360. [15]

Ceccarelli, S., S. Grando, A. Amri, et al. 2001. Decentralized and Participatory Plant Breeding for Marginal Environments. In: Broadening the Genetic Base of Crop Production, ed. H. D. Cooper et al., pp. 115–136. Wallingford: CABI. [2]

Ceccarelli, S., S. Grando, M. Maatougui, et al. 2010. Plant Breeding and Climate Changes. *J. Agric. Sci.* **148**:627–637. [4, 7]

Ceccarelli, S., E. P. Guimarães, and E. Weltizien. 2009. Plant Breeding and Farmer Participation. Rome: FAO. [3]

Celis, C., M. Scurrah, S. Cowgill, et al. 2004. Environmental Biosafety and Transgenic Potato in a Centre of Diversity for This Crop. *Nature* **432**:222–225. [3]

CEPAL. 2014. Los Pueblos Indígenas en América Latina: Avances en el Último Decenio y Retos Pendientes Para la Garantía de Sus Derechos. Santiago: Comisión Económica para América Latina y el Caribe (CEPAL). [12]

Cerdán, C. R., M. C. Rebolledo, G. Soto, B. Rapidel, and F. L. Sinclair. 2012. Local Knowledge of Impacts of Tree Cover on Ecosystem Services in Smallholder Coffee Production Systems. *Agric. Syst.* **110**:119–130. [2, 4]

Chable, V., J. Dawson, R. Bocci, and I. Goldringer. 2014. Seeds for Organic Agriculture: Development of Participatory Plant Breeding and Farmers' Networks in France In: Organic Farming, Prototype for Sustainable Agricultures, ed. S. Bellon and S. Penvern. Dordrecht: Springer. [13]

Challinor, A. J., J. Watson, D. B. Lobell, et al. 2014. A Meta-Analysis of Crop Yield under Climate Change and Adaptation. *Nat. Clim. Chang.* **4**:287–291. [6]

Chambers, K. J., and J. H. Momsen. 2007. From the Kitchen and the Field: Gender and Maize Diversity in the Bajío Region of Mexico. *Singap. J. Trop. Geogr.* **28**:39–56. [8]

Chan, K. M. A., P. Balvanera, K. Benessaiah, et al. 2016. Why Protect Nature? Rethinking Values and the Environment. *PNAS* **113**:1462–1465. [6]

Chase Smith, R., M. Benavides, M. Pariona, and M. Tuesta. 2013. Mapping the Past and the Future: Geomatics and Indigenous Territories in the Peruvian Amazon. *Hum. Organ.* **62**:357–368. [11]

Checkel, J. T. 1999. Why Comply? Constructivism, Social Norms and the Study of International Institutions, Arena Working Papers 99/24. Oslo: ARENA Centre for European Studies. [14]

Chiarolla, C., L. Selim, and M. Schloen. 2012. An Analysis of the Relationship between the Nagoya Protocol and Instruments Related to Genetic Resources for Food and Agriculture and Farmers' Rights. In: The 2010 Nagoya Protocol on Access and Benefit-Sharing in Perspective Implications for International Law and Implementation Challenges, ed. E. Morgera et al., pp. 83–122. Leiden: Brill Publ. [13]

Choudhury, B., M. L. Khan, and S. Dayanandan. 2013. Genetic Structure and Diversity of Indigenous Rice (*Oryza Sativa*) Varieties in the Eastern Himalayan Region of Northeast India. *Springerplus* **2**:1–10. [3]

Chweya, J., and C. J. M. Almekinders. 2000. Supporting the Utilization and Development of Traditional Leafy Vegetables in Africa. In: Encouraging Diversity: The Conservation and Development of Plant Genetic Resources, ed. C. J. M. Almekinders and W. D. Boef, pp. 294–299. London: ITDG. [6]

Chweya, J. A., and P. B. Eyzaguirre, eds. 1999. The Biodiversity of Traditional Leafy Vegetables. Rome: Intl. Plant Genetic Resources Institute. [9]

Civáň, P., H. Craig, C. J. Cox, and T. A. Brown. 2015. Three Geographically Separate Domestications of Asian Rice. *Nat. Plants* 1:15164. [3]

Claessens, L., J. M. Antle, J. J. Stoorvogel, et al. 2012. A Method for Evaluating Climate Change Adaptation Strategies for Small-Scale Farmers Using Survey, Experimental and Modeled Data. *Agric. Syst.* 111:85–95. [6]

Clark, W. C., T. P. Tomich, M. Van Noordwijk, et al. 2016. Boundary Work for Sustainable Development: Natural Resource Management at the Consultative Group on International Agricultural Research (CGIAR). *PNAS* 113:4615–4622. [1]

Cleveland, D., D. Soleri, and S. E. Smith. 1994. Do Folk Crop Varieties Have a Role in Sustainable Agriculture? *Bioscience* 44:740–751. [13]

Clotault, J., A.-C. Thuillet, M. Buiron, et al. 2012. Evolutionary History of Pearl Millet (*Pennisetum Glaucum* [L.] R. Br.) and Selection on Flowering Genes since Its Domestication. *Mol. Biol. Evol.* 29:1199–1212. [3]

Coats, V. C., and M. E. Rumpho. 2014. The Rhizosphere Microbiota of Plant Invaders: An Overview of Recent Advances in the Microbiomics of Invasive Plants. *Front. Microbiol.* 5:368. [2]

Cockrall-King, J. 2012. Food and the City: Urban Agriculture and the New Food Revolution. New York: Prometheus Books. [6, 8]

Cohn, A. S., P. Newton, J. D. B. Gil, et al. 2017. Smallholder Agriculture and Climate Change. *Annu. Rev. Environ. Resour.* 42:347–375. [6]

Coimbra, C. E. A., R. V. Santos, J. R. Welch, et al. 2013. The First National Survey of Indigenous People's Health and Nutrition in Brazil: Rationale, Methodology, and Overview of Results. *BMC Public Health* 13:52. [9]

CONABIO. 2011. Recopilación, Generación, Actualización y Análisis de Información Acerca de la Diversidad Genética de Maíces y Sus Parientes Silvestres en México, Informe de Gestión y Resultados. Mexico City: Comision Nacional para el Conocimiento y Uso de la Biodiversidad. [2]

———. 2013. Bases de Datos de Maíz. Comision Nacional Para el Conocimiento y Uso de la Biodiversidad. Mexico City: Comisión Nacional para el Conocimiento y Uso de la Biodiversidad. [2]

Condori, B., R. J. Hijmans, J. F. Ledent, and R. Quiroz. 2014. Managing Potato Biodiversity to Cope with Frost Risk in the High Andes: A Modeling Perspective. *PLoS One* 9:e81510. [4]

Constitutional Court Republic of Colombia. 2012. Sentence C-1051 de 2012, MP Luis Guillermo Guerrero Pérez: The Court Declares Unconstitutional the Law 1518 of 2012 "by Means of Which Approved the International Convention for the Protection of Plant Varieties." http://www.corteconstitucional.gov.co/RELATORIA/2012/C-1051-12.htm. (accessed July 21, 2017). [12]

Coomes, O. T. 2010. Of Stakes, Stems, and Cuttings: The Importance of Local Seed Systems in Traditional Amazonian Societies. *Prof. Geogr.* 62:323–334. [13]

Coomes, O. T., S. J. McGuire, E. Garine, et al. 2015. Farmer Seed Networks Make a Limited Contribution to Agriculture? Four Common Misconceptions. *Food Pol.* 56:41–50. [2, 11, 13]

Cordell, D., J.-O. Drangert, and S. White. 2009. The Story of Phosphorus: Global Food Security and Food for Thought. *Glob. Environ. Change* 19:292–305. [9]

Corntassel, J. J. 2003. Who Is Indigenous? Peoplehood and Ethnonationalist Approaches to Rearticulating Indigenous Identity. *Nationalism Ethn. Polit.* 9:75–100. [12]

Costanzo, A., and P. Barberi. 2016. Field Scale Functional Agrobiodiversity in Organic Wheat: Effects on Weed Reduction, Disease Susceptibility and Yield. *Eur. J. Agron.* 76:1–16. [4]

Creissen, H. E., T. H. Jorgensen, and J. K. M. Brown. 2016. Increased Yield Stability of Field-Grown Winter Barley (*Hordeum Vulgare* L.) Varietal Mixtures through Ecological Processes. *Crop Protect.* **85**:1–8. [4]

Crosby, A. 1986. Ecological Imperialism: The Biological Expansion of Europe, 900–1900. Cambridge: Cambridge Univ. Press. [8]

Cruz-Garcia, G. S., and P. L. Howard. 2013. I Used to Be Ashamed: The Influence of an Educational Program on Tribal and Non-Tribal Children's Knowledge and Valuation of Wild Food Plants. *Learn. Individ. Differ.* **27**:234–240. [11]

Cruz-Garcia, G. S., and P. C. Struik. 2015. Spatial and Seasonal Diversity of Wild Food Plants in Home Gardens of Northeast Thailand. *Econ. Bot.* **69**:99–113. [10]

Cullather, N. 2010. The Hungry World. Cambridge, MA: Harverd Univ. Press. [6]

da Fonseca, R. R., B. D. Smith, N. Wales, et al. 2015. The Origin and Evolution of Maize in the Southwestern United States. *Nat. Plants* **1**:14003. [3]

Daes, E. I. 1996. Working Paper for the Working Group on Indigenous Populations. New York: UN-ECOSOC, Commission on Human Rights, Sub-Commission on the Prevention of Discrimination and Protection of Minorities, 14th Session. [12]

Dang, A., and J. V. Meenakshi. 2017. The Nutrition Transition and the Intra-Household Double Burden of Malnutrition in India. Tokyo: ADB Institute. [10]

Darby, M. R., and E. Karni. 1973. Free Competition and the Optimal Amount of Fraud. *J. Law Econ.* **16**:67–88. [6]

Da Vià, E. 2012. Seed Diversity, Farmers' Rights, and the Politics of Repeasantization. *Int. J. Sociol. Agricult. Food* **19**:229–242. [13]

Davis, A. S., J. D. Hill, C. A. Chase, A. M. Johanns, and M. Liebman. 2012. Increasing Cropping System Diversity Balances Productivity, Profitability and Environmental Health. *PLoS One* **7**:e47149. [2]

Dawson, J. C., K. M. Murphy, and S. S. Jones. 2008. Decentralized Selection and Participatory Approaches in Plant Breeding for Low-Input Systems. *Euphytica* **160**:143–154. [2]

Dawson, J. C., P. Rivière, J. F. Berthellot, et al. 2016a. Collaborative Plant Breeding for Organic Agricultural Systems in Developed Countries. *Sustainability* **3**:1206–1223. [6]

Dawson, N., A. Martin, and T. Sikor. 2016b. Green Revolution in Sub-Saharan Africa: Implications of Imposed Innovation for the Wellbeing of Rural Smallholders. *World Dev.* **78**:204–208. [8]

De Boef, W. S., M. H. Thijssen, P. Shrestha, et al. 2012. Moving Beyond the Dilemma: Practices That Contribute to the on-Farm Management of Agrobiodiversity. *J. Sustain. Agricult.* **36**:788–809. [15]

De Grenade, R., and G. P. Nabhan. 2013. Baja California Peninsula Oases: An Agro-Biodiversity of Isolation and Integration. *Appl. Geogr.* **41**:24–35. [8]

de Haan, S. 2009. Potato Diversity at Height: Multiple Dimensions of Farmer-Driven *in-Situ* Conservation in the Andes. PhD Thesis. Wageningen Wageningen Univ. [13]

de Haan, S., M. Bonierbale, M. Ghislain, J. Núñez, and G. Trujillo. 2007. Indigenous Biosystematics of Andean Potatoes: Folk Taxonomy, Descriptors, and Nomenclature. *Acta Horticult.* **745**:89–134. [2]

de Haan, S., G. Burgos, R. Ccanto, et al. 2012a. Effect of Production Environment, Genotype and Process on the Mineral Content of Native Bitter Potato Cultivars Converted into White Chuño. *J. Sci. Food Agric.* **92**:2098–2105. [9]

de Haan, S., G. Burgos, R. Liria, M. Bonierbale, and G. Thiele. 2009. The Role of Biodiverse Potatoes in the Human Diet in Central Peru: Nutritional Composition, Dietary Intake and Cultural Connotations. In: Potato Diversity at Height: Multiple Dimensions of Farmer Driven in-Situ Conservation, ed. S. de Haan, pp. 161–182. Wageningen: Wageningen Univ. [9]

de Haan, S., J. Núñez, M. Bonierbale, and M. Ghislain. 2010. Multilevel Agrobiodiversity and Conservation of Andean Potatoes in Central Peru: Species, Morphological, Genetic, and Spatial Diversity. *Mt. Res. Dev.* **30**:222–231. [3, 5]

de Haan, S., J. Núñez, M. Bonierbale, M. Ghislain, and J. Van der Maesen. 2013. A Simple Sequence Repeat (Ssr) Marker Comparison of a Large in- and Ex-Situ Potato Landrace Cultivar Collection from Peru Reaffirms the Complementary Nature of Both Conservation Strategies. *Diversity* **5**:505–521. [3, 14]

de Haan, S., S. Polreich, F. Rodriguez, et al. 2016. A Long-Term Systematic Monitoring Framework for On-Farm Conserved Potato Landrace Diversity. In: Enhancing Crop Genepool Use: Capturing Wild Relative and Landrace Diversity for Crop Improvement, ed. N. Maxted et al., pp. 289–296. Boston: CABI Publishing. [2, 3, 5]

de Haan, S., and R. N. Y. Salazar. 2006. Catálogo de Variedades de Papa Nativa de Huancavelica-Perú. Lima: Centro Internacional de la Papa. [13, 14]

de Haan, S., M. Scurrah, C. Bastos, et al. 2012b. Becoming Wild: Investigating the Putative Origin, Feral Capacity and Ethnobotany of *Araq* Potatoes. In: Xxv Congreso de la Asociación Latinoamericana de la Papa. Uberlandia: ALAP. [2]

de Haan, S., and R. O. Villanueva. 2015. Catálogo de Variedades de Papa Nativa de Chugay, La Libertad, Perú Peru: International Potato Center (CIP). [13, 14]

De Jonge, B. 2011. What Is Fair and Equitable Benefit-Sharing? *J. Agric. Environ. Ethics* **24**:127–146. [14]

De La Cadena, M. 2000. Indigenous Mestizos: The Politics of Race and Culture, Cuzco, Peru, 1919–1991. Durham: Duke Univ. Press. [12]

Delaquis, E., S. de Haan, and K. A. G. Wyckhuys. 2018. On-Farm Diversity Offsets Environmental Pressures in Tropical Agroecosystems: A Synthetic Review for Cassava-Based Systems. *Agricult. Ecosyst. Environ.* **251**:226–235. [8]

Delêtre, M., D. B. McKey, and T. R. Hodkinson. 2011. Marriage Exchanges, Seed Exchanges, and the Dynamics of Manioc Diversity. *PNAS* **108**:18249–18254. [3, 5, 13]

Demeritt, D. 2001. The Construction of Global Warming and the Politics of Science. *Ann. Assoc. Am. Geogr.* **91**:307–337. [7]

Demeulenaere, E. 2014. A Political Ontology of Seeds: The Transformative Frictions of a Farmers' Movement in Europe. *Focaal* **2014**:45–61. [13]

De Mita, S., A.-C. Thuillet, L. Gay, et al. 2013. Detecting Selection Along Environmental Gradients: Analysis of Eight Methods and Their Effectiveness for Outbreeding and Selfing Populations. *Mol. Ecol.* **22**:1383–1399. [3]

Denison, R. F. 2012. Darwinian Agriculture: How Understanding Evolution Can Improve Agriculture. Princeton: Princeton Univ. Press. [2]

De Schutter, O. 2011. How Not to Think of Land-Grabbing: Three Critiques of Large-Scale Investments in Farmland. *J. Peasant Stud.* **38**:249–279. [9]

Desclaux, D., J. M. Nolot, Y. Chiffoleau, E. Gozé, and C. Leclerc. 2008. Changes in the Concept of Genotype X Environment Interactions to Fit Agriculture Diversification and Decentralized Participatory Plant Breeding from a Pluridisciplinary Point of View. *Euphytica* **163**:533–546. [7]

Descola, P., and G. Pálsson, eds. 1996. Nature and Society: Anthropological Perspectives. London: Routledge. [14]

338   *Bibliography*

Deu, M., F. Sagnard, J. Chantereau, et al. 2010. Spatio-Temporal Dynamics of Genetic Diversity in *Sorghum Bicolor* in Niger. *Theor. Appl. Genet.* **120**:1301–1313. [2]

Devaux, A., D. Horton, C. Velasco, et al. 2009. Collective Action for Market Chain Innovation in the Andes. *Food Pol.* **34**:31–38. [15]

Devaux, A., M. Ordinola, and D. Horton, eds. 2011. Innovation for Development: The Papa Andina Experience. Peru: International Potato Center (CIP). [15]

Devaux, A., M. Torero, J. Donovan, and D. E. Horton. 2016. Innovation for Inclusive Value-Chain Development: Successes and Challenges. Washington, D.C.: Intl. Food Policy Research Institute (IFPRI). [5]

de Wit, M. M. 2016. Are We Losing Diversity? Navigating Ecological, Political, and Epistemic Dimensions of Agrobiodiversity Conservation. *Agricult. Human Values* **33**:625–640. [8, 14]

———. 2017. Stealing into the Wild: Conservation Science, Plant Breeding and the Makings of New Seed Enclosures. *J. Peasant Stud.* **44**:169–212. [14]

Dhileepan, K. 1994. Variation in Populations of the Introduced Pollinating Weevil (*Elaeidobius Kamerunicus*)(Coleoptera: Curculionidae) and Its Impact on Fruitset of Oil Palm (*Elaeis Guineensis*) in India. *Bull. Entom. Res. India* **84**:477–485. [2]

Di Falco, S., M. Bezabih, and M. Yesuf. 2010. Seeds for Livelihood: Crop Biodiversity and Food Production in Ethiopia. *Ecol. Econ.* **69**:1695–1702. [8]

Diamond, J., and P. Bellwood. 2003. Farmers and Their Languages: The First Expansions. *Science* **300**:597–603. [3]

Diener, E., and M. E. P. Seligman. 2004. Beyond Money: Toward and Economy of Well-Being. *Psychol. Sci. Public Interest* **5**:1–31. [11]

Dillard, C. J., and J. B. German. 2000. Phytochemicals: Nutraceuticals and Human Health. *J. Sci. Food Agric.* **80**:1744–1756. [11]

Dorst, K. 2015. Frame Innovation: Create New Thinking by Design. Cambridge, MA: MIT Press. [7]

Dove, M. 1996. Center, Periphery, and Biodiversity: A Paradox of Governance a Developmental Challenge. In: Valuing Local Knowledge: Indigenous People and Intellectal Property Rights, ed. S. Brush and D. Stabinsky. Washington, D.C.: Island Press. [13]

Dressler, W. W., and J. R. Bindon. 2000. The Health Consequences of Cultural Consonance: Cultural Dimensions of Lifestyle, Social Support, and Arterial Blood Pressure in an African American Community. *Am. Anthropol.* **102**:244–260. [11]

Duru, M., O. Therond, G. Martin, et al. 2015. How to Implement Biodiversity-Based Agriculture to Enhance Ecosystem Services: A Review. *Agron. Sustain. Dev.* **35**:1259–1281. [7]

Dutfield, G. 2010. Why Traditional Knowledge Is Important in Drug Discovery. *Future Med. Chem.* **2**:1405–1409. [12]

Duvick, D. 1984. Genetic Diversity in Major Farm Crops on the Farm and in Reserve. *Econ. Bot.* **38**:161–178. [7]

Dyer, G. A., A. López-Feldman, A. Yúnez-Naude, and J. E. Taylor. 2014. Genetic Erosion in Maize's Center of Origin. *PNAS* **111**:14094–14099. [2, 3, 5]

Dze, M. 2005. State Policies, Shifting Cultivation and Indigenous Peoples in Ims. *Indigen. Aff.* **2**:30–37. [12]

Eakin, H., H. R. Perales, K. Appendini, and S. Sweeney. 2014. Selling Maize in Mexico: The Persistence of Peasant Farming in an Era of Global Markets. *Dev. Change* **45**:133–155. [8]

Eddleston, M., L. Karalliedde, N. Buckley, et al. 2002. Pesticide Poisoning in the Developing World: A Minimum Pesticides List. *Lancet* **360**:1163–1167. [11]

Eilers, E. J., C. Kremen, S. Smith Greenleaf, A. K. Garber, and A.-M. Klein. 2011. Contribution of Pollinator-Mediated Crops to Nutrients in the Human Food Supply. *PLoS One* **6**:e21363. [9]

Elith, J., and J. R. Leathwick. 2009. Species Distribution Models: Ecological Explanation and Prediction across Space and Time. *Annu. Rev. Ecol. Evol. Syst.* **40**:677–697. [2, 5]

Ellen, R., and S. Platten. 2011. The Social Life of Seeds: The Role of Networks of Relationships in the Dispersal and Cultural Selection of Plant Germplasm. *J. R. Anthropol. Inst.* **17**:563–584. [8]

Ellis, K. E., and M. E. Barbercheck. 2015. Management of Overwintering Cover Crops Influences Floral Resources and Visitation by Native Bees. *Environ. Entomol.* **44**:999–1010. [4]

Ellouze, W., C. Hamel, V. Vujanovic, et al. 2013. Chickpea Genotypes Shape the Soil Microbiome and Affect the Establishment of the Subsequent Durum Wheat Crop in the Semiarid North American Great Plains. *Soil Biol. Biochem.* **63**:129–141. [4]

Ellstrand, N. C. 2003. Dangerous Liaisons? When Cultivated Plants Mate with Their Wild Relatives. Baltimore: Johns Hopkins Univ. Press. [2]

Elzinga, C. L., D. W. Salzer, J. W. Willoughby, and D. P. Gibbs. 2001. Monitoring Plant and Animal Populations. Abingdon, UK: Blackwell. [2]

Empaire, L., and N. Peroni. 2007. Traditional Management of Agrobiodiversity in Brazil: A Case Study of Manioc. *Hum. Ecol.* **2007**:761–768. [13]

Enjalbert, J. N., and J. J. Johnson. 2011. Guide for Producing Dryland Camelina in Eastern Colorado. http://extension.colostate.edu/docs/pubs/crops/00709.pdf (accessed Sept. 11, 2018). [2]

Entman, R. S. 1993. Framing: Toward the Clarification of a Fractured Paradigm. *J. Commun.* **43**:51–58. [7]

Escobar, A. 1995. Encountering Development: The Making and Unmaking of the Third World. Princeton: Princeton Univ. Press. [13]

Escobar, A., and W. Harcourt. 2002. Practices of Difference: Introducing Women and the Politics of Place. *Development* **45**:7–14. [13]

Excoffier, L., I. Dupanloup, E. Huerta-Sánchez, V. C. Sousa, and M. Foll. 2013. Robust Demographic Inference from Genomic and Snp Data. *PLoS Genet.* **9**:e1003905. [3]

Fafchamps, M., and R. Vargas. 2005. Selling at the Farmgate or Travelling to Market. *Am. J. Agric. Econ.* **87**:717–734. [15]

FANTA. 2006. Developing and Validating Simple Indicators of Dietary Quality and Energy Intake of Infants and Young Children in Developing Countries: Summary of Findings from Analysis of 10 Data Sets. Washington, D.C.: Food and Nutrition Technical Assistance Project (FANTA). [10]

FAO. 1996. Rome Declaration on World Food Security and World Food Summit Plan of Action. In: World Food Summit, November 13–17, 1996. Rome: FAO. [9, 11]

———. 1999a. Background Paper 1: Agricultural Biodiversity, Multifunctional Character of Agriculture and Land Conference, Sept. 1999. Maastricht: FAO. [1, 8, 9, 11]

———. 1999b. Women: Users, Preservers and Managers of Agrobiodiversity. Rome: FAO. [9]

———. 2004. What Is Agrobiodiversity? Rome: FAO. [13]

———. 2006. Community Diversity Seed Fairs in Tanzania: Guidelines for Seed Fairs. Rome: FAO. [5]

———. 2009. The International Treaty on Plant Genetic Resources for Food and Agriculture. Rome: FAO. [9]

———. 2010a. Policy on Indigenous and Tribal Peoples. Rome: FAO. [12]

FAO. 2010b. The Second Report on the State of the World's Plant Genetic Resources for Food and Agriculture. Rome: FAO. [1, 3, 5, 9, 10, 15]

———. 2013. Food Outlook: Biennial Report on Global Food Markets, Nov. 2013. Rome: FAO. [9]

———. 2015. Designing Nutrition-Sensitive Agriculture Investments; Checklist and Guidance for Programme Formulation. Food and Agriculture Organization of the United Nations. Rome: FAO. [9]

———. 2017. FAO Cereal Supply and Demand Brief. Rome: FAO. [10]

FAO/INFOODS. 2013. Report on the Nutrition Indicators for Biodiversity: Food Composition and Food Consumption: Global Progress Report 2013. Rome: FAO. [10]

Fargione, J. E., and D. Tilman. 2005. Diversity Decreases Invasion via Both Sampling and Complementarity Effects. *Ecol. Lett.* **8**:604–611. [4]

Feitosa Vasconcelos, A. C., M. Bonatti, S. L. Schlindwein, et al. 2013. Landraces as an Adaptation Strategy to Climate Change for Smallholders in Santa Catarina, Southern Brazil. *Land Use Policy* **34**:250–254. [4]

Felix, D.-T., J. Coello-Coello, and J. Martinez-Castillo. 2014. Wild to Crop Introgression and Genetic Diversity in Lima Bean (*Phaseolus Lunatus* L.) in Traditional Mayan Milpas from Mexico. *Conser. Genet.* **15**:1315–1328. [4]

Ferrand, D., A. Gibson, and H. Scott. 2004. Making Markets Work for the Poor: An Objective and an Approach for Governments and Development Agencies. Woodmead: ComMark Trust. [15]

Fielder, H. V. 2015. Developing Methodologies for the Genetic Conservation of UK Crop Wild Relatives. Ph.D. Thesis. Birmingham: Univ. of Birmingham. [5]

Fierer, N., C. L. Lauber, K. S. Ramirez, et al. 2012. Comparative Metagenomic, Phylogenetic and Physiological Analyses of Soil Microbial Communities across Nitrogen Gradients. *ISME J.* **6**:1007–1017. [4]

Fischer, C. G., and T. Garnett. 2016. Plates, Pyramids, and Planets: Developments in National Healthy and Sustainable Dietary Guidelines: A State of Play Assessment. Rome: FAO and the Food Climate Research Network at the Univ. of Oxford. [9]

Fischer-Kowalski, M., A. Mayer, A. Schaffartzik, and A. Reenberg, eds. 2014. Ester Boserup's Legacy on Sustainability. Heidelberg: Springer. [8]

Fitzpatrick, I., R. Young, M. Perry, and E. Rose. 2017. The Hidden Cost of UK Food. Bristol: Sustainable Food Trust. [1]

Flachs, A. 2015. Persistent Agrobiodiversity on Genetically Modified Cotton Farms in Telangana, India. *J. Ethnobiol.* **35**:406–426. [8]

Foley, J. A., R. Defries, G. P. Asner, et al. 2005. Global Consequences of Land Use. *Science* **309**:570–574. [9]

Foley, J. A., N. Ramankutty, K. A. Brauman, et al. 2011. Solutions for a Cultivated Planet. *Nature* **478**:337–342. [9]

Fonte, S. J., S. J. Vanek, P. Oyarzun, et al. 2012. Pathways to Agroecological Intensification of Soil Fertility Management by Smallholder Farmers in the Andean Highlands. In: Advances in Agronomy, Vol. 116, ed. D. L. Sparks, vol. 116, pp. 125–184. Advances in Agronomy. Burlington: Academic Press. [4]

Foote, J. A., S. P. Murphy, L. R. Wilkens, P. P. Basiotis, and A. Carlson. 2004. Dietary Variety Increases the Probability of Nutrient Adequacy among Adults. *J. Nutr.* **134**:1779–1785. [10]

Forouzanfar, M. H., L. Alexander, H. R. Anderson, et al. 2015. Global, Regional, and National Comparative Risk Assessment of 79 Behavioural, Environmental and Occupational, and Metabolic Risks or Clusters of Risks in 188 Countries, 1990–2013; a Systematic Analysis for the Global Burden of Disease Study 2013. *Lancet* **386**:2287–2323. [10]

Forrest, J. R. K., R. W. Thorp, C. Kremen, and N. M. Williams. 2015. Contrasting Patterns in Species and Functional-Trait Diversity of Bees in an Agricultural Landscape. *J. Appl. Ecol.* **52**:706–715. [4]

Fort, J. 2012. Synthesis between Demic and Cultural Diffusion in the Neolithic Transition in Europe. *PNAS* **109**:18669–18673. [7]

Foucault, M. 1980. Power/Knowledge: Selected Interviews and Other Writings 1972-1977. New York: Pantheon Books. [14]

———. 1994/1996. The Order of Things: An Archaeology of the Human Sciences. New York: Vintage Books, Random House. [13, 14]

Fowler, C., and P. R. Mooney. 1990. Shattering: Food, Politics and the Loss of Genetic Diversity. Tucson: Univ. Arizona Press. [9, 14]

Fowler, G., and T. Hodgkin. 2004. Plant Genetic Resources for Food and Agriculture: Assessing Global Availability. *Annu. Rev. Environ. Resour.* **29**:143–179. [2]

François, O., M. G. B. Blum, M. Jakobsson, and N. A. Rosenberg. 2008. Demographic History of European Populations of *Arabidopsis Thaliana*. *PLoS Genet.* **4**:e1000075. [3]

Frantz, L. A. F., J. G. Schraiber, O. Madsen, et al. 2015. Evidence of Long-Term Gene Flow and Selection during Domestication from Analyses of Eurasian Wild and Domestic Pig Genomes. *Nat. Genet.* **47**:1141–1148. [3]

Frei, M., and K. Becker. 2004. Agro-Biodiversity in Subsistence-Oriented Farming Systems in a Philippine Upland Region: Nutritional Considerations. *Biodivers. Conserv.* **13**:1591–1610. [15]

Friedmann, H., and P. McMichael. 1989. Agriculture and the State System: The Rise and Decline of National Agricultures, 1870 to the Present. *Sociol. Ruralis* **29**:93–117. [11]

Friel, S., D. Gleeson, A.-M. Thow, et al. 2013. A New Generation of Trade Policy: Potential Risks to Diet-Related Health from the Trans Pacific Partnership Agreement. *Global. Health* **9**:1–7. [9]

Frison, E. A., J. Cherfas, and T. Hodgkin. 2011. Agricultural Biodiversity Is Essential for a Sustainable Improvement in Food and Nutrition Security. *Sustainability* **3**:238–253. [11]

Frison, E. A., I. F. Smith, T. Johns, J. Cherfas, and P. B. Eyzaguirre. 2006. Agricultural Biodiversity, Nutrition, and Health: Making a Difference to Hunger and Nutrition in the Developing World. *Food Nutr. Bull.* **27**:167–179. [9, 11]

Fuller, D. Q., E. Kingwell-Banham, L. Lucas, C. Murphy, and C. J. Stevens. 2015. Comparing Pathways to Agriculture. *Archaeol. Int.* **18**:61–66. [7]

Gadgil, M., F. Berkes, and C. Folke. 1993. Indigenous Knowledge for Biodiversity Conservation. *Ambio* **22**:151–156. [12]

Galluzzi, G., P. Eyzaguirre, and V. Negri. 2010. Home Gardens: Neglected Hotspots of Agro-Biodiversity and Cultural Diversity. *Biodivers. Conserv.* **19**:3635–3654. [8]

García, M. E. 2013. The Taste of Conquest: Colonialism, Cosmopolitics, and the Dark Side of Peru's Gastronomic Boom. *J. Lat. Am. Caribb. Anthropol.* **18**:505–524. [8]

Garibaldi, A., and N. Turner. 2004. Cultural Keystone Species: Implications for Ecological Conservation and Restoration. *Ecol. Soc.* **9**:1–18. [9]

Garibaldi, L. A., M. A. Aizen, A. M. Klein, S. A. Cunningham, and L. D. Harder. 2011. Global Growth and Stability of Agricultural Yield Decrease with Pollinator Dependence. *PNAS* **108**:5909–5914. [2, 9]

Garnett, T., M. C. Appleby, A. Balmford, et al. 2013. Sustainable Intensification in Agriculture: Premises and Policies. *Science* **341**:33–34. [8]

Garrett, K. A., and C. C. Mundt. 1999. Epidemiology in Mixed Host Populations. *Phytopathology* **89**:984–990. [2]

Garris, A. J., T. H. Tai, J. Coburn, S. Kresovich, and S. McCouch. 2005. Genetic Structure and Diversity in *Oryza Sativa* L. *Genetics* **169**:1631–1638. [3]

Gatto, E., A. Marino, and G. Signorino. 2013. Biodiversity and Risk Management in Agriculture: What Do We Learn from Cap Reforms? A Farm-Level Analysis. In: Proc. 53rd ERSA Congress on Regional Integration: Europe, the Mediterranean and the World Economy. Palermo: European Regional Science Association. [11]

Gauchan, D., M. Smale, and P. Chaudhary. 2005. Market-Based Incentives for Conserving Diversity on Farms: The Case of Rice Landraces in Central Tarai, Nepal. *Genet. Resour. Crop Evol.* **52**:293–303. [15]

Gaudin, A. C., T. N. Tolhurst, A. P. Ker, et al. 2015. Increasing Crop Diversity Mitigates Weather Variations and Improves Yield Stability. *PLoS One* **10**:e0113261. [6]

Gavin, M., J. Mccarter, A. Mead, et al. 2015. Defining Biocultural Approaches to Conservation. *Trends Ecol. Evol.* **30**:140–145. [12]

GBD Risk Factor Collaborators. 2017. Global Burden of Disease (GBD) 2016 Risk Factor Collaborators. Global, regional, and national comparative risk assessment of 84 behavioural, environmental and occupational, and metabolic risks or clusters of risks, 1990 2016: a systematic analysis for the Global Burden of Disease Study 2016. *Lancet* **390**: 1345–1422. [09]

Gepts, P. 2006. Plant Genetic Resources Conservation and Utilization. *Crop Sci.* **46**:2278–2292. [1, 5]

Gepts, P., T. R. Famula, R. L. Bettinger, et al. 2012. Biodiversity in Agriculture: Domestication, Evolution, and Sustainability. Cambridge: Cambridge Univ. Press. [1, 3]

Gerbault, P., R. G. Allaby, N. Boivin, et al. 2014. Storytelling and Story Testing in Domestication. *PNAS* **111**:6159–6164. [3]

Gibney, M. J., C. G. Forde, D. Mullally, and E. R. Gibney. 2017. Ultra-Processed Foods in Human Health: A Critical Appraisal. *Am. J. Clin. Nutr.* **406**:717–724. [10]

Gilbert, M. E., and N. M. Holbrook. 2011. Limitations to Crop Diversification for Enhancing the Resilience of Rain-Fed Subsistence Agriculture to Drought. CID Working Paper No. 228. Cambridge, MA: Center for Intl. Development, Harvard Univ. [7]

Gilbert, P. R. 2013. Deskilling, Agrodiversity, and the Seed Trade: A View from Contemporary British Allotments. *Agricult. Human Values* **30**:193–217. [8]

Gill, R. J., K. C. R. Baldock, M. J. F. Brown, et al. 2016. Protecting an Ecosystem Service: Approaches to Understanding and Mitigating Threats to Wild Insect Pollinators. In: Ecosystem Services: From Biodiversity to Society, Pt. 2, Vol. 54, ed. G. Woodward and D. A. Bohan, vol. 54, pp. 135–206. London: Academic Press. [4]

Girard, F., and C. Frison, eds. 2018. The Commons, Plant Breeding and Agricultural Research. New York: Routledge. [13]

Giuliani, A., F. Hintermann, W. Rojas, and S. Padulosi, eds. 2012. Biodiversity of Andean Grains: Balancing Market Potential and Sustainable Livelihoods. Rome: Bioversity Intl. [9]

Giuliani, A., F. van Oudenhoven, and S. Mubalieva. 2011. Agricultural Biodiversity in the Tajik Pamirs: A Bridge between Market Development and Food Sovereignty. *Mt. Res. Dev.* **31**:16–26. [15]

Gladek, E., M. Fraser, G. Roemers, et al. 2016. The Global Food System: An Analysis. Amsterdam: Metabolic. [9]

Glaser, B., G. Guggenberger, W. Zech, and M. De Lourdes. 2003. Soil Organic Matter Stability in Amazonian Dark Earths. In: Amazonian Dark Earths: Origins, Properties, Management, ed. J. Lehmann et al. Dordrecht: Kluwer. [6]

Glinwood, R., E. Ahmed, E. Qvarfordt, V. Ninkovic, and J. Pettersson. 2009. Airborne Interactions between Undamaged Plants of Different Cultivars Affect Insect Herbivores and Natural Enemies. *Arthropod Plant Interact.* **3**:215–224. [4]

GLOPAN. 2016. Food Systems and Diets: Facing the Challenges of the 21st Century. London: Global Panel on Agriculture and Food Systems for Nutrition. [9, 10]

Goldringer, I., J. Enjalbert, J. David, et al. 2001. Dynamic Management of Genetics Resources: A 13-Year Experiment on Wheat. Broadening the Genetic Base of Crop Production. Rome: IPGRI/FAO. [13]

Gonsalves, J. 2013. A New Relevance and Better Prospects for Wider Uptake of Social Learning within CGIAR. In: CCAFS Working Paper No. 37. Copenhagen: CGIAR Research Program on Climate Change, Agriculture and Food Security (CCAFS). [5]

Gonzalez, C. G. 2011. Climate Change, Food Security, and Agrobiodiversity: Toward a Just, Resilient, and Sustainable Food System. *Fordham Environ. Law Rev.* **22**:493–522. [15]

González-Esquivel, C. E., M. E. Gavito, M. Astier, et al. 2015. Ecosystem Service Trade-Offs, Perceived Drivers, and Sustainability in Contrasting Agroecosystems in Central Mexico. *Ecol. Soc.* **20**:38. [2, 4]

Gorenflo, L., S. Romaine, R. Mittermeier, and K. Walker-Painemilla. 2012. Co-Occurrence of Linguistic and Biological Diversity in Biodiversity Hotspots and High Biodiversity Wilderness Areas. *PNAS* **109**:8032–8037. [12]

GOS. 2017. A National Food Strategy for Sweden: More Jobs and Sustainable Growth Throughout the Country. Short Version of Government Bill 2016/17:104. Stockholm: Government Offices of Sweden. [1]

Gotor, E., and C. Irungu. 2010. The Impact of Bioversity International's African Leafy Vegetables Programme in Kenya. *Impact Assess. Proj. Apprais.* **28**:41–55. [9]

Gracey, M., and M. King. 2009. Indigenous Health Part 1: Determinants and Disease Patterns. *Lancet* **374**:65–75. [9]

Graddy, T. G. 2013. Regarding Biocultural Heritage: *In Situ* Political Ecology of Agricultural Biodiversity in the Peruvian Andes. *Agricult. Human Values* **30**:587–604. [12]

———. 2014. Situating *in Situ*: A Critical Geography of Agricultural Biodiversity Conservation in the Peruvian Andes and Beyond. *Antipode* **46**:426–454. [13]

Graeub, B. E., M. J. Chappell, H. Wittman, et al. 2016. The State of Family Farms in the World. *World Dev.* **87**:1–15. [6]

Graham, R. D., R. M. Welch, D. A. Saunders, et al. 2007. Nutritious Subsistence Food Systems. *Adv. Agron.* **92**:1–74. [9]

GRAIN. 1999. Union for the Protection of New Varieties of Plants (UPOV) on the War Path. https://www.grain.org/article/entries/257-upov-on-the-war-path. [13]

Gray, C. L. 2009. Rural out-Migration and Smallholder Agriculture in the Southern Ecuadorian Andes. *Popul. Environ.* **30**:193–217. [8]

Gray, C. L., and R. E. Bilsborrow. 2014. Consequences of out-Migration for Land Use in Rural Ecuador. *Land Use Policy* **36**:182–191. [8]

Greenberg, L. 2003. Women in the Garden and Kitchen: The Role of Cuisine in the Conservation of Traditional House Lots among Yucatec Mayan Immigrants. In: Women and Plants: Gender Relations in Biodiversity Management and Conservation, ed. P. L. Howard, pp. 51–65. London: Zed Books. [11]

Greene, S. L., T. C. Hart, and A. Afonin. 1999. Using Geographic Information to Acquire Wild Crop Germplasm for *Ex Situ* Collections: I. Map Development and Field Use. *Crop Sci.* **39**:836–842. [5]

Greenleaf, S. S., and C. Kremen. 2006. Wild Bees Enhance Honey Bees' Pollination of Hybrid Sunflower. *PNAS* **103**:13890–13895. [4]

Grobman, A., W. Salhuana, and R. Sevilla. 1961. The Races of Maize in Peru: Their Origins, Evolution and Classification. Washington, D.C.: National Academy of Sciences. [3]

Gruberg, H., G. Meldrum, S. Padulosi, et al. 2013. Towards a Better Understanding of Custodian Farmers and Their Roles: Insights from a Case Study in Cachilaya, Bolivia. Rome: Bioversity International, Fundación PROINPA. [2]

Gruère, G., L. Nagarajan, and O. King. 2009a. The Role of Collective Action in the Marketing of Underutilized Plant Species: Lessons from a Case Study on Minor Millets in South India. *Food Pol.* **34**:39–45. [15]

Gruère, G., M. Smale, and A. Giuliani. 2009b. Marketing Underutilized Species for the Benefit of the Poor: A Conceptual Framework. In: Agrobiodiversity, Conservation and Economic Development, ed. A. Kontoleon et al., pp. 62–81. London: Routledge. [15]

Guarino, L. 1995. Mapping the Ecogeographic Distribution of Biodiversity. In: Collecting Plant Genetic Diversity, Technical Guidelines, ed. L. Guarino et al., pp. 287–328. Wallingford: CAB International. [5]

Guarino, L., A. Jarvis, R. J. Hijmans, and N. Maxted. 2002. Geographic Information Systems (GIS) and the Conservation and Use of Plant Genetic Resources. In: Managing Plant Genetic Diversity, ed. J. M. M. Engels et al., pp. 387–404. Wallingford: CABI Publishing. [5]

Guitart, D., C. Pickering, and J. Byrne. 2012. Past Results and Future Directions in Urban Community Gardens Research. *Urb. Forest. Urb. Green.* **11**:364–373. [2]

Gustafson, D., A. Gutman, W. Leet, et al. 2016. Seven Food System Metrics of Sustainable Nutrition Security. *Sustainability* **8**:196. [9]

Gutenkunst, R. N., R. D. Hernandez, S. H. Williamson, and C. D. Bustamante. 2009. Inferring the Joint Demographic History of Multiple Populations from Multidimensional Snp Frequency Data. *PLoS Genet.* **5**:e1000695. [3]

Gutiérrez Escobar, L., and E. Fitting. 2016. The Red de Semillas Libres: Contesting Biohegemony in Colombia. *J. Agrar. Change* **16**:711–719. [13]

Gyawali, S., and B. Sthapit. 2006. Participatory Plant Breeding for Enhancing the Use of Local Crop Genetic Diversity to Manage Abiotic Stresses In: Enhancing the Use of Crop Genetic Diversity to Manage Abiotic Stress in Agricultural Production Systems, ed. D. I. Jarvis et al., pp. 72–83. Rome: IPGRI. [4]

Haas, P. M. 1992. Introduction: Epistemic Communities and International Policy Coordination. *Int. Organ.* **46**:1–35. [14]

Hadley, C., and C. L. Patil. 2006. Food Insecurity in Rural Tanzania Is Associated with Maternal Anxiety and Depression. *Am. J. Hum. Biol.* **18**:359–368. [11]

Haichar, F. Z., C. Marol, O. Berge, et al. 2008. Plant Host Habitat and Root Exudates Shape Soil Bacterial Community Structure. *ISME J.* **2**:1221–1230. [2]

Hajjar, R., D. I. Jarvis, and B. Gemmill-Herren. 2008. The Utility of Crop Genetic Diversity in Maintaining Ecosystem Services. *Agricult. Ecosyst. Environ.* **123**:261–270. [4]

Hall, S. J. G., ed. 2004. Livestock Biodiversity: Genetic Resources for the Farming of the Future. Oxford: Blackwell. [2]

Halpert, M. T., and M. J. Chappell. 2017. Prima Facie Reasons to Question Enclosed Intellectual Property Regimes and Favor Open-Source Regimes for Germplasm. *F1000Research* **6**:284. [13]

Hammond, J., S. Fraval, J. van Etten, et al. 2016. The Rural Household Multi-Indicator Survey (Rhomis) for Rapid Characterisation of Households to Inform Climate Smart Agriculture Interventions: Description and Applications in East Africa and Central America. *Agric. Syst.* **151**:225–233. [7]

Handley, L. J., A. M. Lawson, J. Goudet, and F. Balloux. 2007. Going the Distance: Human Population Genetics in a Clinal World. *Trends Genet.* **23**:432–439. [3]

Hanf, K., and A. Underdal. 1998. Domesticating International Commitments: Linking National and International Decicion-Making. In: The Politics of International Environmental Management, ed. A. Underdal, pp. 149–170. Dordrecht: Kluwer Academic Publishers. [14]

Hannah, L., P. R. Roehrdanz, M. Ikegami, et al. 2013. Climate Change, Wine and Conservation. *PNAS* **110**:6907–6912. [5]

Hanotte, O., D. G. Bradley, J. W. Ochieng, et al. 2002. African Pastoralism: Genetic Imprints of Origins and Migrations. *Science* **296**:336–339. [3]

Haraway, D. 2008. When Species Meet. Minneapolis: Univ. of Minnesota Press. [13]

Hardigan, M. A., F. P. E. Laimbeer, L. Newton, et al. 2017. Genome Diversity of Tuber-Bearing Solanum Uncovers Complex Evolutionary History and Targets of Domestication in the Cultivated Potato. *PNAS* **114**:E9999–E10008. [3]

Hardison, P. 2016. La Cogeneración de Servicios Ecosistémicos Por Parte de la Naturaleza y Los Pueblos Indígenas: el Valor de Las Relaciones Bioculturales de Adaptación y el Papel de Los Conocimientos Tradicionales. In: Biodiversidad y Propiedad Intelectual en Disputa, ed. S. Roca, pp. 214–240. Lima: Esan Ediciones. [12]

Hardon-Baars, A. 2000. The Role of Agrobiodiversity in Farm-Household Livelihood and Food Security: A Conceptual Analysis. In: Encouraging Diversity, ed. C. Almekinders, pp. 31–35. London: Intermediate Technology Publications Ltd. [11]

Harlan, H. V., and M. L. Martini. 1936. Problems and Results of Barley Plant Breeding. In: Usda Yearbook of Agriculture, pp. 303–346. Washington, D.C.: U.S. GPO. [14]

Harlan, J. R. 1975. Our Vanishing Genetic Resources. *Science* **188**:618–621. [5, 14]

———. 1992. Crops and Man (2nd ed.). Madison: American Society of Agronomy/ Crop Science Society of America. [14]

Harlan, J. R., and J. M. J. De Wet. 1971. Toward a Rational Classification of Cultivated Plants. *Taxon* **20**:509–517. [3]

Hartigan, J. 2017. Care of the Species: Races of Corn and the Science of Plant Biodiversity. St. Paul: Univ. of Minnesota Press. [13]

Harwood, J. 2009. Peasant Friendly Plant Breeding and the Early Years of the Green Revolution in Mexico. *Agric. Hist.* **83**:384–410. [6]

Haselmair, R., H. Pirker, E. Kuhn, and C. R. Vogl. 2014. Personal Networks: A Tool for Gaining Insight into the Transmission of Knowledge About Food and Medicinal Plants among Tyrolean (Austrian) Migrants in Australia, Brazil and Peru. *J. Ethnobiol. Ethnomed.* **10**:1. [11]

Hawkes, C., S. Friel, T. Lobstein, and T. Lang. 2012. Linking Agricultural Policies with Obesity and Noncommunicable Diseases: A New Perspective for a Globalising World. *Food Pol.* **37**:343–353. [14]

Hayden, C. 2003. When Nature Goes Public: The Making and Unmaking of Bioprospecting in Mexico. Princeton: Princeton Univ. Press. [12]

Heald, P. J., and S. Chapman. 2009. Crop Diversity Report Card for the Twentieth Century: Diversity Bust or Diversity Boom? SSRN. https://ssrn.com/abstract=1462917. (accessed July 12, 2017). [9]

Hedenus, F., S. Wirsenius, and D. J. A. Johansson. 2014. The Importance of Reduced Meat and Dairy Consumption for Meeting Stringent Climate Change Targets. *Clim. Change* **124**:79–91. [9]

Heinrich, M., M. Leonti, S. Nebel, and W. Peschel. 2005. Local Food—Nutraceuticals: An Example of a Multidisciplinary Research Project on Local Knowledge. *J. Physiol. Pharmacol.* **56(Suppl. 1)**:5–22. [11]

Heisey, P., and K. Day-Rubenstein. 2015. Using Crop Genetic Resources to Help Agriculture Adapt to Climate Change: Economics and Policy. USDA: Economic Information Bulletin 139. [7]

Henry, M., N. Cerrutti, P. Aupinel, et al. 2015. Reconciling Laboratory and Field Assessments of Neonicotinoid Toxicity to Honeybees. *Proc. R. Soc. Lond. B Biol. Sci.* **282**:20152110. [4]

Herforth, A. 2010. Promotion of Traditional African Vegetables in Kenya and Tanzania: A Case Study of an Intervention Representing Emerging Imperatives in Global Nutrition. Ph.D., Cornell Univ., Ithaca, NY. [9]

———. 2015. Access to Adequate Nutritious Food: New Indicators to Track Progress and Inform Action. In: The Fight against Hunger and Malnutrition, ed. D. Sahn, pp. 139–164. Oxford: Oxford Univ. Press. [9]

———. 2016. Seeking Indicators of Healthy Diets: It Is Time to Measure Diets Globally, How? Washington, D.C.: Gallup Intl. Association, Swiss Agency for Development Cooperation. [9, 10]

Herforth, A., and S. Ahmed. 2015. The Food Environment, Its Effects on Dietary Consumption, and Potential for Measurement within Agriculture-Nutrition Interventions. *Food Secur.* **7**:505–520. [9]

Hermann, M. 2013. Successes and Pitfalls of Linking Nutritionally Promising Andean Crops to Markets. In: Diversifying Food and Diets: Using Agricultural Biodiversity to Improve Nutrition and Health, ed. J. Fanzo et al., pp. 164–185. Issues in Agricultural Biodiversity, M. Halewood and D. Hunter, series ed. London: Earthscan and Bioversity International. [15]

Hernández, J. 2001. Reclamos de la Identidad: La Formación de Las Organizaciones Indígenas en Oaxaca. Mexico City: Universidad Autónoma Benito Juárez de Oaxaca/Porrúa. [12]

Heslot, N., D. Akdemir, M. E. Sorrells, and J. L. Jannink. 2014. Integrating Environmental Covariates and Crop Modelling into the Genomic Selection Framework to Predict Genotype by Environment Interactions. *Theor. Appl. Genet.* **127**:463–480. [7]

Heun, M., R. Schäfer-Pregl, D. Klawan, et al. 1997. Site of Einkorn Wheat Domestication Identified by DNA Fingerprinting. *Science* **278**:1312–1314. [3]

Hidalgo, C. A. 2015. Why Information Grows: The Evolution of Order, from Atoms to Economies. New York: Basic Books. [7]

Hidalgo, C. A., B. Klinger, A. L. Barabási, and R. Hausmann. 2007. The Product Space Conditions the Development of Nations. *Science* **317**:482–487. [2]

Hijmans, R. J., H. Choe, and J. Perlman. 2016. Spatio-Temporal Patterns of Field Crop Diversity in the United States, 1870-2012. *Agric. Environ. Lett.* **1**:160022. [2]

Hijmans, R. J., K. A. Garrett, Z. Huamán, et al. 2000. Assessing the Geographic Representativeness of Genebank Collections: The Case of Bolivian Wild Potatoes. *Conserv. Biol.* **14**:1755–1765. [2, 5]

Hijmans, R. J., M. Jacobs, J. B. Bamberg, and D. M. Spooner. 2003. Frost Tolerance in Wild Potatoes: Unraveling the Predictivity of Taxonomic, Geographic and Ecological Factors. *Euphytica* **130**:47–59. [5]

Hijmans, R. J., and D. M. Spooner. 2001. Geographic Distribution of Wild Potato Species. *Am. J. Bot.* **88**:2101–2112. [2, 5]

Hijmans, R. J., D. M. Spooner, A. R. Salas, L. Guarino, and J. de la Cruz. 2002. Atlas of Wild Potatoes, vol. 10. Systematic and Ecogeographic Studies on Crop Genepools. Rome: Intl. Plant Genetic Resources Institute (IPGRI). [5]

Hirsch, E. 1995. Landscape: Between Space and Place. In: The Anthropology of Landscape: Perspectives on Space and Place, ed. E. Hirsch and M. O'Hanlon. Oxford: Clarendon Press. [13]

HLPE. 2017. Nutrition and Food Systems: A Report by the High Level Panel of Experts on Food Security and Nutrition of the Committee on World Food Security, Rome. Rome: FAO. [8–10]

Hoddinott, J., and Y. Yohannes. 2002. Dietary Diversity as a Food Security Indicator. In: Fcnd Briefs, Discussion Paper 136. Washington, D.C.: Intl. Food Policy Research Institute. [10]

Holt-Gimenez, E., M. A. Altieri, and P. Rosset. 2006. Ten Reasons Why the Rockefeller and the Bill and Melinda Gates Foundations' Alliance for Another Green Revolution Will Not Solve the Problems of Poverty and Hunger in Sub-Saharan Africa. In: Food First Policy Brief No. 12. Oakland: Institute for Food and Development Policy. [15]

Hooper, D. U., D. E. Bignell, V. K. Brown, et al. 2000. Interactions between Aboveground and Belowground Biodiversity in Terrestrial Ecosystems: Patterns, Mechanisms, and Feedbacks. *Bioscience* **50**:1049–1061. [4]

Hooper, D. U., F. S. Chapin, J. J. Ewel, et al. 2005. Effects of Biodiversity on Ecosystem Functioning: A Consensus of Current Knowledge. *Ecol. Monogr.* **75**:3–35. [2, 4, 10]

Horna, D., S. Timpo, and G. Gruère. 2007. Marketing Underutilized Crops: The Case of the African Garden Egg (*Solanum aethiopicum*) in Ghana. Rome: Global Facilitation Unit for Underutilized Species (GFU): 3. [15]

Horst, M., and B. Gaolach. 2015. The Potential of Local Food Systems in North America: A Review of Foodshed Analyses. *Renew. Agr. Food Syst.* **20**:399–407. [9]

Howard, P. L. 2006. Gender and Social Dynamics in Swidden and Homegardens in Latin America. In: Tropical Homegardens: A Time-Tested Example of Sustainable Agroforestry, ed. B. M. Kumar and P. K. R. Nair, pp. 1–24. Heidelberg: Springer. [11]

Huamán, Z. 2002. Tecnología Disponible Para Reforzar la Conservación "*in Situ*" de Los Cultivares de Papa Tradicionales de Los Andes. *Rev. Elect. Red Mundial Cient. Peruanos* **1**:1–10. [5]

Huaman, Z., A. Salas, R. Gomez, A. Panta, and J. Toledo. 2000. Conservation of Potato Genetic Resources at CIP. In: Potato: Global Research and Development, ed. S. M. P. Khurana et al., pp. 102–112. Shimla, India: Indian Potato Association. [2]

Huang, X., N. Kurata, X. Wei, et al. 2012. A Map of Rice Genome Variation Reveals the Origin of Cultivated Rice. *Nature* **490**:497–501. [3]

Huang, X., M. Wolf, M. W. Ganal, et al. 2007. Did Modern Plant Breeding Lead to Genetic Erosion in European Winter Wheat Varieties? *Crop Sci.* **47**:343–349. [3]

Hufford, M. B., X. Xu, J. van Heerwaarden, et al. 2012. Comparative Population Genomics of Maize Domestication and Improvement. *Nat. Genet.* **44**:808–811. [3]

Hughes, A. R., B. D. Inouye, M. T. J. Johnson, N. Underwood, and M. Vellend. 2008. Ecological Consequences of Genetic Diversity. *Ecol. Lett.* **11**:609–623. [4]

Hunter, D., B. Burlingame, and R. Remans. 2015. Biodiversity and Nutrition. In: Connecting Global Priorities: Biodiversity and Human Health: A State of Knowledge Review, ed. C. Romanelli et al., pp. 97–129. Geneva: WHO and Secretariat of the Convention on Biological Diversity. [10]

Hunter, D., and V. Heywood. 2010. Crop Wild Relatives: A Manual of *in Situ* Conservation. London: Bioversity International, Earthscan Publishers. [2]

Hunter, D., I. Özkan, D. Moura de Oliveira Beltrame, et al. 2016. Enabled or Disabled: Is the Environment Right for Using Biodiversity to Improve Nutrition? *Front. Nutr.* **3**:14. [10]

Ickowitz, A., B. Powell, M. A. Salim, and T. Sunderland. 2014. Dietary Quality and Tree Cover in Africa. *Glob. Environ. Change* **24**:287–294. [9, 10]

Ickowitz, A., D. Rowland, B. Powell, M. A. Salim, and T. Sunderland. 2016. Forests, Trees, and Micronutrient-Rich Food Consumption in Indonesia. *PLoS One* **11**:e0154139. [9]

ICTSD. 2016. Wipo Members Update Draft Text on Protecting Traditional Knowledge. *Bridges* **20**:42. [12]

IFPRI. 2016. Global Nutrition Report, from Promise to Impact: Ending Malnutrition by 2030. Washington, D.C. : Intl. Food Policy Research Institute. [9]

IIED. 2016. Biocultural Heritage: Promoting Resilient Farmer Communities and Local Economies. London: Intl. Institute for Environment and Development. [12]

———. 2017. Protecting Community Rights over Traditional Knowledge: Key Findings and Recommendations 2005-2009. London: Intl. Institute for Environment and Development. [12]

ILO. 2017. Ratifications of C169: Indigenous and Tribal Peoples Convention, 1989 (No. 169): Intl. Labor Organization. [12]

Imbach, P., E. Fung, L. Hannah, et al. 2017. Coupling of Pollination Services and Coffee Suitability under Climate Change. *PNAS* **114**:10438–10442. [4]

Imbruce, V. 2007. Bringing Southeast Asia to the Southeast United States: New Forms of Alternative Agriculture in Homestead, Florida. *Agricult. Human Values* **24**:41–59. [8]

INE. 2013. Principales Resultados del Censo Nacional de Población y Vivienda, CNPV-2012. Bolivia: Instituto Nacional de Estadística, Estado Plurinacional de Bolivia. [12]

INEGI. 2000. XII Censo General de Población y Vivienda: Marco Conceptual, Mexico.
———. 2003. Anuario Estadístico del Estado de Chiapas. Oaxaca de Juárez: Instituto Nacional de Estadística, Geografía e Informática. [12]

Infield, M. 2001. Cultural Values: A Forgotten Strategy for Building Community Support for Protected Areas in Africa. *Conserv. Biol.* **15**:800–802. [12]

Ingold, T. 2000. The Perception of the Environment: Essays in Livelihood, Dwelling and Skill. New York: Routledge. [13]

———. 2011. Being Alive: Essays on Movement, Knowledge, and Description. London: Routledge. [13]

Ingold, T., and T. Kurttila. 2000. Perceiving the Environment in Finnish Lapland. *Body & Society* **6**:183–196. [13]

IPBES. 2018. Summary for Policymakers of the Regional Assessment Report on Biodiversity and Ecosystem Services for Africa of the Intergovernmental Science-Policy Platform on Biodiversity and Ecosystem Service. Bonn: IPBES Secretariat. [9]

IPES-Food. 2017. Unravelling the Food–Health Nexus: Addressing Practices, Political Economy, and Power Relations to Build Healthier Food Systems. The Global Alliance for the Future of Food and IPES-Food. [6, 8]

Ironside, J. 2013. Thinking Outside the Fence: Exploring Culture/Land Relationships, a Case Study on Ratanakiri Province, Cambodia. PhD Dissertation, Univ. of Otago, Dunedin, New Zealand. [8]

Ishizawa, J. 2010. Affirmation of Cultural Diversity: Learning with the Communities in the Central Andes. In: Towards an Alternative Development Paradigm: Indigenous People's Self-Determined Development, ed. V. Tauli-Corpus et al., pp. 205–247. Baguio City: Tebtebba Foundation. [12]

———. 2016. La Centralidad de Las Comunidades Andino-Amazónicas Criadoras de la Biodiversidad: Apuntes Sobre el Concepto de Proyecto *in Situ*. In: Biodiversidad y Propiedad Intelectual en Disputa, ed. S. Roca, pp. 267–285. Lima: Esan Ediciones. [12]

Iskandar, B. S., J. Iskandar, B. Iriwan, and B. Partasasmita. 2018. Traditional Markets and Diversity of Edible Plant Trading: Case Study in Ujung Berung, Bandung, West Java, Indonesia. *Biodiversitas* **19**:437452. [6]

Ives, C. D., and D. Kendal. 2014. The Role of Social Values in the Management of Ecological Systems. *J. Environ. Manage.* **144**:67–72. [6]

Jackson, J. F., and G. R. Clarke. 1991. Gene Flow in an Almond Orchard. *Theor. Appl. Genet.* **82**:169–173. [4]

Jackson, L. E., U. Pascual, and T. Hodgkin. 2007. Utilizing and Conserving Agrobiodiversity in Agricultural Landscapes. *Agricult. Ecosyst. Environ.* **121**:196–210. [1, 8, 13, 15]

Jackson, M., B. Ford-Lloyd, and M. Parry, eds. 2013. Plant Genetic Resources and Climate Change. Wallingford: CABI. [7]

Jacobsen, S. E. 2011. The Situation for Quinoa and Its Production in Southern Bolivia: From Economic Success to Environmental Disaster. Journal of Agronomy and Crop. *Science* **197**:390–399. [8]

Jacobsen, S. E., M. Sørensen, S. M. Pedersen, and J. Weiner. 2015. Using Our Agrobiodiversity: Plant-Based Solutions to Feed the World. *Agronom. Sustain. Devel.* **35**:1217–1235. [1]

Jaeger, M., A. Giuliani, and I. van Loosen. 2017. Markets, Consumer Demand and Agricultural Biodiversity. In: Routledge Handbook of Agricultural Biodiversity, ed. D. Hunter et al. London: Routledge. [15]

Jaenicke, H., and D. Virchow. 2013. Entry Points into a Nutrition-Sensitive Agriculture. *Food Secur.* **5**:679–692. [11]

Jänicke, M. 1995. The Political Systems Capacity for Environmental Policy, FFU Report 1995/6. Berlin: Forschungsstelle für Umweltpolitik, Freie Universität Berlin. [14]

Jansky, S. H., J. Dawson, and D. M. Spooner. 2015. How Do We Address the Disconnect between Genetic and Morphological Diversity in Germplasm Collections? *Am. J. Bot.* **102**:1213–1215. [3]

Janssen, M., Ö. Bodin, J. Anderies, et al. 2006. Toward a Network Perspective of the Study of Resilience in Social-Ecological Systems. *Ecol. Soc.* **11**:ART. 15. [8]

Jardón-Barbolla, L. 2015. De la Evolución Al Valor de Uso, Ida y Vuelta: Exploraciones en la Domesticación y Diversificación de Plantas. *Rev. Interdisc.* **3**:99–129. [2]

Jarvis, A., M. E. Ferguson, D. E. Williams, et al. 2003. Biogeography of Wild Arachis: Assessing Conservation Status and Setting Future Priorities. *Crop Sci.* **43**:1100–1108. [2, 5]

Jarvis, A., A. Lane, and R. J. Hijmans. 2008a. The Effect of Climate Change on Crop Wild Relatives. *Agricult. Ecosyst. Environ.* **126**:13–23. [2, 5, 7]

Jarvis, A., K. Williams, D. E. Williams, et al. 2005. Use of GIS for Optimizing a Collecting Mission for a Rare Wild Pepper (*Capsicum Flexuosum* Sendtn.) in Paraguay. *Genet. Resour. Crop Evol.* **52**:671–682. [5]

Jarvis, D. I., A. H. D. Brown, P. H. Cuong, et al. 2008b. A Global Perspective of the Richness and Evenness of Traditional Crop-Variety Diversity Maintained by Farming Communities. *PNAS* **105**:5326–5331. [3–5]

Jarvis, D. I., T. Hodgkin, B. R. Sthapit, C. Fadda, and I. Lopez-Noriega. 2011. An Heuristic Framework for Identifying Multiple Ways of Supporting the Conservation and Use of Traditional Crop Varieties within the Agricultural Production System. *CRC Crit. Rev. Plant Sci.* **30**:125–176. [4, 14]

Jarvis, D. I., C. Padoch, and H. D. Cooper, eds. 2007. Managing Biodiversity in Agricultural Ecosystems. New York: Columbia Univ. Press. [8]

Jenkins-Smith, H. C., and P. A. Sabatier, eds. 1993. Policy Change and Learning: An Advocacy Coalition Approach. Boulder: Westview Press. [14]

———. 1994. Evaluating the Advocacy Coalition Approach. *J. Public Policy* **14**:175–203. [14]

Jennings, B. H. 1988. Foundations of International Agricultural Research: Science and Politics in Mexican Agriculture. Boulder: Westview Press. [6]

Johns, T., and P. B. Eyzaguirre. 2006. Linking Biodiversity, Diet and Health in Policy and Practice. *Proc. Nutr. Soc.* **65**:182–189. [11]

Johns, T., B. Powell, P. Maundu, and P. B. Eyzaguirre. 2013. Agricultural Biodiversity as a Link between Traditional Food Systems and Contemporary Development, Social Integrity and Ecological Health. *J. Sci. Food Agric.* **93**:3433–3442. [9–11, 15]

Johns, T., and B. R. Sthapit. 2004. Biocultural Diversity in the Sustainability of Developing Country Food Systems. *Food Nutr. Bull.* **25**:143–155. [9]

Johnson, D., F. Martin, J. W. G. Cairney, and I. C. Anderson. 2012. The Importance of Individuals: Intraspecific Diversity of Mycorrhizal Plants and Fungi in Ecosystems. *New Phytol.* **194**:614–628. [4]

Johnson, K. B., A. Jacob, and M. E. Brown. 2013. Forest Cover Associated with Improved Child Health and Nutrition: Evidence from the Malawi Demographic and Health Survey and Satellite Data. *Glob. Health Sci. Pract.* **1**:237–248. [9, 10]

Johnson, M. T. J., M. J. Lajeunesse, and A. A. Agrawal. 2006. Additive and Interactive Effects of Plant Genotypic Diversity on Arthropod Communities and Plant Fitness. *Ecol. Lett.* **9**:24–34. [4]

Johnson, N. C., G. W. T. Wilson, M. A. Bowker, J. A. Wilson, and R. M. Miller. 2010. Resource Limitation Is a Driver of Local Adaptation in Mycorrhizal Symbioses. *PNAS* **107**:2093–2098. [4]

Jones, A. D. 2016. On-Farm Crop Species Richness Is Associated with Household Diet Diversity and Quality in Subsistence- and Market-Oriented Farming Households in Malawi. *J. Nutr.* **147**:86–96. [10]

———. 2017. Critical Review of the Emerging Research Evidence on Agricultural Biodiversity, Diet Diversity, and Nutritional Status in Low- and Middle-Income Countries. *Nutr. Rev.* **75**:769–782. [9, 10]

Jones, A. D., Y. Acharya, and L. P. Galway. 2017. Deforestation and Child Diet Diversity: A Geospatial Analysis of 15 Sub-Saharan African Countries. *Lancet* **389**:S11. [10]

Jones, A. D., L. Hoey, J. Blesh, et al. 2016. A Systematic Review of the Measurement of Sustainable Diets. *Adv. Nutr.* **7**:641–664. [9]

Jones, A. D., A. Shrinivas, and R. Bezner-Kerr. 2014. Farm Production Diversity Is Associated with Greater Household Dietary Diversity in Malawi: Findings from Nationally Representative Data. *Food Pol.* **46**:1–12. [11]

Jones, T. S., E. Allan, S. A. Haerri, et al. 2011. Effects of Genetic Diversity of Grass on Insect Species Diversity at Higher Trophic Levels Are Not Due to Cascading Diversity Effects. *Oikos* **120**:1031–1036. [4]

Jordan, N. R., and A. S. Davis. 2015. Middle-Way Strategies for Sustainable Intensification of Agriculture. *Bioscience* **65**:513–519. [8]

Jost, L. 2006. Entropy and Diversity. *Oikos* **113**:363–375. [2]

Juarez, H., F. Plasencia, and S. de Haan. 2011. Zooming in on the Secret Life of Genetic Resources in Potatoes: High Technology Meets Old-Fashioned Footwork. Esri Conservation Map Book. Redlands, CA [5]

Jump, A., and J. Penuelas. 2005. Running to Stand Still: Adaptation and the Response of Plants to Rapid Climate Change. *Ecol. Lett.* **8**:1010–1020. [2]

Junqueira, A. B., C. J. M. Almekinders, T. J.Stomph, C. R. Clement, and P. C. Struik. 2016. The Role of Amazonian Anthropogenic Soils in Shifting Cultivation: Learning from Farmers' Rationales. *Ecol. Soc.* **21**:12. [6]

Kagoda, F., J. Derera, P. Tongoona, and D. L. Coyne. 2010. Awareness of Plant-Parasitic Nematodes and Preferred Maize Varieties among Smallholder Farmers in East and Southern Uganda: Implications for Assessing Nematode Resistance Breeding Needs in African Maize. *Int. J. Pest Manag.* **56**:217–222. [4]

Kahane, R., T. Hodgkin, H. Jaenicke, et al. 2013. Agrobiodiversity for Food Security, Health and Income. *Agronom. Sustain. Devel.* **33**:671–693. [11, 13]

Kahiluoto, K., J. Kaseva, K. Hakala, et al. 2014. Cultivating Resilience by Empirically Revealing Response Diversity. *Glob. Environ. Change* **25**:186–193. [5]

Kauffman, S. 1993. The Origins of Order: Self Organization and Selection in Evolution. Oxford: Oxford University Press. [7]

Kawa, N. C., C. McCarty, and C. R. Clement. 2013. Manioc Varietal Diversity, Social Networks, and Distribution Constraints in Rural Amazonia. *Curr. Anthropol.* **54**:764–770. [8, 11]

Kehlenbeck, K., E. Asaah, and R. Jamnadass. 2013. Diversity of Indigenous Fruit Trees and Their Contribution to Nutrition and Livelihoods in Sub-Saharan Africa: Examples from Kenya and Cameroon. In: Diversifying Food and Diets Using Agricultural Biodiversity to Improve Nutrition and Health, ed. J. C. Fanzo et al., pp. 257–259. Abingdon, UK: Earthscan from Routledge. [9]

Keleman, A., and J. Hellin. 2009. Specialty Maize Varieties in Mexico: A Case Study in Market-Driven Agro-Biodiversity Conservation. *J. Lat. Amer. Geogr.* **8**:147–174. [2]

Keleman, A., J. Hellin, and M. R. Bellon. 2009. Maize Diversity, Rural Development Policy, and Farmers' Practices: Lessons from Chiapas, Mexico. *Geogr J.* **175**:52–70. [4, 15]

Keller, D. 2009. Deep Ecology. In: Encyclopedia of Environmental Ethics and Philosophy, ed. J. Callicot and R. Frodeman, vol. 1, pp. 206–211. Farmington Hills, MI: Thomson Gale. [12]

Kennedy, C. M., E. Lonsdorf, M. C. Neel, et al. 2013. A Global Quantitative Synthesis of Local and Landscape Effects on Wild Bee Pollinators in Agroecosystems. *Ecol. Lett.* **16**:584–599. [4]

Kennedy, G., D. Hunter, J. Garrett, and S. Padulosi. 2017a. Leveraging Agrobiodiversity to Create Sustainable Food Systems for Healthier Diets. *UNSCN News* **42**:23–31. [10]

Kennedy, G., L. T. K. Warren, C. Termote, R. Charrindière, and J. Y. A. Tung. 2017b. Guidelines on Assessing Biodiverse Foods in Dietary Intake Surveys. Rome: FAO. [10]

Kerssen, T. M. 2015. Food Sovereignty and the Quinoa Boom: Challenges to Sustainable Re-Peasantisation in the Southern Altiplano of Bolivia. *Third World Q.* **36**:489–507. [8]

Khoury, C. K., H. A. Achicanoy, A. D. Bjorkman, et al. 2016. Origins of Food Crops Connect Countries Worldwide. *Proc. R. Soc. Lond. B Biol. Sci.* **283**:20160792. [9, 13]

Khoury, C. K., A. D. Bjorkman, H. Dempewolf, et al. 2014. Increasing Homogeneity in Global Food Supplies and the Implications for Food Security. *PNAS* **111**:4001–4006. [8–11, 14, 15]

Khoury, C. K., N. P. Castañeda-Álvarez, H. A. Achicanoy, et al. 2015a. Crop Wild Relatives of Pigeonpea [*Cajanus Cajan* (L .) Millsp.]: Distributions, *Ex Situ* Conservation Status, and Potential Genetic Resources for Abiotic Stress Tolerance. *Biol. Conserv.* **184**:259–270. [5]

Khoury, C. K., B. Heider, N. P. Castañeda-Álvarez, et al. 2015b. Distributions, *Ex Situ* Conservation Priorities, and Genetic Resource Potential of Crop Wild Relatives of Sweetpotato [*Ipomoea Batatas* (L.) Lam., *I.* Series *Batatas*] *Front. Plant Sci.* **6**:251. [2, 5]

Kirksey, E., and S. Helmreich. 2010. The Emergence of Multispecies Ethnography. *Cult. Anthropol.* **25**:545–576. [14]

Kirmayer, L. J., G. M. Brass, and C. L. Tait. 2000. The Mental Health of Aboriginal Peoples: Transformations of Identity and Community. *Can. J. Psychiatry* **45**:607–616. [11]

Klein, A.-M., B. E. Vaissiere, J. H. Cane, et al. 2007. Importance of Pollinators in Changing Landscapes for World Crops. *Proc. R. Soc. Lond. B Biol. Sci.* **274**:303–313. [2]

Kloppenburg, J. 1988. First the Seed: The Political Economy of Plant Biotechnology, 1492–2000. Cambridge: Cambridge Univ. Press. [13, 14]

———. 2014. Re-Purposing the Master's Tools: The Open Source Seed Initiative and the Struggle for Seed Sovereignty. *J. Peasant Stud.* **41**:1225–1246. [13]

Köberl, M., M. Dita, M. A., C. Staver, and G. Berg. 2015. Agroforestry Leads to Shifts within the Gammaproteobacterial Microbiome of Banana Plants Cultivated in Central America. *Front. Microbiol.* **6**:91. [2, 8]

Kohn, E. 2015. Anthropology of Ontologies. *Annu. Rev. Anthropol.* **44**:311–327. [14]

Kontoleon, A., U. Pascual, and M. Smale, eds. 2008. Agrobiodiversity Conservation and Economic Development. London: Routledge. [15]

Koonan, S. 2014. India's *Sui Generis* System of Plant Variety Protection. Briefing Paper 4, pp. 1–5. Food, Biological Diversity and Intellectual Property. New York: Quaker United Nations Offices. [12]

Kraft, K. H., C. H. Brown, G. P. Nabhan, et al. 2014. Multiple Lines of Evidence for the Origin of Domesticated Chili Pepper, *Capsicum Annuum*, in Mexico. *PNAS* **111**:6165–6170. [2, 5]

Kral, M. J., L. Idlout, J. B. Minore, R. J. Dyck, and L. J. Kirmayer. 2011. Unikkaartuit: Meanings of Well-Being, Unhappiness, Health, and Community Change among Inuit in Nunavut, Canada. *Am. J. Community Psychol.* **48**:426–438. [11]

Kremen, C., and A. Miles. 2012. Ecosystem Services in Biologically Diversified versus Conventional Farming Systems: Benefits, Externalities, and Trade-Offs. *Ecol. Soc.* **17**:40. [4]

Kremen, C., N. M. Williams, and R. W. Thorp. 2002. Crop Pollination from Native Bees at Risk from Agricultural Intensification. *PNAS* **99**:16812–16816. [2]

Kristjanson, P., B. Harvey, M. Van Epp, and P. K. Thornton. 2014. Social Learning and Sustainable Development. *Nat. Clim. Chang.* **4**:5–7. [7]

Kruijssen, F., A. Giuliani, and M. Sudha. 2009a. Marketing Underutilized Crops to Sustain Agrobiodiversity and Improve Livelihoods. *Acta Horticult.* **806**:415–422. [15]

Kruijssen, F., M. Keizer, and A. Giuliani. 2009b. Collective Action for Small-Scale Producers of Agricultural Biodiversity Products. *Food Pol.* **34**:46–52. [15]

Krupke, C. H., G. J. Hunt, B. D. Eitzer, G. Andino, and K. Given. 2012. Multiple Routes of Pesticide Exposure for Honey Bees Living near Agricultural Fields. *PLoS One* **7**:e29268. [4]

Kuhnlein, H. V., B. Erasmus, and B. Spigelski. 2009. Indigenous Peoples' Food Systems. Rome: Food and Agricultural Organization of the United Nations. [11]

Kuhnlein, H. V., B. Erasmus, D. Spigelski, and B. Burlingame, eds. 2013. Indigenous Peoples' Food Systems and Well-Being: Interventions and Policies for Healthy Communities. Rome: FAO/CINE. [11]

Kuhnlein, H. V., S. T. Smitasiri, S. Yesudas, et al. 2006. Documenting Traditional Food Systems of Indigenous Peoples, International Case Studies: Guidelines for Procedures. Toronto: Centre for Indigenous Peoples' Nutrition and Environment, McGill Univ. [9]

Kuhnlein, H. V., R. Soueida, and O. Receveur. 1996. Dietary Nutrient Profiles of Canadian Baffin Island Inuit Differ by Food Source, Season, and Age. *J. Am. Diet Assn.* **96**:155–162. [11]

Kwak, Y.-S., and D. M. Weller. 2013. Take-All of Wheat and Natural Disease Suppression: A Review. *Plant Pathol. J.* **29**:125–135. [4]

Labeyrie, V., M. Deu, A. Barnaud, et al. 2014. Influence of Ethnolinguistic Diversity on the Sorghum Genetic Patterns in Subsistence Farming Systems in Eastern Kenya. *PLoS One* **9**:e92178. [3, 5]

Labeyrie, V., M. Thomas, Z. K. Muthamia, and C. Leclerc. 2016. Seed Exchange Networks, Ethnicity, and Sorghum Diversity. *PNAS* **113**:98–103. [3, 5, 13]

Lachat, C., J. E. Raneri, K. W. Smith, et al. 2018. Dietary Species Richness as a Measure of Food Biodiversity and Nutritional Quality of Diets. *PNAS* **115**:201709194. [10]

Lammerts van Bueren, E. T., S. S. Jones, L. Tamm, et al. 2011. The Need to Breed Crop Varieties Suitable for Organic Farming, Using Wheat, Tomato and Broccoli as Examples: A Review. *NJAS* **58**:193–205. [4]

Lang, T., and M. Heasman. 2004. Food Wars: The Global Battle for Mouths, Minds and Markets. London: Earthscan. [11]

Larsen, C. S. 2006. The Agricultural Revolution as Environmental Catastrophe: Implications for Health and Lifestyle in the Holocene. *Quatern. Int.* **150**:12–20. [2]

Larson, G., U. Albarella, K. Dobney, et al. 2007. Ancient DNA, Pig Domestication, and the Spread of the Neolithic into Europe. *PNAS* **104**:15276–15281. [3]

Latour, B. 2005. Reassembling the Social: An Introduction to Actor-Network-Theory. Oxford: Oxford Univ. Press. [13]

Lavelle, P., D. E. Bignell, M. C. Austen, et al. 2004. Connecting Soil and Sediment Biodiversity: The Role of Scale and Implications for Management. In: Sustaining Biodiversity and Ecosystem Services in Soils and Sediments, ed. D. H. Wall, pp. 193–224. Washington, D.C.: Island Press. [4]

Leclerc, C. 2012. L'adoption de L'agriculture Chez Les Pygmées Baka Du Cameroun: Dynamique Sociale Et Continuité Structurale. Paris: Editions Quae, Maison des Sciences de L'homme. [2]

Leclerc, C., and G. C. d'Eeckenbrugge. 2012. Social Organization of Crop Genetic Diversity: The G × E × S Interaction Model. *Diversity* **4**:1–32. [3, 5, 13, 14]

Legg, J., E. A. Somado, I. Barker, et al. 2014. A Global Alliance Declaring War on Cassava Viruses in Africa. *Food Secur.* **6**:231–248. [2]

Leopold, A. 1966. A Sand County Almanac. Oxford: Oxford Univ. Press. [12]

Lerner, A. M., and K. Appendini. 2011. Dimensions of Peri-Urban Maize Production in the Toluca-Atlacomulco Valley, Mexico. *J. Lat. Amer. Geogr.* **10**:87–106. [8]

Lestrelin, G., J. Bourgoin, B. Bouahom, and J. C. Castella. 2011. Measuring Participation: Case Studies on Village Land Use Planning in Northern Lao Pdr. *Appl. Geogr.* **31**:950–958. [8]

Lewis, D., B. Barham, and K. S. Zimmerer. 2008. Spatial Externalities in Agriculture: Empirical Analysis, Statistical Identification, and Policy Implications for Development and Environment. *World Dev.* **36**:1813–1829. [5]

Li, B., Y.-Y. Li, H.-M. Wu, et al. 2016. Root Exudates Drive Interspecific Facilitation by Enhancing Nodulation and N2 Fixation. *PNAS* **113**:6496–6501. [2]

Li, H., and R. Durbin. 2011. Inference of Human Population History from Individual Whole-Genome Sequences. *Nature* **475**:493–496. [3]

Li, J., E. T. Lammerts van Bueren, J. Jiggins, and C. Leeuwis. 2012. Farmers' Adoption of Maize (*Zea Mays* L.) Hybrids and the Persistence of Landraces in Southwest China: Implications for Policy and Breeding. *Genet. Resour. Crop Evol.* **59**:1147–1160. [4]

Li, L.-F., and K. M. Olsen. 2016. To Have and to Hold: Selection for Seed and Fruit Retention During Crop Domestication. *Curr. Top. Development. Biol.* **119**:63–109. [3]

Lin, B. B. 2011. Resilience in Agriculture through Crop Diversification: Adaptive Management for Environmental Change. *Bioscience* **61**:183–193. [7]

Lipper, L., P. Thornton, B. M. Campbell, et al. 2014. Climate-Smart Agriculture for Food Security. *Nat. Clim. Chang.* **4**:1068–1072. [6]

Litt, J. S., M.-J. Soobader, M. S. Turbin, et al. 2011. The Influence of Social Involvement, Neighborhood Aesthetics, and Community Garden Participation on Fruit and Vegetable Consumption. *Am. J. Public Health* **101**:1466–1473. [11]

Little, D., and P. Edwards. 2003. Integrated Livestock-Fish Farming Systems. Rome: FAO. [2]

Lloyd, K., S. Wright, S. Suchet-Pearson, L. Burarrwanga, and B. Country. 2012. Reframing Development through Collaboration: Towards a Relational Ontology of Connection in Bawaka, North East Arnhem Land. *Third World Q.* **33**:1075–1094. [12]

Loarie, S. R., P. B. Duffy, H. Hamilton, et al. 2009. The Velocity of Climate Change. *Nature* **462**:pages 1052–1055. [7]

Lobell, D. B., W. Schlenker, and J. Costa-Roberts. 2011. Climate Trends and Global Crop Production since 1980. *Science* **333**:616–620. [6]

Lockie, S., and D. Carpenter, eds. 2010. Agriculture, Biodiversity and Markets: Livelihoods and Agroecology in Comparative Perspective. Oxford: Earthscan. [8, 15]

Londo, J. P., Y.-C. Chiang, K.-H. Hung, T.-Y. Chiang, and B. A. Schaal. 2006. Phylogeography of Asian Wild Rice, *Oryza Rufipogon*, Reveals Multiple Independent Domestications of Cultivated Rice, *Oryza Sativa*. *PNAS* **103**:9578–9583. [3]

López-García, D., and G. I. Guzmán-Casado. 2013. Si la Tierra Tiene Sazón:el Conocimiento Tradicional Campesino Como Movilizador de Procesos de Transición Agroecológica. *Agroecologia* **7**:7–20. [11]

Lopez-Raez, J. A. 2016. How Drought and Salinity Affect Arbuscular Mycorrhizal Symbiosis and Strigolactone Biosynthesis? *Planta* **243**:1375–1385. [4]

Loreau, M., S. Naeem, P. Inchausti, et al. 2001. Biodiversity and Ecosystem Functioning: Current Knowledge and Future Challenges. *Science* **294**:804–808. [6]

Louette, D., A. Charrier, and J. Berthaud. 1997. *In Situ* Conservation of Maize in Mexico: Genetic Diversity and Maize Seed Management in a Traditional Community *Econ. Bot.* **51**:20–38. [13]

Louzada, M. L., L. G. Baraldi, E. M. Steele, et al. 2015. Consumption of Ultra-Processed Foods and Obesity in Brazilian Adolescents and Adults. *Prev. Med.* **81**:9–15. [10]

Lovon, M., and A. Mathiassen. 2014. Are the World Food Programme's Food Consumption Groups a Good Proxy for Energy Deficiency? *Food Secur.* **6**:461–470. [10]

Low, J. W., M. Arimond, N. Osman, et al. 2007. A Food-Based Approach Introducing Orange-Fleshed Sweet Potatoes Increased Vitamin a Intake and Serum Retinol Concentrations in Young Children in Rural Mozambique. *J. Nutr.* **137**:1320–1327. [2]

Lowder, S. K., J. Skoet, and T. Raney. 2016. The Number, Size, and Distribution of Farms, Smallholder Farms, and Family Farms Worldwide. *World Dev.* **87**:16–29. [6]

Lubinsky, P., M. Van Dam, and A. Van Dam. 2006. Pollination of *Vanilla* and Evolution in Orchidaceae. *Lindleyana* **75**:926–929. [2]

Macdiarmid, J. I. 2013. Is a Healthy Diet an Environmentally Sustainable Diet? *Proc. Nutr. Soc.* **72**:13–20. [9]

Machaca, M. 2016. El Reto de Las Organizaciones Comunitarias: la Recuperación y Vigorización de Las Sabidurías de Crianza de la Agrobiodiversidad en Un Contexto de Cambio Climático. In: Biodiversidad y Propiedad Intelectual en Disputa, ed. S. Roca, pp. 343–366. Lima: Esan Ediciones. [12]

Maffi, L. 2005. Linguistic, Cultural and Biological Diversity. *Annu. Rev. Anthropol.* **34**:599–617. [12]

Magalhães, M. V., A. R. Baby, V. R. Velasco, D. M. Pereira, and T. M. Kaneko. 2011. Patenting in the Cosmetic Sector: Study of the Use of Herbal Extracts. *Braz. J. Pharm. Sci.* **47**:693–700. [12]

Mallinger, R. E., and C. Gratton. 2015. Species Richness of Wild Bees, but Not the Use of Managed Honeybees, Increases Fruit Set of a Pollinator-Dependent Crop. *J. Appl. Ecol.* **52**:323–330. [4]

Manica, A., F. Prugnolle, and F. Balloux. 2005. Geography Is a Better Determinant of Human Genetic Differentiation Than Ethnicity. *Hum. Genet.* **118**:366–371. [3]

Marchi, E. 2018. Accommodation of Cultural Diversity and Collective Rights at the Crossroads of Conservation Discourses: The Case of Indigenous Communities in Oaxaca, Mexico. PhD Dissertation. Florence: Univ. Degli Studi Firenze. [12]

Mariac, C., I. S. Ousseini, A. K. Alio, et al. 2016. Spatial and Temporal Variation in Selection of Genes Associated with Pearl Millet Varietal Quantitative Traits *in Situ*. *Front Genet.* **7**:130. [3]

Marion Suiseeya, K. R. 2014. Negotiating the Nagoya Protocol: Indigenous Demands for Justice. *Glob. Environ. Polit.* **14**:102–124. [14]

Martin-Prevel, Y., P. Allemand, D. Wiesmann, et al. 2015. Moving Forward on Choosing a Standard Operational Indicator of Women's Dietary Diversity. Rome: FAO. [10]

Martinez, T. N., and N. C. Johnson. 2010. Agricultural Management Influences Propagule Densities and Functioning of Arbuscular Mycorrhizas in Low- and High-Input Agroecosystems in Arid Environments. *Appl. Soil Ecol.* **46**:300–306. [4]

Martinez-Alier, J., L. Temper, D. Del Bene, and A. Scheidel. 2016. Is There a Global Environmental Justice Movement? *J. Peasant Stud.* **43**:731–755. [9]

Masset, E., L. Haddad, A. Cornelius, and J. Isaza-Castro. 2012. Effectiveness of Agricultural Interventions That Aim to Improve Nutritional Status of Children: Systematic Review. *Br. Med. J.* **344**:d8222. [10]

Massey, D. 2005. For Space. London: SAGE Publications. [13]

Mathew, A. G., R. Cissell, and S. Liamthong. 2007. Antibiotic Resistance in Bacteria Associated with Food Animals: A United States Perspective of Livestock Production. *Foodborne Pathog. Dis.* **4**:115–133. [11]

Mathur, P. N. 2013. Empowering Farmers to Use Plant Genetic Diversity for Adapting to Climate Change. In: A Road Map for Implementing the Multilateral System of Access and Benefit-Sharing in India, ed. M. Halewood et al. Rome: Bioversity Intl. [6]

Mato, D. 2016. Indigenous People in Latin America: Movements and Universities, Achievements, Challenges, and Intercultural Conflicts. *J. Intercult. Stud.* **37**:211–233. [12]

Matson, P. A., and P. M. Vitousek. 2006. Agricultural Intensification: Will Land Spared from Farming Be Land Spared for Nature? *Conserv. Biol.* **20**:709–710. [9]

Matsuda, M. 2013. Upland Farming Systems Coping with Uncertain Rainfall in the Central Dry Zone of Myanmar: How Stable Is Indigenous Multiple Cropping under Semi-Arid Conditions? *Hum. Ecol.* **41**:927–936. [7]

Matsuoka, Y., Y. Vigouroux, M. M. Goodman, et al. 2002. A Single Domestication for Maize Shown by Multilocus Microsatellite Genotyping. *PNAS* **99**:6080–6084. [3]

Maundu, P., M. Yasuyuki, E. Towett, J. A. Ombonya na, and E. Obel-Lawson, eds. 2011. Mboga Za Watu Wa Pwani: Kilifi Utamaduni Conservation Group. Rome: Bioversity Intl. [9]

Maxted, N., and E. Dulloo. 2016. Enhancing Crop Genepool Use: Capturing Wild Relative and Landrace Diversity for Crop Improvement. Oxfordshire: CABI. [13]

Maxted, N., B. V. Ford-Lloyd, and J. G. Hawkes. 1997. Complementary Conservation Strategies. In: Plant Genetic Conservation: The *in Situ* Approach, ed. N. Maxted et al., pp. 15–39. Dordrecht: Kluwer Academic Publishers. [2]

Maxted, N., and S. Kell. 2009. Establishment of a Global Network for the in-Situ Conservation of Crop Wild Relatives: Status and Needs. Background Study Paper No. 39. Rome: FAO. [3]

Maxted, N., P. Mabuza-Diamini, H. Moss, et al. 2004. Systematic and Ecogeographic Studies on Crop Genepools 11: An Ecogeographic Study African *Vigna*. Rome: Intl. Plant Genetic Resources Institute (IPGRI). [5]

McCord, P. F., M. Cox, M. Schmitt-Harsh, and T. Evans. 2015. Crop Diversification as a Smallholder Livelihood Strategy within Semi-Arid Agricultural Systems near Mount Kenya. *Land Use Policy* **42**:738–750. [8]

McGuire, S. J. 2008. Securing Access to Seed: Social Relations and Sorghum Seed Exchange in Eastern Ethiopia. *Hum. Ecol.* **36**:217–229. [8, 13]

McGuire, S. J., and L. Sperling. 2008. Leveraging Farmers' Strategies for Coping with Stress: Seed Aid in Ethiopia. *Glob. Environ. Change* **18**:679–688. [8]

———. 2013. Making Seed Systems More Resilient to Stress. *Glob. Environ. Change* **23**:644–653. [13]

———. 2016. Seed Systems Smallholder Farmers Use. *Food Secur.* **8**:179–195. [13]

McKey, D., T. R. Cavagnaro, J. Cliff, and R. Gleadow. 2010a. Chemical Ecology in Coupled Human and Natural Systems: People, Manioc, Multitrophic Interactions and Global Change. *Chemoecol.* **20**:109–133. [8]

McKey, D., M. Elias, B. Pujol, and A. Duputié. 2010b. The Evolutionary Ecology of Clonally Propagated Domesticated Plants. *New Phytol.* **186**:318–332. [2, 3]

———. 2012. Ecological Approaches to Crop Domestication. In: Biodiversity in Agriculture: Domestication, Evolution and Sustainability, ed. P. Gepts et al., pp. 377–406. Cambridge: Cambridge Univ. Press. [2]

McLeod, A. 2011. World Livestock 2011: Livestock in Food Security. Rome: FAO. [9]

McMichael, P. 2009. A Food Regime Analysis of the World Food Crisis. *Agricult. Human Values* **26**:281–295. [11]

———. 2013. Value-Chain Agriculture and Debt Relations: Contradictory Outcomes. *Third World Q.* **34**:671–690. [8]

Mekbib, F. 2008. Genetic Erosion of Sorghum in the Centre of Diversity, Ethiopia. *Genet. Resour. Crop Evol.* **55**:351–364. [9]

Meles, K. 2011. Integrated GIS and Survey Approaches in Assessing Diversity and Sustainability in Agricultural Landscapes in Tigray, Northern Ethiopia. *J. Drylands* **4**:267–282. [5]

Mercer, K. L., A. Martínez-Vásquez, and H. R. Perales. 2008. Asymmetrical Local Adaptation of Maize Landraces Along an Altitudinal Gradient. *Evol. Appl.* **1**:489–500. [2, 4, 13]

Mercer, K. L., and H. R. Perales. 2010. Evolutionary Response of Landraces to Climate Change in Centers of Crop Diversity. *Evol. Appl.* **3**:480–493. [2]

Mercer, K. L., H. R. Perales, and J. D. Wainwright. 2012. Climate Change and the Transgenic Adaptation Strategy: Smallholder Livelihoods, Climate Justice, and Maize Landraces in Mexico. *Glob. Environ. Change* **22**:495–504. [4]

Meyer, R. S., E. D. Ashley, and R. J. Helen. 2012. Patterns and Processes in Crop Domestication: An Historical Review and Quantitative Analysis of 203 Global Food Crops. *New Phytol.* **196**:29–48. [3]

Meyer, R. S., and M. D. Purugganan. 2013. Evolution of Crop Species: Genetics of Domestication and Diversification. *Nat. Rev. Genet.* **14**:840–852. [3]

Midega, C. A. O., J. Pickett, A. Hooper, J. Pittchar, and Z. R. Khan. 2016. Maize Landraces Are Less Affected by Striga Hermonthica Relative to Hybrids in Western Kenya. *Weed. Technol.* **30**:21–28. [4]

Midler, E., U. Pascual, A. G. Drucker, U. Narloch, and J. L. Soto. 2015. Unraveling the Effects of Payments for Ecosystem Services on Motivations for Collective Action. *Ecol. Econ.* **120**:394–405. [2]

Mijatović, D., F. Van Oudenhoven, P. B. Eyzaguirre, and T. Hodgkin. 2013. The Role of Agricultural Biodiversity in Strengthening Resilience to Climate Change: Towards an Analytical Framework. *Int. J. Agricult. Sustainabil.* **11**:95–107. [10]

Mikkelsen, C. 2014. The Indigenous World 2014. Copenhagen: Intl. Work Group for Indigenous Affairs (IWGIA). [12]

Milla, R., C. P. Osborne, M. M. Turcotte, and C. Violle. 2015. Plant Domestication through an Ecological Lens. *Trends Ecol. Evol.* **30**:463–469. [2]

Milligan, C., A. Gatrell, and A. Bingley. 2004. Cultivating Health: Therapeutic Landscapes and Older People in Northern England. *Soc. Sci. Med.* **58**:1781–1793. [11]

MINAGRI. 2017. Catálogo de Variedades de Papa Nativa del Sureste del Departamento de Junín, Perú. Lima: Intl. Potato Center (CIP). [2, 5]

Mitchell, L., E. Brook, J. E. Lee, C. Buizert, and T. Sowers. 2013. Constraints on the Late Holocene Anthropogenic Contribution to the Atmospheric Methane Budget. *Science* **342**:964–966. [7]

Money, P. 1996. Viewpoint of Non-Governmental Organizations. In: Agrobiodiversity and Farmers' Rights: Proceedings of a Technical Consultation on an Implementation Framework for Farmers' Rights, ed. M. S. Swaminathan, pp. 40–43. Madras: Swaminathan Research Foundation. [12]

Monteiro, C. A., G. Cannon, and R. B. Levy. 2016. Nova: The Star Shines Bright. *World Nutri.* **7**:28–38. [10]

Monteros, A. R. 2011. Potato Landraces: Description and Dynamics in Three Areas of Ecuador. PhD Thesis thesis, Wageningen Agricultural University, Wageningen. [3]

Montesano, V., D. Negro, G. Sarli, G. Logozzo, and P. S. Zeuli. 2012. Landraces in Inland Areas of the Basilicata Region, Italy: Monitoring and Perspectives for on Farm Conservation. *Genet. Resour. Crop Evol.* **59**:701–716. [2]

Mooney, P. R. 1983. The Law of the Seed: Another Development and Plant Genetic Resources. In: Development Dialogue, 1–2. Uppsala: Dag Hammarskjöld Foundation. [14]

Moore, C., and R. D. Raymond. 2006. Back by Popular Demand: The Benefits of Traditional Vegetables. Rome: Intl. Plant Genetic Resources Institute. [9]

Morandin, L. A., and C. Kremen. 2013. Hedgerow Restoration Promotes Pollinator Populations and Exports Native Bees to Adjacent Fields. *Ecol. Appl.* **23**:829–839. [4]

Morueta-Holme, N., K. Engemann, P. Sandoval-Acuña, et al. 2015. Strong Upslope Shifts in Chimborazo's Vegetation over Two Centuries since Humboldt. *PNAS* **112**:12741–12745. [5]

Moscoe, L. J., R. Blas, D. H. Masi, M. H. Masi, and E. Emshwiller. 2017. Genetic Basis for Folk Classification of Oca (*Oxalis Tuberosa* Molina; Oxalidaceae): Implications for Research and Conservation of Clonally Propagated Crops. *Genet. Resour. Crop Evol.* **64**:867–887. [3, 9]

Moss, R. H., J. A. Edmonds, K. A. Hibbard, et al. 2010. The Next Generation of Scenarios for Climate Change Research and Assessment. *Nature* **463**:747–756. [6]

Mossakowski, K. N. 2003. Coping with Perceived Discrimination: Does Ethnic Identity Protect Mental Health? *J. Health Soc. Behav.* **44**:318–331. [11]

Mouser, B., E. Nuitjen, F. Okry, and P. Richards. 2012. Commodity and Anti-Commodity: Linked Histories of Slavery, Emancipation, and Red and White Rice at Sierra Leone. In: Commodities of Empire Working Paper No. 19. Wageningen: Univ. of Wageningen. [6]

Mucioki, M., G. M. Hickey, L. Muhammad, and T. Johns. 2016. Supporting Farmer Participation in Formal Seed Systems: Lessons from Tharaka, Kenya. *Dev. Pract.* **26**:137–148. [9]

Mueller, B. 2014. Seeds: Grown, Governed, and Contested, or the Ontic in Political Anthropology. *Focaal* **69**:3–11. [13]

Mukanga, M., J. Derera, P. Tongoona, and M. D. Laing. 2011. Farmers' Perceptions and Management of Maize Ear Rots and Their Implications for Breeding for Resistance. *Afr. J. Agric. Res.* **6**:4544–4554. [4]

Mulumba, J. W., R. Nankya, J. Adokorach, et al. 2012. A Risk-Minimizing Argument for Traditional Crop Varietal Diversity Use to Reduce Pest and Disease Damage in Agricultural Ecosystems of Uganda. *Agricult. Ecosyst. Environ.* **157**:70–86. [4]

Mundt, C. C. 2002. Use of Multiline Cultivars and Cultivar Mixtures for Disease Management. *Annu. Rev. Phytopathol.* **40**:381–410. [4]

Murphy, K. M., D. Bazile, J. Kellogg, and M. Rahmanian. 2016. Development of a Worldwide Consortium on Evolutionary Participatory Breeding in Quinoa. *Front. Plant Sci.* **7**:608. [4, 6]

Murphy, K. M., J. C. Dawson, and S. S. Jones. 2008. Relationship among Phenotypic Growth Traits, Yield and Weed Suppression in Spring Wheat Landraces and Modern Cultivars. *Field Crops Res.* **105**:107–115. [4]

Mwongera, C., J. Boyard-Micheau, C. Baron, and C. Leclerc. 2014. Social Process of Adaptation to Environmental Changes: How Eastern African Societies Intervene between Crops and Climate. *Weather Clim. Soc.* **6**:341–353. [2, 7]

Myers, N. 1997. Environmental Refugees. *Popul. Environ.* **19**:167–182. [9]

Myers, N., R. A. Mittermeier, C. G. Mittermeier, G. A. da Fonseca, and J. Kent. 2000. Biodiversity Hotspots for Conservation Priorities. *Nature* **403**:853–858. [5]

Naess, A. 1973. The Shallow and the Deep, Long-Range Ecology Movement: A Summary. *Inquiry* **16**:95–100. [12]

NAFRI. 2016. Lao Pdr National Agro-Biodiversity Programme and Action Plan II (2015-2025). Vientiane: National Agriculture and Forestry Research Institute. [1]

Nakagome, S., G. Alkorta-Aranburu, R. Amato, et al. 2016. Estimating the Ages of Selection Signals from Different Epochs in Human History. *Mol. Biol. Evol.* **33**:657–669. [3]

Narloch, U., A. G. Drucker, and U. Pascual. 2011a. Payments for Agrobiodiversity Conservation Services for Sustained on-Farm Utilization of Plant and Animal Genetic Resources. *Ecol. Econ.* **70**:1837–1845. [14, 15]

———. 2017. What Role for Cooperation in Conservation Tenders? Paying Farmer Groups in the High Andes. *Land Use Policy* **63**:659–671. [14]

Narloch, U., U. Pascual, and A. Drucker. 2011b. Cost-Effectiveness Targeting under Multiple Conservation Goals and Equity Considerations in the Andes. *Environ. Conserv.* **38**:417–425. [2]

Narloch, U., U. Pascual, and A. G. Drucker. 2013. How to Achieve Fairness in Payments for Ecosystem Services? Insights from Agrobiodiversity Conservation Auctions. *Land Use Policy* **35**:107–118. [14]

Nazarea, V. D. 2005a. Cultural Memory and Biodiversity. Tuscon: Univ. of Arizona Press. [13, 14]

———. 2005b. Heirloom Seeds and Their Keepers: Marginality and Memory in the Conservation of Biological Diversity. Tucson: Univ. of Arizona Press. [14]

———. 2006. Local Knowledge and Memory in Biodiversity Conservation. *Annu. Rev. Anthropol.* **35**:317–335. [13, 14]

Negin, J., R. Remans, S. Karuti, and J. C. Fanzo. 2009. Integrating a Broader Notion of Food Security and Gender Empowerment into the African Green Revolution. *Food Secur.* **1**:351–360. [9]

Nelson, A., and K. M. Chomitz. 2011. Effectiveness of Strict vs. Multiple Use Protected Areas in Reducing Tropical Forest Fires: A Global Analysis Using Matching Methods. *PLoS One* **6**:e22722. [9]

Nemogá, G. 2013. Estudio de Caso en Perú: Registro de Conocimientos Colectivos Asociados a la Biodiversidad. In: Seis Estudios de Caso en América Latina y el Caribe: Acceso a Recursos Genéticos y Distribución de Beneficios, ed. M. Rios and A. Mora, pp. 105–116. Quito: UICN-PNUMA/GEF-ABS-LAC. [12]

———. 2016. Biocultural Diversity: Innovating in Research for Conservation. *Acta Biol. Columb.* **21**:311–319. [12, 14]

———. 2018. Designing Biocultural Protocols with the Embera People of Colombia. *Landscape* **7**:20–24. [12]

Nendel, C., K. C. Kersebaum, W. Mirschel, and K. O. Wenkel. 2014. Testing Farm Management Options as Climate Change Adaptation Strategies Using the Monica Model. *Eur. J. Agron.* **52**:47–56. [6]

Nevo, E. 1998. Genetic Diversity in Wild Cereals: Regional and Local Studies and Their Bearing on Conservation *Ex Situ* and *in Situ*. *Genet. Resour. Crop Evol.* **45**:355–370. [2]

Ninkovic, V., I. Dahlin, A. Vucetic, et al. 2013. Volatile Exchange between Undamaged Plants: A New Mechanism Affecting Insect Orientation in Intercropping. *PLoS One* **8**:e69431. [4]

Noack, A.-L., and N. R. M. Pouw. 2015. A Blind Spot in Food and Nutrition Security: Where Culture and Social Change Shape the Local Food Plate. *Agricult. Human Values* **32**:169–182. [11]

Nuijten, E. 2010. Gender and Management of Crop Diversity in the Gambia. *J. Polit. Ecol.* **17**:42–58. [6]

Oberlack, C., L. Tejada, P. Messerli, S. Rist, and M. Giger. 2016. Sustainable Livelihoods in the Global Land Rush? Archetypes of Livelihood Vulnerability and Sustainability Potentials. *Glob. Environ. Change* **41**:153–171. [8]

Obregon-Tito, A. J., R. Y. Tito, J. Metcalf, et al. 2015. Subsistence Strategies in Traditional Societies Distinguish Gut Microbiomes. *Nat. Commun.* **6**:6505. [9]

O'Brien, K. L., and R. M. Leichenko. 2000. Double Exposure: Assessing the Impacts of Climate Change within the Context of Economic Globalization. *Glob. Environ. Change* **10**:221–232. [6]

Ofstehage, A. 2012. The Construction of an Alternative Quinoa Economy: Balancing Solidarity, Household Needs, and Profit in San Agustín, Bolivia. *Agricult. Human Values* **29**:441–454. [8]

Oka, H. 1969. A Note on the Design of Germplasm Presentation Work in Grain Crops. *SABRAO News.* **1**:127–134. [5]

Okonya, J. S., R. O. M. Mwanga, K. Syndikus, and J. Kroschel. 2014. Insect Pests of Sweet Potato in Uganda: Farmers' Perceptions of Their Importance and Control Practices. *Springerplus* **3**:303. [4]

Oliveira, H. R., M. G. Campana, H. Jones, et al. 2012. Tetraploid Wheat Landraces in the Mediterranean Basin: Taxonomy, Evolution and Genetic Diversity. *PLoS One* **7**:e37063. [5]

O'Neill, B. C., E. Kriegler, K. Riahi, et al. 2014. A New Scenario Framework for Climate Change Research: The Concept of Shared Socioeconomic Pathways. *Clim. Change* **122**:387–400. [6]

Ordinola, M., T. Bernet, and K. Manrique. 2007. T'ikapapa: Linking Urban Consumers and Small-Scale Andean Producers with Potato Diversity. Lima: International Potato Center. [2, 5]

Orindi, V. A., and A. Ochieng. 2005. Case Study 5: Kenya Seed Fairs as a Drought Recovery Strategy in Kenya. *IDS Bull.* **36**:87–102. [4]

Orozco-Ramirez, Q., H. R. Perales, and R. J. Hijmans. 2017. Geographical Distribution and Diversity of Maize (*Zea Mays* L. Subsp. Mays) Races in Mexico. *Genet. Resour. Crop Evol.* **64**:855–865. [2, 5]

Orozco-Ramírez, Q., J. Ross-Ibarra, A. Santacruz-Varela, and S. B. Brush. 2016. Maize Diversity Associated with Social Origin and Environmental Variation in Southern Mexico. *Heredity* **116**:477–484. [8, 13]

Ortiz, R. 2011. Agrobiodiversity Management for Climate Change. In: Agrobiodiversity Management for Food Security: A Critical Review, ed. J. Lenné and D. Wood, pp. 189–211. Wallingford: CAB International. [4]

Ostrom, E. 1990. Governing the Commons: The Evolution of Institutions for Collective Action. Cambridge: Cambridge University Press. [11]

Otero, G. 2012. The Neoliberal Food Regime in Latin America: State, Agribusiness Transnational Corporations and Biotechnology. *Rev. Can. Etudes Dev.* **33**:282–294. [11]

Oumar, I., C. Mariac, J.-L. Pham, and Y. Vigouroux. 2008. Phylogeny and Origin of Pearl Millet (*Pennisetum Glaucum* [L.] R. Br) as Revealed by Microsatellite Loci. *Theor. Appl. Genet.* **117**:489–497. [3]

Overmars, K. P., C. J. Schulp, R. Alkemade, et al. 2014. Developing a Methodology for a Species-Based and Spatially Explicit Indicator for Biodiversity on Agricultural Land in the EU. *Ecol. Indic.* **37**:186–198. [8]

Ozkan, H., A. Brandolini, R. Schäfer-Pregl, and F. Salamini. 2002. Aflp Analysis of a Collection of Tetraploid Wheats Indicates the Origin of Emmer and Hard Wheat Domestication in Southeast Turkey. *Mol. Biol. Evol.* **19**:1797–1801. [3]

Pacicco, L., M. Bodesmo, R. Torricelli, and V. Negri. 2018. A Methodological Approach to Identify Agro-Biodiversity Hotspots for Priority in Situ Conservation of Plant Genetic Resources. *PLoS One* **13**:e0197709. [5]

Padmanabhan, M. A. 2007. The Making and Unmaking of Gendered Crops in Northern Ghana. *Singap. J. Trop. Geogr.* **28**:57–70. [14]

Padulosi, S., N. Bergamini, and T. Lawrence, eds. 2011a. On-Farm Conservation of Neglected and Underutilized Species: Status, Trends and Novel Approaches to Cope with Climate Change. Rome: Bioversity International. [15]

Padulosi, S., V. Heywood, D. Hunter, and A. Jarvis. 2011b. Underutilized Species and Climate Change: Current Status and Outlook. In: Crop Adaptation to Climate Change, ed. S. S. Yadav et al., pp. 507–521. New York: Wiley. [7]

Paine, R. T. 1995. A Conversation on Refining the Concept of Keystone Species. *Conserv. Biol.* **9**:962–964. [9]

Pallante, G., A. G. Drucker, and S. Sthapit. 2016. Assessing the Potential for Niche Market Development to Contribute to Farmers' Livelihoods and Agrobiodiversity Conservation: Insights from the Finger Millet Case Study in Nepal. *Ecol. Econ.* **130**:92–105. [4]

Parra, F., A. Casas, J. M. Peñaloza-Ramírez, et al. 2010. Evolution under Domestication: Ongoing Artificial Selection and Divergence of Wild and Managed *Stenocereus Pruinosus* (Cactaceae) Populations in the Tehuacán Valley, Mexico. *Ann. Bot.* **106**:483–496. [3]

Parra-Quijano, M., J. M. Iriondo, and E. Torres. 2011. Improving Representativeness of Genebank Collections through Species Distribution Models, Gap Analysis and Ecogeographical Maps. *Biodivers. Conserv.* **21**:79–96. [5]

Parsa, S. 2010. Native Herbivore Becomes Key Pest after Dismantlement of a Traditional Farming System. *Am. Entomol.* **56**:242–251. [2]

Parsa, S., R. Ccanto, and J. A. Rosenheim. 2011. Resource Concentration Dilutes a Key Pest in Indigenous Potato Agriculture. *Ecol. Appl.* **21**:539–546. [2, 4]

Pascual, U., and C. Perrings. 2007. Developing Incentives and Economic Mechanisms for *in Situ* Biodiversity Conservation in Agricultural Landscapes. *Agricult. Ecosyst. Environ.* **121**:256–268. [15]

Pati, R. N., S. Shukla, and L. Chanza, eds. 2014. Traditional Environmental Knowledge and Biodiversity. New Delhi: Sarup Book Publishers [12]

Pautasso, M., G. Aistara, A. Barnaud, et al. 2013. Seed Exchange Networks for Agrobiodiversity Conservation: A Review. *Agronom. Sustain. Devel.* **33**:151–175. [8, 13, 14]

Peiffer, J. A., A. Spor, O. Koren, et al. 2013. Diversity and Heritability of the Maize Rhizosphere Microbiome under Field Conditions. *PNAS* **110**:6548–6553. [4]

Perales, H. R., B. Benz, and S. B. Brush. 2005. Maize Diversity and Ethnolinguistic Diversity in Chiapas, Mexico. *PNAS* **102**:949–954. [2, 13]

Perales, H. R., and D. Golicher. 2014. Mapping the Diversity of Maize Races in Mexico. *PLoS One* **9**:e114657. [2, 5, 9, 13, 14]

Perales, H. R., and C. J. M. Hernández. 2005. Diversidad del Maíz en Chiapas. In: Diversidad Biológica en Chiapas, ed. M. González-Espinosa et al., pp. 419–440. Mexico: Plaza y Valdez y Ecosur. [12]

Pereira, H. M., and H. D. Cooper. 2006. Towards the Global Monitoring of Biodiversity Change. *Trends Ecol. Evol.* **21**:123–129. [8]

Perkins, J. M., S. V. Subramanian, and N. A. Christakis. 2015. Social Networks and Health: A Systematic Review of Sociocentric Network Studies in Low- and Middle-Income Countries. *Soc. Sci. Med.* **125**:60–78. [11]

Perreault, T. 2005. Why Chacras (Swidden Gardens) Persist: Agrobiodiversity, Food
Security, and Cultural Identity in the Ecuadorian Amazon. *Hum. Organ.* **64**:327–
339. [13]

Perrier, X., E. De Langhe, M. Donohue, et al. 2011. Multidisciplinary Perspectives on
Banana (*Musa* Spp.) Domestication. *PNAS* **108**:11311–11318. [3]

Perrings, C., S. Baumgärtner, W. A. Brock, et al. 2009. The Economics of Biodiversity and Ecosystem Services. In: Biodiversity, Ecosystem Functioning, and Human
Wellbeing: An Ecological and Economic Perspective, ed. S. Naeem et al., pp. 230–
247. Oxford: Oxford Univ. Press. [15]

Perrings, C., L. Jackson, K. Bawa, et al. 2006. Biodiversity in Agricultural Landscapes:
Saving Natural Capital without Losing Interest. *Conserv. Biol.* **20**:263–264. [1]

Peschard, K. 2014. Farmers' Rights and Food Sovereignty: Critical Insights from India.
J. Peasant Stud. **41**:1085–1108. [14]

Pestalozzi, H. 2000. Sectoral Fallow Systems and the Management of Soil Fertility:
The Rationality of Indigenous Knowledge in the High Andes of Bolivia. *Mt. Res.
Dev.* **20**:64–71. [2, 4]

Pettis, J. S., E. M. Lichtenberg, M. Andree, et al. 2013. Crop Pollination Exposes Honey
Bees to Pesticides Which Alters Their Susceptibility to the Gut Pathogen Nosema
Ceranae. *PLoS One* **8**:e70182. [4]

Phalan, B., M. Onia, A. Balmford, and R. E. Green. 2011. Reconciling Food Production
and Biodiversity Conservation: Land Sharing and Land Sparing Compared. *Science*
333:1289–1291. [9]

Phillips, S. L., and M. S. Wolfe. 2005. Evolutionary Plant Breeding for Low Input Systems. *J. Agric. Sci.* **143**:245–254. [4]

Pierce, J., D. G. Martin , and J. T. Murphy. 2011. Relational Place-Making: The Networked Politics of Place. *Trans. Inst. Br. Geogr.* **36**:54–70. [13]

Pieroni, A., S. Nebel, R. F. Santoro, and M. Heinrich. 2005. Food for Two Seasons: Culinary Uses of Non-Cultivated Local Vegetables and Mushrooms in a South Italian
Village. *Int. J. Food Sci. Nutr.* **56**:245–272. [11]

Pierotti, R. 2011. Indigenous Knowledge, Ecology, and Evolutionary Biology. New
York: Routledge. [12]

Pimentel, D., M. McNair, L. Buck, M. Pimentel, and J. Kamil. 1997. The Value of Forests to World Food Security. *Hum. Ecol.* **25**:91–120. [9]

Pingali, P. 2015. Agricultural Policy and Nutrition Outcomes: Getting Beyond the Pre-
occupation with Staple Grains. *Food Secur.* **7**:583–591. [7]

Pinna, S. 2017. Sowing Landscapes: Social and Ecological Aspects of Food Production
in Peri-Urban Spatial Planning Initiatives-a Study from the Madrid Area. *Future
Food* **5**:34–45. [8]

Pinstrup-Andersen, P. 2009. Food Security: Definition and Measurement. *Food Secur.*
1:5–7. [11]

Pionetti, C. 2006. Seed Diversity in the Drylands: Women and Farming in South India.
Gatekeeper Series 126. Stockholm: International Institute for Environment and Development. [13]

Pirro, C., and I. Anguelovski. 2017. Farming the Urban Fringes of Barcelona: Competing Visions of Nature and the Contestation of a Partial Sustainability Fix. *Geoforum*
82:53–65. [8]

Pistorius, R. 1997. Scientists, Plants and Politics: A History of the Plant Genetic Resources Movement. Rome: Intl. Plant Genetic Resources Institute (IPGRI). [5]

Plenderleith, K. 1999. Traditional Agriculture and Soil Management. In: Cultural and Spiritual Values of Biodiversity, ed. D. A. Posey, pp. 285–324. London: UNEP-ITP. [12]

Pollan, M. 2006. The Omnivore's Dilemma: A Natural History of Four Meals. London: Penguin. [6]

Polreich, S., S. de Haan, H. Juárez, et al. 2014. An Interdisciplinary Monitoring Network of Diversity Hotspots of Long-Term *in Situ* Conservation of Potato Landraces: Bridging the Gap between Increasing Knowledge and Decreasing Resources. In: Proc. Tropentag 2014: Intl. Research on Food Security, Natural Resource Management and Rural Development. Lima: Intl. Potato Center (CIP). [5]

Poot-Pool, W. S., H. van der Wal, S. Flores–Guido, J. M. Pat-Fernández, and L. Esparza–Olguín. 2015. Home Garden Agrobiodiversity Differentiates Along a Rural: Periurban Gradient in Campeche, México. *Econ. Bot.* **69**:203–217. [8]

Popkin, B. 2004. The Nutrition Transition: An Overview of World Patterns of Change. *Nutr. Rev.* **62**:S140 – S143. [9]

Porter, J. R., L. Xie, A. J. Challinor, et al. 2014. Food Security and Food Production Systems. In: Climate Change 2014: Impacts, Adaptation and Vulnerability, ed. Working Group II Contribution to the IPCC 5th Assessment Report, pp. 485–533. Cambridge: Cambridge Univ. Press. [6, 7]

Posey, D. A., ed. 1999a. Cultural and Spiritual Values of Biodiversity. Nairobi: United Nations Environment Programme (UNEP). [9]

———. 1999b. Introduction: Culture and Nature: The Inextricable Link. In: Cultural and Spiritual Values of Biodiversity, ed. D. A. Posey, pp. 3–18. London: UNEP-ITP. [12]

Posey, D. A., J. Frechione, J. Eddins, et al. 1984. Ethnoecology as Applied Anthropology in Amazonian Development. *Hum. Org.* **43**:95–107. [12]

Poudel, R., A. Jumpponen, D. C. Schlatter, et al. 2016. Analytical and Theoretical Plant Pathology Microbiome Networks: A Systems Framework for Identifying Candidate Microbial Assemblages for Disease Management. Analytic. *Phytopathology* **106**:1083–1096. [13]

Powell, B. 2012. Biodiversity and Human Nutrition in a Landscape Mosaic of Farms and Forests in the East Usambara Mountains, Tanzania. Ph.D., McGill Univ., Montreal. [9]

Powell, B., A. Ickowitz, S. McMullin, et al. 2013. The Role of Forests, Trees and Wild Biodiversity for Improved Nutrition-Sensitivity of Food and Agriculture Systems. Expert Background Paper for the International Conference on Nutrition 2. Rome: FAO and the World Health Organization. [9]

Powell, B., S. H. Thilsted, A. Ickowitz, et al. 2015. Improving Diets with Wild and Cultivated Biodiversity from across the Landscape. *Food Secur.* **7**:535–554. [9–11]

Power, A. G. 2010. Ecosystem Services and Agriculture: Tradeoffs and Synergies. *Philos. Trans. R. Soc. Lond. B Biol. Sci.* **365**:2959–2971. [4]

Prain, G., and M. Dubbeling. 2011. Urban Agriculture: A Sustainable Solution to Alleviating Urban Poverty, Addressing the Food Crisis, and Adapting to Climate Change: Case Studies of the Cities of Accra, Nairobi, Lima, and Bangalore. Leusden: RUAF Foundation. [9]

Prain, G., N. Karanja, and D. Lee-Smith, eds. 2010. African Urban Harvests: Agriculture in the Cities of Cameroon, Kenya and Uganda. New York: Springer. [8]

Prugnolle, F., A. Manica, and F. Balloux. 2005. Geography Predicts Neutral Genetic Diversity of Human Populations. *Curr. Biol.* **15**:R159–160. [3]

Purugganan, M. D., and D. Q. Fuller. 2009. The Nature of Selection during Plant Domestication. *Nature* **457**:843–848. [2]

Putt, E. D. 1997. Early History of Sunflower. In: Sunflower Technology and Production, ed. A. A. Schneiter, vol. 35, pp. 1–20. Agronomy. Madison: American Society of Agronomy, Inc., Crop Science Society of America, Inc., Soil Science Society of America, Inc. [9]

Quiggin, J. C., and J. K. Horowitz. 2003. Costs of Adjustment to Climate Change. *Aust. J. Agric. Resour. Econ.* **47**:429–446. [7]

Rabbi, I. Y., P. A. Kulakow, J. A. Manu-Aduening, et al. 2015. Tracking Crop Varieties Using Genotypingby-Sequencing Markers: A Case Study Using Cassava (*Manihot Esculenta* Crantz). *BMC Genet.* **16**:115. [5]

Rabitz, F. 2017. Access without Benefit-Sharing: Design, Effectiveness and Reform of the FAO Seed Treaty. *Int. J. Commons* **11**:621–640. [13]

Radel, C., B. Schmook, J. McEvoy, C. Méndez, and P. Petrzelka. 2012. Labour Migration and Gendered Agricultural Relations: The Feminization of Agriculture in the Ejidal Sector of Calakmul, Mexico. *J. Agrar. Change* **7**:98–119. [6]

Ramanna, A. 2003. India's Plant Variety and Farmers' Rights Legislation: Potential Impact on Stakeholder Access to Genetic Resources. Eptd Discussion Paper No. 96. Washington, D.C.: Environment and Production Technology Division (EPTD). [12]

Ramirez, A. S., L. K. Diaz Rios, Z. Valdez, E. Estrada, and A. Ruiz. 2017. Bringing Produce to the People: Implementing a Social Marketing Food Access Intervention in Rural Food Deserts. *J. Nutr. Educ. Behav.* **49**:166–174. [9]

Ramírez-Villegas, J., C. K. Khoury, A. Jarvis, D. G. Debouck, and L. Guarino. 2010. A Gap Analysis Methodology for Collecting Crop Genepools: A Case Study with Phaseolus Beans. *PLoS One* **5**:e13497. [5]

Raneri, J. E., and G. Kennedy. 2017. Agricultural Biodiversity for Healthy Diets and Food Systems. In: Routledge Handbook of Agricultural Biodiversity, ed. D. Hunter et al. New York: Routledge. [10]

Rangan, H., E. A. Alpers, T. Denham, C. A. Kull, and J. Carney. 2015. Food Traditions and Landscape Histories of the Indian Ocean World: Theoretical and Methodological Reflections. *Environ. Hist. Camb.* **21**:135–157. [8]

Rao, E. J., and M. Qaim. 2011. Supermarkets, Farm Household Income, and Poverty: Insights from Kenya. *World Dev.* **39**:784–796. [6]

Ratnadass, A., P. Fernandes, J. Avelino, and R. Habib. 2012. Plant Species Diversity for Sustainable Management of Crop Pests and Diseases in Agroecosystems: A Review. *Agronom. Sustain. Devel.* **32**:273–303. [4]

Ravera, F., U. Pascual, A. Drucker, et al. 2019. Gendered agrobiodiversity management and everyday adaptation practices in two marginal rural areas of India. *Agricul. Human Values*, in press. [11]

Reardon, T., K. Chen, B. Minten, and L. Adriano. 2012. The Quiet Revolution in Staple Food Value Chains: Enter the Dragon, the Elephant and the Tiger. Mandaluyong City: Asian Development Bank and International Food Policy Research Institute. [14]

Rebelo, A. G. 1994. Iterative Selection Procedures: Centres of Endemism and Optimal Placement of Reserves. *Strelitzia* **1**:231–257. [5]

Reenberg, A., I. Maman, and P. Oksen. 2013. Twenty Years of Land Use and Livelihood Changes in Se-Niger: Obsolete and Short-Sighted Adaptation to Climatic and Demographic Pressures? *J. Arid Environ.* **94**:47–58. [8]

Reichman, J. H., T. Dedeurwaerdere, and P. F. Uhlir. 2016. Governing Digitally Integrated Genetic Resources, Data, and Literature: Global Intellectual Property Strategies for a Redesigned Microbial Research Commons. New York: Cambridge Univ. Press. [15]

Renting, H., and H. Wiskerke. 2010. New Emerging Roles for Public Institutions and Civil Society in the Promotion of Sustainable Local Agro-Food Systems. In: Proc. 9th European IFSA Symposium: Transitions Towards Sustainable Agriculture from Farmers to Agro-Food Systems, ed. I. Darnhofer and M. Grötzer, pp. 1902–1912. Vienna: Vienna. [11]

Reyes-García, V., T. Huanca, V. Vadez, W. Leonard, and D. Wilkie. 2006. Cultural, Practical, and Economic Value of Wild Plants: A Quantitative Study in the Bolivian Amazon. *Econ. Bot.* **60**:62–74. [1]

Reyes-García V., J. Luis Molina, L. Calvet-Mir, et al. 2013. *Tertius Gaudens*: Germplasm Exchange Networks and Agroecological Knowledge among Home Gardeners in the Iberian Peninsula. *J. Ethnobiol. Ethnomed.* **9**:53. [11]

Reyes-García, V., G. Menendez-Baceta, L. Aceituno-Mata, et al. 2015. From Famine Foods to Delicatessen: Interpreting Trends in the Consumption and Gathering of Wild Edible Plants through Their Connection to Cultural Ecosystem Services. *Ecol. Econ.* **12**:303–311. [8, 11]

Ribot, J. C., and N. L. Peluso. 2003. A Theory of Access. *Rural Sociol.* **68**:153–181. [8]

Ricciardi, V. 2015. Social Seed Networks: Identifying Central Farmers for Equitable Seed Access. *Agric. Syst.* **139**:110–121. [11]

Richards, M. P., R. J. Schulting, and R. E. Hedges. 2003. Archaeology: Sharp Shift in Diet at Onset of Neolithic. *Nature* **425**:366–366. [2]

Richards, P., and G. Ruivenkamp. 1997. Seeds and Survival: Crop Genetic Resources in War and Reconstruction in Africa. Rome: International Plant Genetic Resources Institute (IPGRI). [2, 8]

Ríos, D., M. Ghislain, F. Rodríguez, and D. M. Spooner. 2007. What Is the Origin of the European Potato? Evidence from Canary Island Landraces. *Crop Sci.* **47**:1271–1280. [3, 5]

Roa, C., R. S. Hamilton, P. Wenzl, and W. Powell. 2016. Plant Genetic Resources: Needs, Rights, and Opportunities. *Trends Plant Sci.* **21**:633–636. [13]

Rockström, J., W. Steffen, K. Noone, et al. 2009. A Safe Operating Space for Humanity. *Nature* **461**:472–475. [9]

Rodríguez, P., C. Sanjuanelo, L. Ñústez, C. Eduardo, and L. P. Moreno-Fonseca. 2016. Growth and Phenology of Three Andean Potato Varieties (*Solanum tuberosum* L.) under Water Stress. *Agronom. Colomb.* **34**:141–154. [9]

Rogelj, J., M. den Elzen, N. Höhne, et al. 2016. Paris Agreement Climate Proposals Need a Boost to Keep Warming Well Below 2°C. *Nature* **534**:631–639. [6]

Rook, A. J., B. Dumont, J. Isselstein, et al. 2004. Matching Type of Livestock to Desired Biodiversity Outcomes in Pastures–a Review. *Biol. Conserv.* **119**:137–150. [2]

Rosendal, K. 2000. The Convention on Biological Diversity and Developing Countries. Dordrecht: Kluwer Academic Publishers. [14]

Rostami, R., A. Koocheki, P. R. Moghaddam, and M. N. Mahallati. 2016. Effect of Landscape Structure on Agrobiodiversity in Western Iran (Gilan-E Gharb). *Agroecol. Sust. Food* **40**:660–692. [5]

Roullier, C., L. Benoit, D. McKey, and V. Lebot. 2013a. Historical Collections Reveal Patterns of Diffusion of Sweet Potato in Oceania Obscured by Modern Plant Movements and Recombination. *PNAS* **110**:2205–2210. [3, 8]

Roullier, C., R. Kambouo, J. Paofa, D. McKey, and V. Lebot. 2013b. On the Origin of Sweet Potato (*Ipomoea Batatas* (L.) Lam) Genetic Diversity in New Guinea, a Secondary Centre of Diversity. *Heredity* **110**:594–604. [2]

Rowland, D., A. Ickowitz, B. Powell, R. Nasi, and T. Sunderland. 2016. Forest Foods and Healthy Diets: Quantifying the Contributions. *Environ. Conserv.* **44**:1–13. [9]

Roy, S., B. C. Marndi, B. Mawkhlieng, et al. 2016. Genetic Diversity and Structure in Hill Rice (*Oryza Sativa* L.) Landraces from the North-Eastern Himalayas of India. *BMC Genet.* **17**:107. [13]

Ruddiman, W. F. 2013. The Anthropocene. *Annu. Rev. Earth Planet. Sci.* **41**:45–68. [7]

Ruel, M. T. 2003. Operationalizing Dietary Diversity: A Review of Measurement Issues and Research Priorities. *J. Nutr.* **133**:3911S–3926S. [11]

Ruiz, M. 2011. Seeking Benefit Sharing through a Defensive Approach: The Experience of the National Commission for the Prevention of Biopiracy. In: The Custodians of Biodiversity: Sharing Access to and Benefits of Genetic Resources, ed. M. Ruiz and R. Vernooy, pp. 43–52. London: Routledge. [12]

Salas, M. 2005. Seeds-Songs: Reflection of Swidden Agriculture, Agro-Biodiversity and Food Sovereignty. *Indigen. Aff.* **2**:14–21. [12]

Salazar, R., N. P. Louwaars, and B. Visser. 2007. Protecting Farmers' New Varieties: New Approaches to Rights on Collective Innovations in Plant Genetic Resources. *World Dev.* **35**:1515–1528. [13, 14]

Salick, J. 2012. Indigenous Peoples Conserving, Managing, and Creating Biodiversity. Biodiversity in Agriculture: Domestication, Evolution and Sustainability, P. Gepts et al., series ed. Cambridge: Cambridge Univ. Press. [3]

Salter, B., and C. Salter. 2017. Controlling New Knowledge: Genomic Science, Governance and the Politics of Bioinformatics. *Soc. Stud. Sci.* **47**:263–287. [14]

Sangabriel-Conde, W., S. Negrete-Yankelevich, I. Eduardo Maldonado-Mendoza, and D. Trejo-Aguilar. 2014. Native Maize Landraces from Los Tuxtlas, Mexico Show Varying Mycorrhizal Dependency for P Uptake. *Biol. Fertil. Soils* **50**:405–414. [4]

Santilli, J. 2012. Agrobiodiversity and the Law: Regulating Genetic Resources, Food Security and Cultural Diversity. New York: Routledge. [13]

Sardinas, H. S., and C. Kremen. 2015. Pollination Services from Field-Scale Agricultural Diversification May Be Context-Dependent. *Agricult. Ecosyst. Environ.* **207**:17–25. [4]

Sarikamis, G., J. Marquez, R. MacCormack, et al. 2006. High Glucosinolate Broccoli: A Delivery System for Sulforaphane. *Mol. Breeding* **18**:219–228. [2]

Särkinen, T., P. Gonzáles, and S. Knapp. 2013. Distribution Models and Species Discovery: The Story of a New *Solanum* Species from the Peruvian Andes. *PhytoKeys* **20**:1–20. [5]

Sawyer, S., and E. Terence-Gomez. 2013. On Indigenous Identity and a Language of Rights. In: The Politics of Resource Extraction: Indigenous Peoples, Multinational Corporations and the State, ed. S. Sawyer and E. Terence-Gomez, pp. 9–32. Hampshire: Palgrave Macmillan. [12]

Scarcelli, N., S. Tostain, Y. Vigouroux, et al. 2006. Farmers' Use of Wild Relative and Sexual Reproduction in a Vegetatively Propagated Crop: The Case of Yam in Benin. *Mol. Ecol.* **15**:2421–2431. [3]

Schaafsma, A., V. Limay-Rios, Y. Xue, J. Smith, and T. Baute. 2016. Field-Scale Examination of Neonicotinoid Insecticide Persistence in Soil as a Result of Seed Treatment Use in Commercial Maize (Corn) Fields in Southwestern Ontario. *Environ. Toxicol. Chem.* **35**:295–302. [4]

Schiffels, S., and R. Durbin. 2014. Inferring Human Population Size and Separation History from Multiple Genome Sequences. *Nat. Genet.* **46**:919–925. [3]

Schiller, J. M., M. B. Chanpengxay, B. Linguist, and S. Appa Rao, eds. 2006. Rice in Laos. Los Baños: International Rice Research Institute (IIRI), Australian Centre for International Agricultural Research (ACIAR). [2]

Schlegel, S. A., and H. A. Guthrie. 1973. Diet and the Tiruray Shift from Swidden to Plow Farming. *Ecol. Food Nutr.* **2**:181–191. [9]

Schmidhuber, J., and F. N. Tubiello. 2007. Global Food Security under Climate Change. *PNAS* **104**:19703–19708. [11]

Schmidt, J. E., T. M. Bowles, and A. C. M. Gaudin. 2016. Using Ancient Traits to Convert Soil Health into Crop Yield: Impact of Selection on Maize Root and Rhizosphere Function. *Front. Plant Sci.* **7**:373. [4]

Schroth, G., U. Krauss, L. Gasparotto, J. A. D. Aguilar, and K. Vohland. 2000. Pests and Diseases in Agroforestry Systems of the Humid Tropics. *Agroforest. Syst.* **50**:199–241. [4]

Schwartz, S. J., J. B. Unger, B. L. Zamboanga, and J. Szapocznik. 2010. Rethinking the Concept of Acculturation Implications for Theory and Research. *Am. Psychol.* **65**:237–251. [11]

Scott, M. P., J. W. Edwards, C. P. Bell, J. R. Schussler, and J. S. Smith. 2006. Grain Composition and Amino Acid Content in Maize Cultivars Representing 80 Years of Commercial Maize Varieties. *Maydica* **51**:417–423. [2]

Scrimshaw, N. S., and J. P. San Giovanni. 1997. Synergism of Nutrition, Infection, and Immunity: An Overview. *Am. J. Clin. Nutr.* **66**:464S–477S. [10]

Scurrah, M., C. Celis-Gamboa, S. Chumbiauca, A. Salas, and R. G. Visser. 2008. Hybridization between Wild and Cultivated Potato Species in the Peruvian Andes and Biosafety Implications for Deployment of Gm Potatoes. *Euphytica* **164**:881–892. [3, 4]

Scurrah, M., S. de Haan, and T. Winge. 2013. Cataloguing Potato Varieties and Traditional Knowledge from the Andean Highlands of Huancavelica, Peru. In: Realizing Farmers' Rights to Crop Genetic Resources: Success Stories and Best Practices, ed. R. Andersen and T. Winge, pp. 65–79. London: Routledge Press. [2, 3, 14]

Scurrah, M., E. Fernandez-Baca, R. Ccanto, et al. 1999. Learning About Biodiversity in Peru. *ILEIA News.* **15**:26–28. [2]

Seimon, T. A., A. Seimon, P. Daszak, et al. 2007. Upward Range Extension of Andean Anurans and Chytridiomycosis to Extreme Elevations in Response to Tropical Deglaciation. *Glob. Chang. Biol.* **13**:288–299. [5]

Seo, S. N. 2012. Decision Making under Climate Risks: An Analysis of Sub-Saharan Farmers' Adaptation Behaviors. *Weather Clim. Soc.* **4**:285–299. [7]

Serra, P., D. Saurí, and L. Salvati. 2017. Peri-Urban Agriculture in Barcelona: Outlining Landscape Dynamics *Vis À Vis* Socio-Environmental Functions. *Landscape Res.* **43**:613–131. [8]

Serrasolses, G., L. Calvet-Mir, E. Carrió, et al. 2016. A Matter of Taste: Local Explanations for the Consumption of Wild Food Plants in the Catalan Pyrenees and the Balearic Islands. *Econ. Bot.* **70**:176–189. [9]

Sessitsch, A., and B. Mitter. 2015. 21st Century Agriculture: Integration of Plant Microbiomes for Improved Crop Production and Food Security. *Microb. Biotechnol.* **8**:32–33. [8]

Severson, K. 2016. A Rush of Americans, Seeking Gold in Cuban Soil. *New York Times* June 20, 2016. [8]

Shaver, I., A. Chain-Guadarrama, K. A. Cleary, et al. 2015. Coupled Social and Ecological Outcomes of Agricultural Intensification in Costa Rica and the Future of Biodiversity Conservation in Tropical Agricultural Regions. *Glob. Environ. Change* **32**:74–86. [8]

Shea, K. M. 2003. Antibiotic Resistance: What Is the Impact of Agricultural Uses of Antibiotics on Children's Health? *Pediatrics* **112**:253–258. [11]

Shen, S., G. Xu, D. Li, et al. 2017. Agrobiodiversity and in Situ Conservation in Ethnic Minority Communities of Xishuangbanna in Yunnan Province, Southwest China. *J. Ethnobiol. Ethnomed.* **13**:28. [13]

Shepherd, C. J. 2010. Mobilizing Local Knowledge and Asserting Culture: The Cultural Politics of *in Situ* Conservation of Agricultural Biodiversity. *Curr. Anthropol.* **51**:629–654. [12]

―――. 2017. Andean Cultural Affirmation and Cultural Integration in Context: Reflections on Indigenous Knowledge for the in Situ Conservation of Agrobiodiversity. In: Indigenous Knowledge: Enhancing Its Contribution to Natural Resources Management, ed. P. Sillitoe, pp. 130–146. Wallingford: CABI Publishing. [5]

Sherwood, S., A. Deaconu, and M. Paredes. 2017. 250 Thousand Families Campaign: The Existence of Flavor and Taste In: Food, Agriculture and Social Change: The Everyday Vitality of Latin America, ed. S. Sherwood et al. London: Routledge. [9]

Sherwood, S., S. van Bommel, and M. Paredes. 2016. Self-Organization and the Bypass: Re-Imagining Institutions for More Sustainable Development in Agriculture and Food. *Agriculture* **6**:1–19. [14]

Shewayrga, H., D. R. Jordan, and I. D. Godwin. 2008. Genetic Erosion and Changes in Distribution of Sorghum Landraces in North-Eastern Ethiopia. *Plant Genet. Resour.* **6**:1–10. [9]

Shillington, L. 2008. Being(S) in Relation at Home: Socio-Natures of Patio Gardens in Managua, Nicaragua. *Soc. Cult. Geogr.* **9**:755–776. [11]

Shoffner, A. V., and J. F. Tooker. 2013. The Potential of Genotypically Diverse Cultivar Mixtures to Moderate Aphid Populations in Wheat (*Triticum Aestivum* L.). *Arthropod Plant Interact.* **7**:33–43. [4]

Sibhatu, K. T., V. V. Krisha, and M. Qaim. 2015. Production Diversity and Dietary Diversity in Smallholder Farm Households. *PNAS* **112**:10657–10662. [6]

Sietz, D., J. C. Ordoñez, M. T. J. Kok, et al. 2017. Nested Archetypes of Vulnerability in African Drylands: Where Lies Potential for Sustainable Agricultural Intensification? *Environ. Res. Lett.* **12**:095006. [8]

Sileshi, G. W., P. Nyeko, P. O. Y. Nkunika, et al. 2009. Integrating Ethno-Ecological and Scientific Knowledge of Termites for Sustainable Termite Management and Human Welfare in Africa. *Ecol. Soc.* **14**:48. [2, 4]

Sissoko, S., S. Doumbia, M. Vaksmann, et al. 2008. Accounting for Farmer Knowledge in Varietal Choice in a Plant Breeding Programme. *Cah. Agric.* **17**:128–133. [2]

Skarbø, K. 2014. The Cooked Is the Kept: Factors Shaping the Maintenance of Agro-Biodiversity in the Andes. *Hum. Ecol.* **42**:711–726. [14]

―――. 2015. From Lost Crop to Lucrative Commodity: Conservation Implications of the Quinoa Renaissance. *Hum. Organ.* **74**:86–99. [8]

Skarbø, K., and K. VanderMolen. 2015. Maize Migration: Key Crop Expands to Higher Altitudes under Climate Change in the Andes. *Clim. Dev.* **8**:245–255. [5]

Smale, M., M. R. Bellon, D. I. Jarvis, and B. R. Sthapit. 2004. Economic Concepts for Designing Policies to Conserve Crop Genetic Resources on Farms. *Genet. Resour. Crop Evol.* **51**:121–135. [15]

Smale, M., L. Diakité, and N. Keita. 2012. Millet transactions in market fairs, millet diversity and farmer welfare in Mali. Environ. Devel. Econ. 17: 523–546. [15]

Smith, K. P., and N. A. Christakis. 2008. Social Networks and Health. *Annu. Rev. Sociol.* **34**:405–429. [11]

Snapp, S. S., M. J. Blackie, R. A. Gilbert, R. Bezner-Kerr, and G. Y. Kanyama-Phiri. 2010. Biodiversity Can Support a Greener Revolution in Africa. *PNAS* **107**:20840–20845. [2]

Sonnenburg, J. L., and F. Backhed. 2016. Diet-Microbiota Interactions as Moderators of Human Metabolism. *Nature* **535**:56–64. [9]

Sperling, L. 2001. The Effect of the Civil War on Rwanda's Bean Seed Systems and Unusual Bean Diversity. *Biodivers. Conserv.* **10**:989–1009. [2, 3]

Sperling, L., and S. J. McGuire. 2010. Persistent Myths About Emergency Seed Aid. *Food Pol.* **35**:195–201. [8]

Spooner, D. M., T. Gavrilenko, S. H. Jansky, et al. 2010. Ecogeography of Ploidy Variation in Cultivated Potato (*Solanum* Sect. *Petota*). *Am. J. Bot.* **97**:2049–2060. [5]

Spooner, D. M., K. McLean, G. Ramsay, R. Waugh, and G. J. Bryan. 2005. A Single Domestication for Potato Based on Multilocus Amplified Fragment Length Polymorphism Genotyping. *PNAS* **102**:14694–14699. [3]

Stahl, P. W. 2015. Interpreting Interfluvial Landscape Transformations in the Pre-Columbian Amazon. *Holocene* **25**:1598–1603. [4]

Stanner, W. 1969. The Kitui Kamba Market, 1938–39. *Ethnology* **8**:125–138. [9]

Steward, A. M., and D. d. M. Lima. 2017. We Also Preserve: Quilombola Defense of Traditional Plant Management Practices against Preservationist Bias in Mumbuca, Minas Gerais, Brazil. *J. Ethnobiol.* **37**:141–165. [6]

Sthapit, S., G. Meldrum, S. Padulosi, and N. Bergamini, eds. 2015. Strengthening the Role of Custodian Farmers in the National Conservation Programme of Nepal. Rome: LI-BIRD, Bioversity International. [2]

Stone, G. D. 2007. Agricultural Deskilling and the Spread of Genetically Modified Cotton in Warangal. *Curr. Anthropol.* **48**:67–103. [2]

———. 2016. Towards a General Theory of Agricultural Knowledge Production: Environmental, Social and Didactic Learning. *Cult. Agricult. Food Environ.* **38**:5–17. [6]

Stone, G. D., A. Flachs, and C. Diepenbrock. 2013. Rythms of the Herd: Long Term Dynamics in Seed Choice by Indian Farmers. *Technol. Soc.* **36**:26–38. [7]

Stone, G. D., and D. Glover. 2017. Disembedding Grain: Golden Rice, the Green Revolution, and Heirloom Seeds in the Philippines. *Agricult. Human Values* **34**:87–102. [13]

Strange, S. 1988. States and Markets. London: Pinter. [14]

Strathern, M. 1996. Cutting the Network. *J. R. Anthropol. Inst.* **2**:517–535. [13]

Struik, P. C., T. W. Kuyper, L. Brussaard, and C. Leeuwis. 2014. Deconstructing and Unpacking Scientific Controversies in Intensification and Sustainability: Why the Tensions in Concepts and Values? *Curr. Opin. Environ. Sustain.* **8**:80–88. [7]

Sturz, A. V., B. G. Matheson, W. Arsenault, J. Kimpinski, and B. R. Christie. 2001. Weeds as a Source of Plant Growth Promoting Rhizobacteria in Agricultural Soils. *Can. J. Microbiol.* **47**:1013–1024. [4]

Sullivan, S. N. 2004. Plant Genetic Resources and the Law: Past, Present, and Future Plant Physiology. *Plant Physiol.* **135**:10–15. [15]

Suma, T. R., and K. Großmann. 2017. Exclusions in Inclusive Programs: State-Sponsored Sustainable Development Initiatives Amongst the Kurichya in Kerala, India. *Agric. Hum. Val.* **34**:995–106. [14]

Sunderlin, W. D., J. Hatcher, and M. Liddle. 2008. From Exclusion to Ownership? Challenges and Opportunities in Advancing Forest Tenure Reform. Washington, D.C.: Rights and Resources Initiative. [9]

Suneson, C. A. 1969. Evolutionary Plant Breeding. *Crop Sci.* **9**:119–121. [2]

Surowiecki, J. 2004. The Wisdom of Crowds. New York: Doubleday. [7]

Swaminathan, M. S., ed. 1996. Agrobiodiversity and Farmers' Rights: Proceedings of a Technical Consultation on an Implementation Framework for Farmers' Rights. Madras: Swaminathan Research Foundation. [12]

Swiderska, K. 2006. Banishing the Biopirates: A New Approach to Protecting Traditional Knowledge. Gatekeeper Series 129, International Institute for Environment and Development, Sustainable Agriculture and Rural Livelihoods Programme. London: IIED. [12]

Swiderska, K., A. Argumedo, Y. Song, et al. 2009. Protecting Community Rights over Traditional Knowledge: Implications of Customary Laws and Practices: Key Findings and Recommendations 2005-2009: London: IIED. [12]

Swift, M. J., A. M. N. Izac, and M. van Noordwijk. 2004. Biodiversity and Ecosystem Services in Agricultural Landscapes: Are We Asking the Right Questions? *Agricult. Ecosyst. Environ.* **104**:113–134. [2, 4]

Swindale, A., and P. Bilinsky. 2006. Household Dietary Diversity Score (Hdds) for Measurement of Household Food Access: Indicator Guide. Washington, D.C.: Food and Nutrition Technical Assistance Project (FANTA). [10]

Syfert, M. M. 2016. Crop Wild Relatives of the Brinjal Eggplant (*Solanum Melongena*: Solanaceae): Poorly Represented in Genebanks and Many Species at Risk of Extinction. *Am. J. Bot.* **103**:1–17. [5]

Szoboszlay, M., J. Lambers, J. Chappell, et al. 2015. Comparison of Root System Architecture and Rhizosphere Microbial Communities of Balsas Teosinte and Domesticated Corn Cultivars. *Soil Biol. Biochem.* **80**:34–44. [4]

Tansey, J., and T. O'Riordan. 1999. Cultural Theory and Risk: A Review. *Health Risk Soc.* **1**:71–90. [6]

Tapia, M. E. 2000. Mountain Agrobiodiversity in Peru. *Mt. Res. Dev.* **20**:220–225. [13]

Tapia, M. E., and A. Rosas. 1993. Seed Fairs in the Andes: A Strategy for Local Conservation of Plant Genetic Resources. In: Cultivating Knowledge: Genetic Diversity, Farmer Experimentation and Crop Research, ed. W. de Boef et al., pp. 111–118. London: Intermediate Technology Publications. [2, 5]

Taylor, J. R., and S. T. Lovell. 2014. Urban Home Food Gardens in the Global North: Research Traditions and Future Directions. *Agricult. Human Values* **31**:285–305. [8]

Temudo, M. P. 2011. Planting Knowledge, Harvesting Agro-Biodiversity: A Case Study of Southern Guinea-Bissau Rice Farming. *Hum. Ecol.* **39**:309–321. [8]

ten Kate, K., and S. A. Laird. 1999. The Commercial Use of Biodiversity: Access to Genetic Resources and Benefit-Sharing. London: Earthscan Publications. [12]

Teshome, A., J. K. Torrance, B. Baum, et al. 1999. Traditional Farmers' Knowledge of Sorghum (*Sorghum Bicolor* [Poaceae]) Landrace Storability in Ethiopia. *Econ. Bot.* **53**:69–78. [4]

Thiele, G. 1999. Informal Potato Seed Systems in the Andes: Why Are They Important and What Should We Do with Them? *World Dev.* **21**:83–99. [2]

Thomann, M., E. Imbert, R. C. Engstrand, and P.-O. Cheptou. 2015. Contemporary Evolution of Plant Reproductive Strategies under Global Change Is Revealed by Stored Seeds. *J. Evol. Biol.* **28**:766–778. [6]

Thomas, M., J. C. Dawson, I. Goldringer, and C. Bonneuil. 2011. Seed Exchanges, a Key to Analyze Crop Diversity Dynamics in Farmer-Led on-Farm Conservation. *Genet. Resour. Crop Evol.* **58**:321–338. [13]

Thomas, M., E. Demeulenaere, J. C. Dawson, et al. 2012. On-Farm Dynamic Management of Genetic Diversity: The Impact of Seed Diffusions and Seed Saving Practices on a Population-Variety of Bread Wheat. *Evol. Appl.* **5**:779–795. [2, 3, 13]

Thornton, P. K., and M. Herrero. 2015. Adapting to Climate Change in the Mixed Crop and Livestock Farming Systems in Sub-Saharan Africa. *Nat. Clim. Chang.* **5**:830–836. [7]

Thrupp, L. A. 2000. Linking Agricultural Biodiversity and Food Security: The Valuable Role of Agrobiodiversity for Sustainable Agriculture. *Int. Aff.* **76**:265–281. [11, 13, 15]

Tiffen, M., M. Mortimore, and F. Gichuki. 1994. More People, Less Erosion: Environmental Recovery in Kenya. New York: John Wiley. [8]

Tilman, D. 1999. Global Environmental Impacts of Agricultural Expansion: The Need for Sustainable and Efficient Practices. *PNAS* **96**:5995–6000. [9, 11]

Tixier, P., P.-F. Duyck, F.-X. Cote, G. Caron-Lormier, and E. Malezieux. 2013. Food Web-Based Simulation for Agroecology. *Agronom. Sustain. Devel.* **33**:663–670. [4]

Tobin, D., R. Bates, M. Brennan, and T. Gill. 2018. Peru Potato Potential: Biodiversity Conservation and Value Chain Development. *Renew. Agr. Food Syst.* **33**:19–32. [5, 8]

Toledo, V., E. Boege, and N. Barrera-Bassols. 2010. The Biocultural Heritage of Mexico: An Overview. *Landscape* **2**:22–31. [12]

Toledo, V. M., D. Garrido, and N. Barrera-Bassols. 2015. The Struggle for Life: Socio-Environmental Conflicts in Mexico. *Lat. Am. Perspect.* **42**:133–147. [9]

Tooker, J. F., and S. D. Frank. 2012. Genotypically Diverse Cultivar Mixtures for Insect Pest Management and Increased Crop Yields. *J. Appl. Ecol.* **49**:974–985. [2, 4]

Trinh, L. N., J. W. Watson, N. N. Hue, et al. 2003. Agrobiodiversity Conservation and Development in Vietnamese Home Gardens. *Agricult. Ecosyst. Environ.* **97**:317–344. [8]

Tscharntke, T., Y. Clough, T. C. Wanger, et al. 2012. Global Food Security, Biodiversity Conservation and the Future of Agricultural Intensification. *Biol. Conserv.* **151**:53–59. [8]

Tsing, A. 2012. Unruly Edges: Mushrooms as Companion Species. *Environ. Human.* **1**:141–154. [13]

Tubiello, F. N., J. F. Soussana, and S. M. Howden. 2007. Crop and Pasture Response to Climate Change. *PNAS* **104**:19686–19690. [6]

Turner, K. L., and I. J. Davidson-Hunt. 2016. Tensions and Synergies in the Central Valley of Tarija, Bolivia: Commercial Viticulture and Agrobiodiversity in Smallholder Farming Systems. *Agroecol. Sust. Food* **40**:518–552. [8]

Turner, W., S. Spector, N. Gardiner, et al. 2003. Remote Sensing for Biodiversity Science and Conservation. *Trends Ecol. Evol.* **18**:306–314. [5]

Tuxill, J. 2005. Agrarian Change and Crop Diversity in Mayan Milpas of Yucatan, Mexico: Implications for on-Farm Conservation. PhD dissertation, Yale University, New Haven. [4]

Tuxill, J., L. A. Reyes, L. L. Moreno, V. C. Uicab, and D. I. Jarvis. 2010. All Maize Is Not Equal: Maize Variety Choices and Mayan Foodways in Rural Yucatan, Mexico. In: Pre-Columbian Foodways: Interdisciplinary Approaches to Food, Culture, and Markets in Ancient Mesoamerica, ed. J. Staller and M. Carrasco, pp. 467–486. New York: Springer. [4]

Uchino, B. N. 2009. Understanding the Links between Social Support and Physical Health: A Life-Span Perspective with Emphasis on the Separability of Perceived and Received Support. *Perspect. Psychol. Sci.* **4**:236–255. [11]

UN. 1993. Convention on Biological Diversity. New York: United Nations. [9]

———. 2004. Access and Benefit-Sharing as Related to Genetic Resources (Article 15). In: Convention on Biological Diversity (CBD). New York: United Nations. [12]

———. 2008. United Nations Declaration on the Rights of Indigenous Peoples. New York: United Nations. [12]

———. 2014a. Additional Information Received on Use of the Term "Indigenous Peoples and Local Communities". Convention on Biological Diversity (CBD). New York: United Nations. [12]

Bibliography

UN. 2014b. The Concept of Local Communities: Background Paper Prepared by the Secretariat of the International Expert Group Meeting for Local Community Representatives. In: Convention on Biological Diversity (CBD). New York: United Nations. [12]

———. 2015. Transforming Our World: The 2030 Agenda for Sustainable Development. New York: United Nations. [9]

———. 2016. 244 Million International Migrants Living Abroad Worldwide, New UN Statistics Reveal. UN News. New York: UN. [8]

———. 2017a. New Urban Agenda. New York/Quito: UN. [9]

———. 2017b. Parties to the Nagoya Protocol. In: Convention on Biological Diversity (CBD). New York: United Nations. [12]

UNICEF. 2015. Unicef's Approach to Scaling up Nutrition for Mothers and Their Children. In: Discussion Paper. New York: United Nations Children's Fund. [10]

Ureta, M. S., A. D. Carrera, M. A. Cantamutto, and M. M. Poverene. 2008. Gene Flow among Wild and Cultivated Sunflower, *Helianthus Annuus,* in Argentina. *Agricult. Ecosyst. Environ.* **123**:343–349. [2]

USDA. 2014. Farmers Marketing: Direct Sales through Markets, Roadside Stands, and Other Means up 8 Percent since 2007. In: 2012 Census of Agriculture: Highlights, ACH 12-7. Washington, D.C.: GPO. [6]

Vadi, V. S. 2011. When Cultures Collide: Foreign Direct Investment, Natural Resources and Indigenous Heritage in International Investment Law. *Columbia Human Rights Law. Rev.* **42**:797–889. [9]

Valente, T. W. 2010. Social Networks and Health : Models, Methods, and Applications. New York: Oxford University Press. [11]

Van Andel, T. R., and M. C. Fundiko. 2017. The Trade in African Medicinal Plants in Matonge-Ixelles, Brussels (Belgium). *Econ. Bot.* **70**:1–11. [8]

Van Andel, T. R., R. S. Meyer, S. A. Aflitos, et al. 2016. Tracing Ancestor Rice of Suriname Maroons Back to Its African Origin. *Nat. Plants* **2**:16149. [8]

Vanderbroek, I., and M. J. Balick. 2012. Globalization and Loss of Plant Knowledge: Challenging the Paradigm. *PLoS One* **7**:e37643. [8]

Vandermeer, J., M. van Noordwijk, J. Anderson, C. Ong, and I. Perfecto. 1998. Global Change and Multi-Species Agroecosystems: Concepts and Issues. *Agricult. Ecosyst. Environ.* **67**:1–22. [1]

van de Wouw, M., C. Kik, T. Van Hintum, R. Van Treuren, and B. Visser. 2009. Genetic Erosion in Crops: Concept, Research Results and Challenges. *Plant Genet. Resour.* **8**:1–15. [3, 9]

van de Wouw, M., T. van Hintum, C. Kik, R. van Treuren, and B. Visser. 2010. Genetic Diversity Trends in Twentieth Century Crop Cultivars: A Meta Analysis. *Theor. Appl. Genet.* **120**:1241–1252. [9]

van Dooren, T. 2008. Inventing Seed: The Nature(S) of Intellectual Property in Plants. *Environ. Plan. D* **26**:676–697. [13]

Van Dusen, M. E., and J. E. Taylor. 2005. Missing Markets and Crop Diversity: Evidence from Mexico. *Environ. Dev. Econ.* **10**:513–531. [8, 15]

van Etten, J. 2006. Molding Maize: The Shaping of a Crop Diversity Landscape in the Western Highlands of Guatemala. *J. His. Geogr.* **32**:689–711. [13]

———. 2011. Crowdsourcing Crop Improvement in Sub-Saharan Africa: A Proposal for a Scalable and Inclusive Approach to Food Security. *IDS Bull.* **42**:102–110. [2, 6]

van Etten, J., M. R. Fuentes López, L. G. Molina Monterroso, and K. M. Ponciano Samayoa. 2008. Genetic Diversity of Maize (*Zea Mays* L. Ssp. Mays) in Communities of the Western Highlands of Guatemala: Geographical Patterns and Processes. *Genet. Resour. Crop Evol.* **55**:303–317. [3]

van Etten, J., and R. J. Hijmans. 2010. A Geospatial Modeling Approach Integrating Archaeobotany and Genetics to Trace the Origin and Dispersal of Domesticated Plants. *PLoS One* **5**:e12060. [2, 3, 5]

van Heerwaarden, J., J. Doebley, W. H. Briggs, et al. 2011. Genetic Signals of Origin, Spread, and Introgression in a Large Sample of Maize Landraces. *PNAS* **108**:1088–1092. [2, 3]

van Heerwaarden, J., J. Hellin, R. F. Visser, and F. A. van Eeuwijk. 2009. Estimating Maize Genetic Erosion in Modernized Smallholder Agriculture. *Theor. Appl. Genet.* **119**:875–888. [4]

van Heerwaarden, J., F. A. Van Eeuwijk, and J. Ross-Ibarra. 2010. Genetic Diversity in a Crop Metapopulation. *Heredity* **104**:28–39. [8]

van Lenteren, J. 2000. A Greenhouse without Pesticides: Fact or Fantasy? *Crop Protect.* **19**:375–384. [7]

van Oudenhoven, A. P. E., K. Petz, R. Alkemade, L. Hein, and R. S. de Groot. 2012. Framework for Systematic Indicator Selection to Assess Effects of Land Management on Ecosystem Services. *Ecol. Indic.* **21**:110–122. [9]

van Vuuren, D. P., E. Kriegler, B. C. O'Neill, et al. 2014. A New Scenario Framework for Climate Change Research: Scenario Matrix Architecture. *Clim. Change* **122**:373–386. [6]

van Zonneveld, M., X. Scheldeman, P. Escribano, et al. 2012. Mapping Genetic Diversity of Cherimoya (*Annona Cherimola* Mill.): Application of Spatial Analysis for Conservation and Use of Plant Genetic Resources. *PLoS One* **7**:e29845. [5]

van Zonneveld, M., E. Thomas, G. Galluzzi, and X. Scheldeman. 2011. Mapping the Ecogeographic Distribution of Biodiversity and GIS Tools for Plant Germplasm Collectors. In: Collecting Plant Genetic Diversity: Technical Guidelines. 2011 Update, ed. L. Guarino et al. Rome Bioversity Intl. [5]

Vavilov, N. I. 1926a. Studies on the Origin of Cultivated Plants. Leningrad: Institute of Applied Botany and Plant Improvement. [14]

———. 1926b. Tzentry Proiskhozhdeniya Kulturnykh Rastenii (the Centers of Origin of Cultivated Plants). *Bull. Appl. Bot. Plant Breed.* **16**:1–248. [9]

Vellema, W., S. Desiere, and M. D'Haese. 2016. Verifying Validity of the Household Dietary Diversity Score. *Food Nutr. Bull.* **37**:27–41. [10]

Verger, E., M. C. Dop, and Y. Martin-Prével. 2016. Not All Dietary Diversity Scores Can Legitimately Be Interpreted as Proxies of Diet Quality. *Public Health Nutr.* **20**:2067–2068. [10]

Vernooy, R., P. Shrestha, and B. Sthapit. 2015. Community Seed Banks: Origin, Evolution and Prospect. London: Bioversity International, Earthscan Publishers. [2]

Vernooy, R., B. Sthapit, G. Otieno, P. Shrestha, and A. Gupta. 2017. The Roles of Community Seed Banks in Climate Change Adaption. *Dev. Pract.* **27**:316–327. [5]

Verweij, M., M. Douglas, R. Ellis, et al. 2006. Clumsy Solution for a Complex World: The Case of Climate Change. *Public Adm.* **84**:817–843. [7]

Veteto, J. R., and K. Skarbø. 2009. Sowing the Seeds: Anthropological Contributions to Agrobiodiversity Studies. *Cult. Agricult.* **31**:73–87. [14]

Vigouroux, Y., A. Barnaud, N. Scarcelli, and A. C. Thuillet. 2011a. Biodiversity, Evolution and Adaptation of Cultivated Crops. *C. R. Biol.* **334**:450–457. [8]

Vigouroux, Y., J. C. Glaubitz, Y. Matsuoka, et al. 2008. Population Structure and Genetic Diversity of New World Maize Races Assessed by DNA Microsatellites. *Am. J. Bot.* **95**:1240–1253. [3, 5]

Vigouroux, Y., C. Mariac, S. De Mita, et al. 2011b. Selection for Earlier Flowering Crop Associated with Climatic Variations in the Sahel. *PLoS One* **6**:e19563. [2, 3, 6]

Villarreal, A. 2014. Ethnic Identification and Its Consequences for Measuring Inequality in Mexico. *Am. Sociol. Rev.* **79**:775–806. [12]

Vincenti, B., P. Eyzaguirre, and T. Johns. 2008. The Nutritional Role of Forest Plant Foods for Rural Communities. In: Human Health and Forests: A Global Overview of Issues, Practice and Policy, ed. C. J. P. Colfer, pp. 63–93. London: Earthscan. [11]

Violon, C., M. Thomas, and E. Garine. 2016. Good Year, Bad Year: Changing Strategies, Changing Networks? A Two-Year Study on Seed Acquisition in Northern Cameroon. *Ecol. Soc.* **21**:34. [2]

Visser, B., H. Mbozi, P. Kasasa, et al. 2019. Options for Scaling up Commuinjity Support in Genetic Divesrity Management. In: Farmer Participation in Plant Breeding Programs: The State of the Art, ed. O. Westengen and T. Winge. Abingdon, UK: Earthscan, in press. [14]

von Hippel, E. 1994. "Sticky Information" and the Locus of Problem Solving: Implications for Innovation *Manag. Sci.* **40**:429–439. [13]

Wale, E., A. G. Drucker, and K. K. Zander. 2011. The Economics of Managing Crop Diversity On-Farm: Case Studies from the Genetic Resources Policy Initiative. London: Earthscan Publishing. [5]

Walsh-Dilley, M. 2013. Negotiating Hybridity in Highland Bolivia: Indigenous Moral Economy and the Expanding Market for Quinoa. *J. Peasant Stud.* **40**:659–682. [8]

Walzer, M. 1992. The Civil Society Argument. In: Dimensions of Radical Democracy: Pluralism, Citizenship and Community, ed. C. Mouffe, pp. 89–107. New York: Verso. [14]

Warmuth, V., A. Eriksson, M. A. Bower, et al. 2012. Reconstructing the Origin and Spread of Horse Domestication in the Eurasian Steppe. *PNAS* **109**:8202–8206. [3]

Watson, J. E. M., D. F. Shanahan, M. Di Marco, et al. 2016. Catastrophic Declines in Wilderness Areas Undermine Global Environment Targets. *Curr. Biol.* **26**:1–6. [5]

Weisenburger, D. D. 1993. Human Health-Effects of Agrichemical Use. *Hum. Pathol.* **24**:571–576. [11]

Weismantel, M. J. 1988. Food, Gender, and Poverty in the Ecuadorian Andes. Illinois: Waveland Press. [14]

Wellhausen, E. J., L. M. Roberts, and E. Hernandez. 1952. Races of Maize in Mexico: Their Origin, Characteristics, and Distribution. Cambridge, MA: Harvard University Press. [3]

Westengen, O. T., M. A. Okongo, L. Onek, et al. 2014. Ethnolinguistic Structuring of Sorghum Genetic Diversity in Africa and the Role of Local Seed Systems. *PNAS* **111**:14100–14105. [3]

Westhoek, H., J. Ingram, S. van Berkum, and M. Hajer. 2014. A Food System Approach for the Identification of Opportunities to Increase Resource Use Efficiency. In: Proceedings of the 9th International Conference on Life Cycle Assessment in the Agri-Food Sector, pp. 1505–1511. San Francisco: American Center for Life Cycle Assessment. [15]

Whitaker, R. C., S. M. Phillips, and S. M. Orzol. 2006. Food Insecurity and the Risks of Depression and Anxiety in Mothers and Behavior Problems in Their Preschool-Aged Children. *Pediatrics* **118**:E859–E868. [11]

WHO. 1946. Preamble to the Constitution of the World Health Organization as Adopted by the International Health Conference, New York, 19-22 June, 1946. In: Official Records of the World Health Organization, p. 100. New York: World Health Organization. [11]

―――. 2003. Annex B: Example of a Research Agreement Concluded between Cine and an Indigenous Community in Canada. In: Indigenous Peoples & Participatory Health Research. Geneva WHO. [9]

WHO/CBD. 2015. Connecting Global Priorities: Biodiversity and Human Health: A State of Knowledge Review. New York: World Health Organization and Convention on Biological Diversity. [9]

Wiesmann, D., L. Bassett, T. Benson, and J. Hoddinott. 2009. Validation of the World Food Programme's Food Consumption Score and Alternative Indicators of Household Food Security. Washington, D.C.: Intl. Food Policy Research Institute. [10]

Wilkinson, R. G. 2000. Mind the Gap: Hierarchies, Health, and Human Evolution. London: Weidenfeld & Nicolson. [11]

Will, M. 2008. Promoting Value Chains of Neglected and Underutilized Species for Pro-Poor Growth and Biodiversity Conservation: Guidelines and Good Practices. Rome: Global Facilitation Unit for Underutilized Species. [15]

Williams, D. E. 1993. *Lycianthes Moziniana* (Solanaceae): An Underutilized Mexican Food Plant with New Crop Potential. *Econ. Bot.* **47**:387–400. [2]

Williams, D. E., and E. Hernández-Xolocotzi. 1996. El Auspicio de Arvenses en Tlaxcala: Un Estudio del Proceso de Domesticación en Marcha *Agrociencia* **30**:215–221. [2]

Williams, L. L. T., J. Germov, S. Fuller, and M. Freij. 2015. A Taste of Ethical Food Consumption at a Slow Food Festival. *Appetite* **91**:321–328. [14]

Winfree, R., N. M. Williams, H. R. Gaines, J. S. Ascher, and C. Kremen. 2008. Wild Bee Pollinators Provide the Majority of Crop Visitation across Land-Use Gradients in New Jersey and Pennsylvania, USA. *J. Appl. Ecol.* **45**:793–802. [4]

Winge, T. 2016. Linking Access and Benefit-Sharing for Crop Genetic Resources to Climate Change Adaptation. *Plant Genet. Resour.* **14**:11–27. [13]

Winkel, T., H. D. Bertero, P. Bommel, et al. 2012. The Sustainability of Quinoa Production in Southern Bolivia: From Misrepresentations to Questionable Solutions. Comments on Jacobsen. *J. Agron. Crop Sci.* **198**:314–319. [8]

Winkel, T., P. Bommel, M. Chevarría-Lazo, et al. 2016. Panarchy of an Indigenous Agroecosystem in the Globalized Market: The Quinoa Production in the Bolivian Altiplano. *Glob. Environ. Change* **39**:195–204. [8]

Wise, R. M., I. Fazey, M. S. Smith, et al. 2014. Reconceptualising Adaptation to Climate Change as Part of Pathways of Change and Response. *Glob. Environ. Change* **28**:325–336. [7]

Wittenmyer, L., and W. Merbach. 2005. Plant Responses to Drought and Phosphorus Deficiency: Contribution of Phytohormones in Root-Related Processes. *J. Plant Nutr. Soil Sci.* **168**:531–540. [4]

Wolff, F. 2004. Legal Factors Driving Agrobiodiversity Loss. *Environ. Law Network Int.* **1**:1–11. [11]

Wood, S. A., D. S. Karp, F. DeClerck, et al. 2015a. Functional Traits in Agriculture: Agrobiodiversity and Ecosystem Services. *Trends Ecol. Evol.* **30**:531–539. [4]

Wood, T. J., J. M. Holland, and D. Goulson. 2015b. Pollinator-Friendly Management Does Not Increase the Diversity of Farmland Bees and Wasps. *Biol. Conserv.* **187**:120–126. [4]

Woodley, E., and L. Maffi. 2012. Biocultural Diversity Conservation: A Global Source-book. London: Routledge. [12]

World Bank. 1991. Operational Directive on Indigenous Peoples. In: Operational Manual OD 4.20. World Bank. [12]

———. 2005. Operational Policy on Indigenous Peoples. In: Operational Manual Op 4.10. World Bank. [12]

World Bank Group. 2014. Learning from World Bank History: Agriculture and Food-Based for Addressing Malnutrition, Agriculture and Environmental Services Discussion Paper No. 10. Washington, D.C.: World Bank. [9]

World Food Programme. 2008. Food Consumption Analysis Technical Guidance Sheet: Calculation and Use of the Food Consumption Score in Food Security Analysis. Rome: World Food Programme, Vulnerability Analysis and Mapping Branch. [10]

Wright, P. J., R. E. Falloon, and D. Hedderley. 2015. Different Vegetable Crop Rotations Affect Soil Microbial Communities and Soilborne Diseases of Potato and Onion: Literature Review and a Long-Term Field Evaluation. *New Zeal. J. Crop Hort.* **43**:85–110. [4]

Wright, S. I., I. V. Bi, S. G. Schroeder, et al. 2005. The Effects of Artificial Selection on the Maize Genome. *Science* **308**:1310–1314. [3]

Yang, Q., J. Carrillo, H. Jin, et al. 2013. Plant–Soil Biota Interactions of an Invasive Species in Its Native and Introduced Ranges: Implications for Invasion Success. *Soil Biol. Biochem.* **65**:78–85. [2]

Zeven, A. C. 1999. The Traditional Inexplicable Replacement of Seed and Seed Ware of Landraces and Cultivars: A Review. *Euphytica* **110**:181–191. [13]

Zimmerer, K. S. 1991a. Labor Shortages and Crop Diversity in the Southern Peruvian Sierra. *Geogr. Rev.* **81**:414–432. [8]

———. 1991b. Wetland Production and Smallholder Persistence: Agricultural Change in a Highland Peruvian Region. *Ann. Assoc. Am. Geogr.* **81**:443–463. [2]

———. 1997. Changing Fortunes: Biodiversity and Peasant Livelihood in the Peruvian Andes. Berkeley: Univ. of California Press. [8, 12, 14, 15]

———. 1998. The Ecogeography of Andean Potatoes. *Bioscience* **48**:445–454. [8]

———. 2001. Report on Geography and the New Ethnobiology. *Geogr. Rev.* **91**:725–734. [13]

———. 2003a. Geographies of Seed Networks for Food Plants (Potato, Ulluco) and Approaches to Agrobiodiversity Conservation in the Andean Countries. *Soc. Nat. Resour.* **16**:583–601. [8, 13, 14]

———. 2003b. Just Small Potatoes (and Ulluco)? The Use of Seed-Size Variation in Native Commercialized Agriculture and Agrobiodiversity Conservation among Peruvian Farmers. *Agricult. Human Values* **20**:107–123. [11]

———. 2010. Biological Diversity in Agriculture and Global Change. *Annu. Rev. Environ. Resour.* **35**:137–166. [6, 8, 11, 13, 14]

———. 2012. The Indigenous Andean Concept of Kawsay, the Politics of Knowledge and Development, and the Borderlands of Environmental Sustainability in Latin America. *PMLA* **127**:600–606. [12]

———. 2013. The Compatibility of Agricultural Intensification in a Global Hotspot of Smallholder Agrobiodiversity (Bolivia). *PNAS* **110**:2769–2774. [2, 5, 8, 14]

———. 2014. Conserving Agrobiodiversity Amid Global Change, Migration, and Nontraditional Livelihood Networks: The Dynamic Uses of Cultural Landscape Knowledge. *Ecol. Soc.* **19**:1–15. [4, 6, 8, 9, 13, 15]

———. 2015a. Time for Change: The Legacy of a Euro-Andean Model of Landscape versus the Need for Landscape Connectivity. *Landsc. Urban Plan.* **139**:104–116. [13]

————. 2015b. Understanding Agrobiodiversity and the Rise of Resilience: Analytic Category, Boundary Concept, or Meta-Level Transition? *Resilience* **3**:183–198. [1]

————. 2017a. A Search for Food Sovereignty: Seeding Post-Conflict Landscapes. *Re-Vista: Harvard Rev. Lat. Am.* **26**:32–34. [8, 13]

————. 2017b. "Territorial Ordering" (*Ordenamiento Territorial*): A "New" Mode of Land Use Planning and a Multi-Scale Idea for Urban-Rural Integration and Their Implications for Next-Generation Conservation. In: The Urban and the Territorial: Housing in Mérida, ed. D. E. Davis et al., pp. 65–74. Cambridge, MA: Harvard Graduate School of Design. [8]

Zimmerer, K. S., J. A. Carney, and S. J. Vanek. 2015. Sustainable Smallholder Intensification in Global Change? Pivotal Spatial Interactions, Gendered Livelihoods, and Agrobiodiversity. *Curr. Opin. Environ. Sustain.* **14**:49–60. [4, 6, 8, 11, 13–15]

Zimmerer, K. S., H. Córdova-Aguilar, R. Mata Olmo, Y. Jiménez Olivencia, and S. J. Vanek. 2017. Mountain Ecology, Remoteness, and the Rise of Agrobiodiversity: Tracing the Geographic Spaces of Human–Environment Knowledge. *Ann. Amer. Assoc. Geogr.* **107**:441–455. [8]

Zimmerer, K. S., and S. de Haan. 2017. Agrobiodiversity and a Sustainable Food Future. *Nat. Plants* **3**:17047. [1, 8, 14]

Zimmerer, K. S., and D. S. Douches. 1991. Geographical Approaches to Crop Conservation: The Partitioning of Genetic Diversity in Andean Potatoes. *Econ. Bot.* **45**:176–189. [2, 3]

Zimmerer, K. S., E. F. Lambin, and S. J. Vanek. 2018. Smallholder Telecoupling and Potential Sustainability. *Ecol. Soc.* **23**:30. [6, 8]

Zimmerer, K. S., and H. L. Rojas Vaca. 2016. Fine-Grain Spatial Patterning and Dynamics of Land Use and Agrobiodiversity Amid Global Changes in the Bolivian Andes. *Reg. Environ. Change* **16**:2199–2214. [5]

Zimmerer, K. S., and S. J. Vanek. 2016. Toward the Integrated Framework Analysis of Linkages among Agrobiodiversity, Livelihood Diversification, Ecological Systems, and Sustainability Amid Global Change. *Land* **5**:10. [2, 4, 8, 15]

Subject Index

Further Titles in the Strüngmann Forum Report Series[1]

Better Than Conscious? Decision Making, the Human Mind, and Implications For Institutions
edited by Christoph Engel and Wolf Singer, ISBN 978-0-262-19580-5

Clouds in the Perturbed Climate System: Their Relationship to Energy Balance, Atmospheric Dynamics, and Precipitation
edited by Jost Heintzenberg and Robert J. Charlson, ISBN 978-0-262-01287-4

Biological Foundations and Origin of Syntax
edited by Derek Bickerton and Eörs Szathmáry, ISBN 978-0-262-01356-7

Linkages of Sustainability
edited by Thomas E. Graedel and Ester van der Voet, ISBN 978-0-262-01358-1

Dynamic Coordination in the Brain: From Neurons to Mind
edited by Christoph von der Malsburg, William A. Phillips and Wolf Singer, ISBN 978-0-262-01471-7

Disease Eradication in the 21st Century: Implications for Global Health
edited by Stephen L. Cochi and Walter R. Dowdle, ISBN 978-0-262-01673-5

Animal Thinking: Contemporary Issues in Comparative Cognition
edited by Randolf Menzel and Julia Fischer, ISBN 978-0-262-01663-6

Cognitive Search: Evolution, Algorithms, and the Brain
edited by Peter M. Todd, Thomas T. Hills and Trevor W. Robbins, ISBN 978-0-262-01809-8

Evolution and the Mechanisms of Decision Making
edited by Peter Hammerstein and Jeffrey R. Stevens, ISBN 978-0-262-01808-1

Language, Music, and the Brain: A Mysterious Relationship
edited by Michael A. Arbib, ISBN 978-0-262-01962-0

Cultural Evolution: Society, Technology, Language, and Religion
edited by Peter J. Richerson and Morten H. Christiansen, ISBN 978-0-262-01975-0

Schizophrenia: Evolution and Synthesis
edited by Steven M. Silverstein, Bita Moghaddam and Til Wykes, ISBN 978-0-262-01962-0

Rethinking Global Land Use in an Urban Era
edited by Karen C. Seto and Anette Reenberg, ISBN 978-0-262-02690-1

Trace Metals and Infectious Diseases
edited by Jerome O. Nriagu and Eric P. Skaar, ISBN 978-0-262-02919-3

Translational Neuroscience: Toward New Therapies
edited by Karoly Nikolich and Steven E. Hyman, ISBN: 978-0-262-029865

---

[1] available at https://mitpress.mit.edu/books/series/str%C3%BCngmann-forum-reports-0

The Pragmatic Turn: Toward Action-Oriented Views in Cognitive Science
edited by Andreas K. Engel, Karl J. Friston and Danica Kragic
ISBN: 978-0-262-03432-6

Complexity and Evolution: Toward a New Synthesis for Economics
edited by David S. Wilson and Alan Kirman, ISBN: 978-0-262-035385

Computational Psychiatry: New Perspectives on Mental Illness
edited by A. David Redish and Joshua A. Gordon, ISBN: 978-0-262-035422

Investors and Exploiters in Ecology and Economics: Principles and Applications
edited by Luc-Alain Giraldeau, Philipp Heeb and Michael Kosfeld
Hardcover: ISBN: 978-0-262-036122

The Cultural Nature of Attachment: Contextualizing Relationships and Development
edited by Heidi Keller and Kim A. Bard
Hardcover: ISBN: 978-0-262-036900

Emergent Brain Dynamics: Prebirth to Adolescence
edited by April A. Benasich and Urs Ribary, ISBN: 978-0-262-038638

Rethinking Environmentalism: Linking Justice, Sustainability, and Diversity
edited by Sharachchandra Lele, Eduardo S. Brondizio, John Byrne, Georgina M.
Mace and Joan Martinez-Alier, ISBN: 978-0-262-038966

Printed in the United States
by Baker & Taylor Publisher Services